AF520487

Recent Trends & Advances in Food Science & Post Harvest Technology

Recent Trends & Advances in Food Science & Post Harvest Technology

Prof. Ivi Chakraborty
Professor
Department of Post Harvest Technology
Faculty of Horticulture, Mohanpur, Nadia, West Bengal

Dr. Riadh Ilahy
Associate Professor
Department of Horticulture
National Agricultural Research Institute of Tunisia

Dr. Balaji Vikram
Assistant Professor
Department of Horticulture
SHUATS, Prayagraj, India

Ms. Sujayasree O.J.
Research Scholar
Division of Post Harvest Technology
ICAR-Indian Institute of Horticultural Research, Bengaluru

Mr. Arghya Mani
Ph.D. Research Scholar
Department of Post Harvest Technology
Bidhan Chandra Krishi Vishwavidyalaya (BCKV),
Mohanpur, Nadia, India

SATISH SERIAL PUBLISHING HOUSE
403, Express Tower, Commercial Complex
Azadpur, Delhi-110033 (India)
Phone : 011-27672852, Fax : 91-11-27672046
e-mail : info@satishserial.com, hkjain1975@yahoo.com
Website : www.satishserial.com

Published by :

SATISH SERIAL PUBLISHING HOUSE

403, Express Tower, Commercial Complex, Azadpur, Delhi-110033 (INDIA)
Phone : 011-27672852 Fax : 91-11-27672046
E-mail : info@satishserial.com, hkjain1975@yahoo.com

ISBN 978-93-88020-53-4

Composed, Designed & Printed in India

About the Editors

Dr. Ivi Chakraborty (Date of Birth: 21.01.1969) is working as a Professor in the Department of Post Harvest Technology, Faculty of Horticulture, Mohanpur, Nadia, West Bengal. She has more than 20 years of teaching and research experience. She has published more than 55 research papers, 04 book chapters and number of technical bulletins in processing and value addition aspects. She has been handled almost 8 (eight) research projects as Principle and Co-Principle Investigator. At present, she is associated with one national and one international project. 05 students have been awarded the Ph.D. degree and 12 students completed M.Sc. degree in Post Harvest Technology under her guidance. She is the recipient of 'Eminent Scientist Award (2017) in the 'Inspiring Women Award Ceremony' in the field of Agriculture & Allied Sciences to be held at Rai Umanath Bali Auditorium, Lucknow, Uttar Pradesh on June 5th, 2017 (World Environment Day). She has also been received the prestigious 'Dr. J.S. Pruthi Award, 2016' from AFST(I), CSIR-CFTRI, India in the 26th ICFoST (Indian Convention of Food Scientists & Technologists) being held at CSIR-IICT, Hyderabad, TS, India on 7th December, 2017 for contribution in fruits and vegetables preservation. Other than this, she has also been received 06 nos. of 'Best poster awards' in different national and international conferences.

Dr. Riadh Ilahy (Date of Birth: 29.03.1979) at Kheireddine, Tunis, Tunisia. Dr. Riadh Ilahy obtained his baccalaureate (experimental sciences) in 1998 from the technical secondary school, le Kram, then he started his higher education (biology/geology) at the Faculty of Science, Bizerte, Tunis, Tunisia until he finished his master degree and Ph.D. (Phytochemistry) in 2011. He finished his Ph.D. after collaboration with Professor Marcello Salvatore Lenucci and Professor Giuseppe Dalessandro from the University of Salento. He is currently working as an Associate Professor at the National Agricultural Research Institute of Tunisia, Department of Horticulture. He is specialized in breeding for higher nutritional quality crops, mainly tomato and potato. He has published 3 Chapter of Books, 17 Communications/posters and more than 33 scientific articles. He acted as an external examiner, paper reviewer for many international indexed and impacted Journals. He obtained the Amitom Young Researcher prize in 2014 from the HM and Clause Company for his research on chloroplast/chromoplsat of high-pigment tomato cvs. He has been bearing 8 H-index, 8 I10 index with 419 citations in "Google Scholar" and an RG score of 15. He has keen research Interest in Horticultural crop-breeding, Phytochemistry, Natural antioxidants, Food chemistry and Post-harvest Technology.

Dr. Balaji Vikram has been serving in the teaching profession for more than 6 years. He got his Bachelor's degree in Agriculture in the year 2007, Master's in Agriculture Science with specialization in Horticulture, in 2009 from the prestigious Allahabad Agricultural Institute, Allahabad. Later on, he completed his Doctoral Research and was conferred Ph. D. in the year 2016 from SHIATS, Allahabad. He is also qualified ASRB (NET) in the field of Horticulture (Fruit Science). Dr. Balaji Vikram started his carrier as a Teaching Assistant from 2013 and joined Assistant Professor on July-2018, in the Department of Horticulture at SHUATS, Allahabad (U.P.), India. Besides contributing 100 publications, which include research papers, books and book chapters, extension articles, etc., he also delivered scientific guest lecture (Expert) at district level training programme (Krishak Goshthi, FSIP and Seasonal Mela's) conducted by Krishi Vibhag and Udyan Vibhag, U.P. He is invited regularly as horticulture expert at "Doordarshan", Allahabad. He also participated actively in NSS during the year 2008 to 2016. He received several awards Young Scientist Award, 2012 from Bioved Research Society, Allahabad. (U.P.) and also Distinguished Service Award, 2016 from Society of Biological Sciences and Rural Development, Allahabad (U.P.), India. Dr. Vikram has membership of different National and International Societies and attended so many International/ National Symposia, conferences, etc. He is always willing to provide his services to the farming community, social service, educational society and to help the students.

Ms. Sujayasree O.J. is a Research Scholar in the Division of Post Harvest Technology, ICAR-Indian Institute of Horticultural Research, Bengaluru, Outreach Campus of ICAR-Indian Agricultural Research Institute, New Delhi, India. She has completed B.Sc. Horticulture in the year of 2014, from UAHS, Shimoga and M.Sc. Post Harvest Technology (Horticulture) in the year 2016 from UHS, Bagalkot. She has attended 20 National/International symposiums or conferences/seminars and 3 trainings/workshops. She has done training on agricultural extension system management and Internship focused on food processing sectors at National Institute of Agricultural Extension

Management (MANAGE), Hyderabad. She has published 2 books on Farmer's Handbook on Food Processing Technologies and Directory of Direct Marketing Agri-business Companies, respectively. She has more than 25 research articles and two technical bulletins. She is also one of the sub-editor in *Research Journal of Chemical and Environmental Sciences.* Also, secured *Young Scholar Award, 2016,* from Society for World Environment Food and Technology, New Delhi. Besides that she got awarded with *Certificate of Merit* during her Bachelor degree for her excellence in academics.

Mr. Arghya Mani (Date of Birth: 07-12-1993) is a Ph.D. Research Scholar in the Department of Post Harvest Technology, Bidhan Chandra Krishi Vishwavidyalaya (BCKV), Mohanpur, India. He has completed B.Sc. Horticulture and M.Sc. Pomology and Post Harvest Technology in the year 2015 and 2017, respectively. He has qualified ICAR-NET in Fruit Science. He has attended 10 National & International conferences/seminars/symposium, 3 workshops, 3 training programs. He has industrial training exposure of one month in "Overseas Agro-product Pvt. Ltd". As a 1st author he has got 1 book, 10 popular articles, 10 research papers, 15 book chapters, 7 review papers published at both national and international level. Besides that he has successfully represented NGO's to discriminate knowledge related to horticulture technology and food processing at farmer's level. He is reviewer of "*Journal of Agricultural and bio-chemical Sciences*" and "*International Journal of Agricultural Sciences*". He is the advisory board member of "*Cambridge Scholar Publishing*". He got awarded with "*Certificate of Merit*" for his academic excellence during his bachelor's degree. Beside that he has been awarded with "*Best Oral presentation*" three times in different events. He has also been awarded "*Young Professional Award*" twice in different seminars. He got awarded "*Young Scientist Award*" by SIRI society as well. Recently, he was felicitated with "*Gold Medal*" for 'Best Master Degree Thesis' during 6th convocation of Uttar Banga Krishi Vishwavidyalaya. He is the founder of Horticulture based website "*Hortipedia India*". He is also the founder and Editor-in-chief of online magazine "***Agriculture & Food: e-Newsletter***". At present, he is working on diverse aspects, related to minor and underutilized fruits. He is actively doing research to ensure doubling of farmers' income by innovative processing technology.

Bidhan Chandra Krishi Viswavidyalaya
P.O. Krishi Viswavidyalaya, Mohanpur 741252
District: Nadia, West Bengal, India
Website: www.bckv.edu.in

☎ (+91) 03473-222666
(+91) 033-25879772
Fax : (+91) (03473)-222275
Email : ddpatra@rediffmail.com
bckvvc@gmail.com
Cell : +91-9830071278

Dr. DD Patra, Ph D (IARI, New Delhi), FNA Sc., FISSS, FNAAS
Vice-Chancellor

No. VC/BCKV/ 114/97
Date: 11/07/2018

Foreword

Post-harvest Technology, Food Processing, Food Technology and Dairy Technology sectors are indispensable in India for safeguarding the available perishable produce from distribution losses. Processing is the brightest wing of post harvest technology that includes value addition and product development. Post Harvest Management and Processing is the need of the hour in order to mitigate post harvest losses and also to improve the economic background of the country by converting the farmers to an agro-processer and ultimately an agro-entrepreneur. The development of Post Harvest Technology and Food Processing in the country would undoubtedly improve the socio-economic status of the small and marginal growers and would also support as an enterprise to large farmers and entrepreneurs. By adoption of the technology, the farmers can not only be benefitted by selling their produce during off-season but can also add value to them and earn upto 10 times of their normal earning. Enhanced demand of processed food products coupled with varied lifestyles have made widespread adoption of Post Harvest Technology very crucial. Further, this will generate employment in these sectors.

In this direction, the efforts of the editorial team in compiling the book series **"Advances in Food Sciences and Post Harvest Technology"** are appreciable. The contents of the book are contributed by authors of renowned institutes from all over the world. It would be helpful for the students studying the courses in B.Sc., B.Tech., M.Sc., M.Tech. and Ph.D. It would be helpful for the researchers in this discipline to generate new ideas and also, be helpful for those entrepreneurs, who sought innovations in their venture.

I believe that the book would be helpful in dissemination of recent knowledge and information to the readers. The content of the book is quite relevant and exhaustive with creative blend of knowledge and skills. Illustrations are quite impressive. Descriptions and explanations generate greater interest to provide advanced applications to the students and practitioners.

D.D. Patra

Preface

Food preservation has been proficient in different parts of the world since time immemorial. Like many other industrial sectors, technological progress in food preservation achieved its strength and momentum only after the outbreak of the First World War, when supply of large quantities of food need to be arranged for the soldiers in the warfront. The 2nd World War added another strong impetus to the growth of food industry. Although, the industry initiated in a problematic war situation, now it is considered to be of perennial importance. It also assures a stable market to the growers as well as consumers at different corners of the world and skips the situation of market glut, extreme fall of price and wastage due to spoilage. However, fruits and vegetables preservation industry is still at its beginning in India. Unfortunately, most of the available research is scattered in scientific papers, reports and technical publications. Our country India is thriving for improved processing technology. The highly perishable horticultural crops need to have enhanced processing technologies. Latest processing information have to be disseminated on a country-wide scale through proper training and demonstration so that full advantage of it can be availed by the interested entrepreneurs in this industry.The recent trend in the changing eating pattern of our population, mainly due to migration from village to cities and increasing number of working women has flexed a large canvas for the food technologists to act. People are much more health conscious and facing acute shortage of time leading to inclination towards convenient food like ready to eat and ready to drink and ready to cook food, specialized menus like snacks customized for various ages and physical requirement, etc. are coming as hot cake. The challenges offered to develop innovative and authentic food, those are being grabbed by new generation entrepreneurs. Under this perspective, the book "**Recent Trends and Advances in Food Science and Harvest Technlogy**" forms a strong basis for students, researchers in the food processing field on which they may further explore and improve food standards.

Editors

Acknowledgement

Books are incomplete without author's dedication. The first thanks goes to all the authors of the book, who are from pioneer institutes of India and abroad. They have not only ensured a very good quality chapter submission but most of them got published with minimum corrections. Their effort to get the book published is sincerely acknowledged. A special thanks to the almighty God without grace of whom, such publication initiative would not have been possible. Besides that those persons who helped us to spread our call for book chapter is also acknowledged. The more diverse the distribution of authors, the greater is the quality of publication. A special thanks to Arghya Mani, Shuvo Saha, Paritosh Halder, Subhankar Adhikary, Venkata Satish Kuchi, D. Sridhar, Ravi Kiran, Raghupati B., Prabhugouda Patil, Nityamanjari Mishra, E. Rambabu, Saheb Pal, Tanmoy Sarkar, Dutta Reddy, Antima Gupta, Siddharood Maragal, Vartika Singh, Manisha momin, Mukta Sinha, Sandeep Singh, Pritesh Halder, Balaji Vikram, Niyati Jain, Bhojaraj Belakud, K. Prasad, Nirmal Kumar Meena, Radha Kushwaha and Nilam Das for collection of chapters and helping us in every aspect. A special thanks to Prof. Surajit Mitra, Prof. Ivi Chakraborty, Prof. Ashis Kumar Banik, Prof. V.B Rajwade, Prof. Swapan Kumar Ghosh and Prof. Prodyut Kumar Paul for their motivation and inspiration. Last but not the least, we want to thank our publisher for trusting our capabilities, accepting our content and publishing our book.

Editors

List of Contribution

Arghya Mani
Research Scholar
Department of Post Harvest Technology, BCKV, Mohanpur, Nadia, India

Kiran Thakur
Associate Professor
School of Food Science and Engineering, Hefei University of Technology, Hefei-230009, People's Republic of China

Zhao-Jun Wei
Professor
School of Food Science and Engineering
Hefei University of Technology, Hefei-230009, People's Republic of China

H. Chikh-Rouhou
Centre Régional des Recherches en Horticulture et Agriculture Biologique (CRRHAB/IRESA). BP 57, 4042 Chott-Mariem, Sousse, Tunisia

R. Ilahy
Institute National de Recherche Agronomique de Tunis (INRAT/IRESA)

R. Sta-Baba
Rue Hédikarray 2049 Ariana, Tunisia, Neha Mishra
Centre of Food Technology, University of Allahabad, Allahabad-211002

M.R. Manikantan
Biochemistry and Post Harvest Technology Division,
ICAR-Central Plantation Crops Research Institute
Kasaragod-671 124, Kerala, India

R. Pandiselvam
Agricultural Engineering College & Research Institute

N. Varadharaju
Tamil Nadu Agricultural University
Coimbatore-641 003, India

Devinder Kaur
Centre of Food Technology, University of Allahabad, Allahabad-211002

Neetu Mishra
Centre of Food Technology, University of Allahabad, Allahabad-211002

Suchetana Purkayastha
Lecturer
Department of Food Processing Technology

Yogendra Singh
Department of Fruit Science, CH&F, Jhalawar, (AU, Kota), Rajasthan-326023

Nirmal Kumar Meena
Department of Fruit Science
CH&F, Jhalawar, (AU, Kota), Rajasthan-326023

Kalpana Choudhary
Department of Fruit Science
CH&F, Jhalawar, (AU, Kota),
Rajasthan-326023

Suresh Gurjar
Department of Fruit Science
CH&F, Jhalawar, (AU, Kota),
Rajasthan-326023

Narendra Verma
Department of Horticulture,
BAU, Birsa, Ranchi

Tapas Sarkar
Food Safety Officer,
Ministry of Health,
Government of West Bengal

Nilanjana Dutta
Assistant Professor
CUTM, Bhubaneshwar

P. Dutta
Professor
Department of Fruit Science,
BCKV, Mohanpur, Nadia, India

Mrityunjay Ghosh
Professor
Department of Agronomy
Bidhan Chandra Krishi
Vishwavidyalaya
Mohanpur, Nadia, West Bengal

Sumit Chakravarty
Department of Forestry
Uttar Banga Krishi Vishwavidyalaya
Pundibari, Cooch, Behar,
West Bengal, India

Vineeta
Department of Forestry
Uttar Banga Krishi
Vishwavidyalaya, Pundibari,
Cooch Behar, West Bengal, India

Gopal Shukla
Department of Forestry
Uttar Banga Krishi
Vishwavidyalaya, Pundibari,
Cooch, Behar, West Bengal, India

N.A. Pala
Department of Forestry
Uttar Banga Krishi
Vishwavidyalaya, Pundibari,
Cooch, Behar, West Bengal, India

Abha Manohar K.
Department of Forestry
Uttar Banga Krishi
Vishwavidyalaya, Pundibari,
Cooch Behar, West Bengal, India

Prakash Rai
Department of Forestry
Uttar Banga Krishi
Vishwavidyalaya, Pundibari,
Cooch, Behar, West Bengal, India

Anju Puri
Baring Union Christian College,
Batala, Punjab

C. Lalmuanpuia
Dept of Food Science and
Technology, Punjab Agricultural
University, Ludhiana,
Punjab-141004

Baljit Singh
Dept of Food Science and
Technology Punjab Agricultural
University, Ludhiana,
Punjab-141004

Savita Sharma
Dept of Food Science and
Technology
Punjab Agricultural University
Ludhiana, Punjab-141004

Tanuja Priya, B.
Assistant Professor
Dr. Y.S.R. Horticultural University
V.R. Gudem, West Godavari, Andhra Pradesh-534101

Syed Sadarunissa
Associate Professor
Dr. Y.S.R. Horticultural University
V.R. Gudem, West Godavari, Andhra Pradesh-534101

Rama Devi P.
Associate Professor
Dr. Y.S.R. Horticultural University
V.R. Gudem, West Godavari, Andhra Pradesh-534101

Rama Krishna
Associate Dean
Dr. Y.S.R. Horticultural University
V.R. Gudem, West Godavari, Andhra Pradesh-534101

Sireesha, K.
Assistant Professor
Dr. Y.S.R. Horticultural University
V.R. Gudem, West Godavari, Andhra Pradesh-534101

Hradesh Rajput
Warner College of Dairy Technology
Sam Higginbottom University of Agriculture Technology and Science, Allahabad

Deepika Goswami
Central Institute of Post Harvest Engineering and Technology, Ludhiana, Punjab

Sandeep GM Nigam
Warner College of Dairy Technology, Sam Higginbottom University of Agriculture Technology & Science, Allahabad

Rekha Rani
Warner College of Dairy Technology, Sam Higginbottom University of Agriculture Technology & Science, Allahabad

Pratistha Srivastava
Warner College of Dairy Technology, Sam Higginbottom University of Agriculture Technology and Science, Allahabad

Shalini Yerukala
Professor Jayashankar Telangana State Agricultural University Hyderabad-500030,Telangana, India

Sumalatha Nallabeema
Department of Plant Pathology, College of Agriculture, Rajendranagar

Suresh Veera
Department of Plant Pathology, College of Agriculture, Rajendra Nagar

Niyati Jain
Department of Horticulture, SHUATS, Allahabad, India

Vijay Bahadur
Professor and Head
Department of Post Harvest Technology, BCKV, Mohanpur, Nadia

K. Rama Krishna
Division of Food Science and Post Harvest Technology, IARI, New Delhi-110012

Gajanan Gundewadi
Division of Food Science and Post Harvest Technology, IARI, New Delhi-110012

Pushpendar Kumar
Division of Food Science and Post Harvest Technology, IARI, New Delhi-110012

Hirak Banerjee
Department of Agronomy
Bidhan Chandra Krishi Vishwavidyalaya, Mohanpur, Nadia, West Bengal, India

Sukamal Sarkar
Research Scholar
Department of Agronomy
Bidhan Chandra Krishi Vishwavidyalaya, Mohanpur, Nadia, West Bengal, India

Krishnendu Ray
Subject Matter Specialist (Agronomy)
Sasya Shyamala Krishi Vigyan Kendra, West Bengal, India

Lalita Rana
Department of Agronomy
Bidhan Chandra Krishi Vishwavidyalaya, Mohanpur, Nadia, West Bengal, India

Sourav Garai
Research Scholar
Department of Agronomy
Bidhan Chandra Krishi Vishwavidyalaya, Mohanpur, Nadia, West Bengal, India

Brati Acharya
Research Scholar
Department of Vegetable Science
Bidhan Chandra Krishi Viswavidyalaya, Mohanpur, Nadia-741252, West Bengal

B. Lalramhlimi
Research Scholar
Department of Vegetable Science
Bidhan Chandra Krishi Vishwavidyalaya, Mohanpur, Nadia- 741252, West Bengal

Subhramalya Dutta
Research Scholar
Department of Vegetable Science
Bidhan Chandra Krishi Vishwavidyalaya, Mohanpur, Nadia- 741252, West Bengal

Ivi Chakraborty
Professor, Department of Post Harvest Technology, Bidhan Chandra Krishi Vishwavidyalaya, Mohanpur, Nadia

Rajib Nath
Professor, Department of Agronomy, Bidhan Chandra Krishi Vishwavidyalaya, Mohanpur, Nadia, West Bengal, India

Bishal Mukherjee
Department of Agronomy
Bidhan Chandra Krishi Vishwavidyalaya, Mohanpur, Nadia, West Bengal, India

Subhrajyoti Chatterjee
Research Scholar
Department of Vegetable Science
Bidhan Chandra Krishi Vishwavidyalaya, Mohanpur, Nadia- 741252, West Bengal

Debmala Mukherjee
Master Degree Student
Department of Vegetable Science
Bidhan Chandra Krishi Vishwavidyalaya, Mohanpur, Nadia- 741252, West Bengal

Partha Choudhuri
Assosiate Professor
Department of Vegetable Science
Bidhan Chandra Krishi Vishwavidyalaya, Mohanpur, Nadia- 741252, West Bengal

Anirban Roy
Research Scholar
Department of Genetics and Plant Breeding, BCKV, Mohanpur

Paramita Roy
Master Degree Student
Department of Vegetable Science
BCKV, Mohanpur

Tanmoy Sarkar
Research Scholar
Department of Fruit science
BCKV, Mohanpur

K. Shunmugapriya
Department of Food Science and Nutrition
Home Science College and Research Institute, Madurai

P. Vennila
Professor
Post Harvest Technology Centre
Tamil Nadu Agricultural University, Coimbatore

Venkata Satish Kuchi
Department of Post Harvest Technology, College of Horticulture
Dr. Y.S.R. Horticultural University, Anantharajupeta, Kadapa District,
Andhra Pradesh-516105, India

Ch. Sai Ratna Sharavani
Department of Floriculture and Landscaping
College of Horticulture
Dr. Y.S.R. Horticultural University, Anantharajupeta, Kadapa District,
Andhra Pradesh-516105, India

Contents

Chapter 1

Post Harvest Management, Processing and Value Addition of Rice

Mrityunjay Ghosh

Professor, Department of Agronomy, Bidhan Chandra Krishi Vishwavidyalaya, Mohanpur, Nadia, West Bengal, India

ABSTRACT

Rice is a major staple food of more than half of global population. It is consumed in different forms like table rice, parched rice (mudi), puffed rice (khai), flattened rice (chuda), etc. along with several preparations. The mature paddy crop in the field undergoes several operations in sequence (harvesting, threshing, drying, parboiling, milling, value-adding processing, etc.) before human consumption. Although manual harvesting is a common practice in Asian and African countries, while combine harvester is mostly used in developed nations. Paddy grain needs to be dried to 12-14% moisture either by sun or mechanical drying for safe storage and milling. Paddy grains are stored at three levels: rural storage at producers level, medium storage at traders or millers level, and bulk storage at government level. The basic principle of parboiling includes soaking of paddy, steaming of soaked paddy and drying of steamed paddy. Milling of paddy has been developed from indigenous method to modern rice mill, which improves the milling quality and saves both, time and labour. The byproducts of rice mill are husk, bran and broken rice, where husk is used as boiler fuel and bran for extraction of oil. Cooking of rice varies in types, methods and time. Rice kernel expands length-wise as well as in volume during cooking, and aromatic rice emits pleasant flavour.

1. Introduction

Rice is a major cereal crop of India and most of the Asian countries. Post harvest management is an important arena of paddy cultivation, which directly influences the quality of rice. Processed and value added products of rice are also getting enormous importance as well. This chapter deals with various post harvest operations and some processed products of rice.

1.1. Harvesting

Harvesting is the process of collecting the mature paddy crop from the field. Harvesting is usually done manually by hand with a sickle at 4-6 cm above ground level in most of the Asian and African countries in the world. In developed countries, harvesting and threshing operations mostly become mechanical, resulting in saving of labour and minimization of losses. Thus, mechanical harvesting of paddy, using combined harvester has been advocated in recent times in our country.

As a thumb rule, harvesting of rice is done at about one month after 50% flowering of the crop. The pre-harvest condition and optimum time of harvesting can be maintained by the following ways:

i) The water in the field should be drained out at 10-15 days before the expected time of harvest. It hastens the time to maturity and improves harvesting condition.

ii) When 80-85% panicles in the field and 80-90% grains of each panicle turn yellow or specific varietal colour, and the grains have 20-22% moisture content, these conditions indicate the optimum time of harvesting of paddy.

iii) If some paddy grains are dehulled by hand at the field and the rice husks are clean and firm, the maturity of the crop can be confirmed.

Sometimes, the farmers place the dehulled rice kernel between teeth, and press it to understand the condition of the maturity of paddy.

Harvesting of paddy at proper time ensures the maximum quantity of paddy and better quality rice. Too early harvesting results in more green or immature grains, when dried, they form shrivelled rice of lower weight, which are prone to breakage during milling operation. But harvesting too late may result in loss of grains owing to shattering followed by eaten by birds and rodents. Over-maturity and over-drying in the field lead to development of sun-cracks in grains, greater breakage during milling, and less head rice recovery.

The harvested plants are kept in field for 3-4 days to dry under the sun. The simplest and most common method is to lay the cut plants in swaths or loose bundles in the field. Then they are tied in bundles manually and transported to the threshing floor or yard of the farmers either by farm labour or bullock-cart or tractor-driven carrier.

In some places, the partially dried plants are gathered after a few days and arranged in layers or heaps for a second drying. The farmers try to protect the grains from sun cracking by arranging the cut plants in rows with the panicles covered by the straw of another row. The cut plants are often piled in large stacks, and as a result suffer from lack of air movement, which leads to sprouting, discolouration and microbial damage.

Combine harvester

The modern combine harvester is a versatile machine-combining three separate operations: harvesting, threshing, and winnowing into a single process. The separated straw, left lying on the field, comprises the stems and any remaining leaves of the crop with limited nutrients left in it: the straw is then either chopped, spread on the field and ploughed back in or baled for bedding and limited-feed for livestock. Combine harvester is important labour saving equipment in paddy cultivation.

1.2. Threshing

Threshing is a process of detachment of paddy grains from the panicle, which involves either one or two actions of rubbing or impact or stripping. It is mostly done manually by hand beating or mechanically in recent times. Manual threshing methods include:

a) Foot threshing or trampling by bullock: When paddy is threshed by trampling with men or animals, it involves rubbing action, but it is not an efficient method of threshing.

b) Beating against solid object: Threshing is commonly done by beating small, hand-held bundles of paddy stalks against a solid object, such as stone, log, ladder or slated bamboo platform, where impact method is followed. About 4-6 blows are required to remove all the grains from the sheaves or stalks.

c) Pedal thresher or treadle thresher: Threshing is done by holding paddy bundles against a rotating drum, having forked spikes or rasp bars, where impulse stripping action occurs ordinarily with impact method.

Threshing of paddy is now mechanized by the use of small stationary machine thresher or large tractor mounted units. Threshing is done in the floor or at the yard of farming house or on the threshing floor in the farm. The threshing ground available in villages is muddy. If any polythene sheet or other cover is not used, a lot of grains stick to the soil causing difficulty in marketability of grains.

1.3. Cleaning / Winnowing

Threshed grains contain chaff, straw, other crop or weed seeds, earth, stones, etc. The simple cleaning method is winnowing, which uses wind or a fan to remove the light elements from the grain. Mechanical winnowers that incorporate a fan and several super imposed reciprocating sieves or screens are used in many countries. These can be manually powered or motorized and have capacities from 100 kg/hour to 2-3 tonnes/hour. When combined harvesters are used, there is a trend toward using large capacity centralized seed cleaners. There are normally equipped with a series of vibrating sieves and are capable of 10-30 tonnes /hour.

1.4. Drying

Drying of harvested paddy is necessary to prepare the grain for storage and milling, maintain quality and minimize losses. It inhibits biological deterioration and prevents mould growth, fermentation, sprouting and discolouration of the grain. Paddy is required to be dried to 12-14% moisture for safe storage and milling. Most of the paddy crop in India is sundried upto about 16-18% moisture in the field. Newly threshed grain with moisture content above 25% must be dried within a day, 21-24% within 2 days and below 21% within 3 days to maintain quality and minimize grain deterioration. Actually, delay in drying results in qualitative losses of grain.

Drying of paddy is done by two methods

a) **Sun-drying**: Sun-drying is a traditional process of drying, which is practiced by farmers from the beginning of rice cultivation. A major quantity of paddy grown in India is dried under sun till today. Small farmers prefer sun or shade drying with natural air, because they can use family labour and avoid fuel and equipment costs.

Small quantities of threshed wet paddy may be spread on woven mats or left in winnowing baskets for further drying. Large lots are spread on drying floors of different types, which may be made of beaten earth or concrete and even on public road. Drying on beaten earth takes a long time and paddy is likely to become soiled. Concrete floors crack and rough surface causes a certain amount of paddy breakage during handling. Besides, the temperature of concrete floor or metal road may be high, causing additional damage.

Layers of grain vary in depth according to the moisture content, air temperature, quantity to be dried, and rain-hazards. High-moisture paddy depths range within 2-5 cm, and low-moisture paddy depths range from 8-10 cm. As the paddy dries, it is raked or otherwise stirred once in an hour by human labour to expose new surfaces. During rainy periods, the paddy must be gathered up or otherwise sheltered until the weather is again favourable. Continuous drying reduces the drying time but causes higher breakage than when drying is slowed down by tempering or holding in between drying steps. Grains, being hygroscopic, even after drying under sun in the day, absorb moisture in night causing drying-wetting cycle. It results in internal stresses, leading to cracks. Drying under shade requires longer time but improves milling quality.

The effectiveness of sun-drying depends on temperature and relative humidity of atmospheric air, speed of wind and condition of grain. Moreover, losses occur due to scattering, winds, rodents, etc. and the method is unhygienic.

Sun-drying has the following features:

i) The uncontrollable non-uniform drying results in cracks in kernels and lower milling recovery due to more brokens.

ii) Large number of unskilled labours are required with no mechanical energy.

iii) During sun-drying, about 1-2% paddy is lost due to birds, rats, etc.

b) **Mechanical drying:** It is the process of drying grain by ventilating natural air or heated air to accomplish moisture from it. When hot air is forced through the grain, the hot air evaporates the moisture from the grain surface, increases the temperature of grain and carries the evaporated moisture from the grain.

Tempering is a process of storing grain in between drying process. During the tempering period, moisture is equalized within the grain and when it is subjected to further drying, the rate of drying is increased. Besides, tempering decreases the moisture and temperature stresses in the kernels and reduces the possibility of developing cracks during the drying process.

Heated air convection type dryers are commonly used for drying of paddy, which include (i) Bag dryer, (ii) Bin or Batch dryer, (iii) Continuous flow dryer and (iv) Rotary drum dryer (Bhole *et al.*, 1985).

i) **Bag dryer:** This type of drier consists of a large floor area with the opening over which the bags of paddy are placed. The openings are covered with a perforated material to support the bags and allow air movement through the bags. The air is heated upto 45°C and that is blown in the holes by a burner and air blower assembly. The bags should be turned once during drying operation for exposing both sides to heated air. It is most suitable for drying smaller quantity of grains, but it requires large number of unskilled labours for handling the bags.

ii) **Bin or Batch dryer :** It consists of a bin or container to hold grains with an air blower and a burner for providing hot air. The bin or the container is fitted with a perforated bottom, however, godown or room with a perforated floor may also be used. There should be air distribution system throughout the room, where a fan with heating unit is connected to the duct system to supply required hot air. The depth of grain is usually limited upto 1.8 metres due to high power requirement to force air through deeper grain mass.

iii) **Continuous flow dryer :** It is usually of two types: Mixing and Non-mixing. Both types of dryers having high drying capacities are used for commercial drying of paddy. Grains are fed at the top to flow through the dryer by gravity, where rate of discharge is mechanically regulated at the bottom. Drying is accomplished by forcing heated air through the grain, as it flows downwards.

For more efficient and economic operation, continuous flow dryer should be coupled with conveying equipment, tempering bin and godown to store dried paddy. A batch of paddy should not be dried at a stretch by continuously re-circulating it through the dryer. But drying operation should be coupled with tempering of grains that normally varies from 8 to 16 hours or more.

iv) **Rotating drum dryer :** It consists of a large drum having 3-6 metres length and 1-2 metres diameter placed on a slight inclination, where the drum rotates about its axis. The inside of the drum is equipped with the helical straight edge ribs to provide the mixing and the forward movement of the paddy during operation. The hot air blown at the lower end of the drum and the paddy flows counter current to the direction of air flow. It is normally used for drying parboiled paddy.

Advantages

There are some advantages of mechanical drying:

i) The drying operation can be controlled more efficiently and the desirable moisture content can be attained accurately.

ii) Drying can be done in fair or foul weather.

1.5. Ageing

Ageing is a natural and spontaneous phenomenon that modifies processing, cooking, eating and nutritional quality due to changes physical and chemical characteristics, and affects the commercial value of rice. Ageing probably begins in the grain during grain desiccation prior to harvest and is accelerated by heat treatment during drying. The rice grain undergoes changes in properties during the first 3-4 months after harvest, particularly if kept above 15^{0}C, regardless of whether it is stored in rough, brown or milled form.

Ageing or storage results in higher total head rice yields because of lower grain breakage on milling. Water absorption and volume expansion of milled rice during cooking decrease, and soluble starch and protein in cooking gruel decrease. The cooked rice is more flaky with desirable hardness. Aroma in grain is lost to some extent with ageing, while seed viability decreases progressively during storage.

Ageing results in less water-imbibing rate, but the total water uptake upon cooling is generally more in old rice than in fresh rice sample. New rice has a very fragile cell-wall structure, which does not stand well the strain of cooking, hence all cell contents get dispersed into the cooking water, making it pasty. Natural ageing hardens these cell-walls sufficiently to cause a smaller loss of rice solids into the cooking water. Ageing decreases the surface whiteness of milled raw rice, while colour change of parboiled rice during storage for about one year is negligible.

1.6. Storage

Storage is an inevitable step in the normal pathway of grain from harvest to consumption. Storage of food grains insures against fluctuation in production, stabilizes prices, restores confidence among consumers, strengthens public distribution system and replenishes the depleted stocks.

Farmers need small storage facilities to hold their produce or food grains or seeds, while co-operatives or traders need medium, sized storage structures for storing rice for a certain period before disposal for sale. The government of a country needs warehouses either for public distribution system or for buffer stock to supply milled rice to the people being affected by flood or vagaries of nature. Thus, paddy grains are stored at three different levels, *viz.* rural storage at the producers' level, medium storage at traders or millers level and bulk storage at the government level. The bulk storage usually requires modern structures like warehouses and facilities for periodic inspection, processing, treatment of grains, etc.

Rice storage facilities vary in many forms depending on the quantity of grains to be stored, the purpose of storage and the location of the store. The requirements for a good storage system include: (i) prevention of reentry of moisture in the grain after drying, (ii) protection from insects, rodents and bird, and (iii) ease of cleaning, loading, inspection, disinfection and unloading (Chandrashekharan, *et al.* 2008).

Generally, paddy is stored in two ways

i) **Storage in bags or containers**: It requires much labour but with minimum investment or permanent structures and equipments. It is a short-time storage, which may not require control measures against stored grain insects.

ii) **Loose storage**: It is usually done for long-bin storage condition. It requires permanent structures and greater care. The apparent advantage of storing grains in bulk are greater storage capacity per unit volume of space, less difficulty in loading and unloading, and no investment in purchasing bags or bins. The lower infestation by insects in bulk storage system is due to conditioning of grains and smaller surface contact with external conditions of atmosphere. Thus, grain in bulk can be kept for a year or more.

Rural storage : Storage of paddy at the village level represents the most vulnerable stage, wherein loss of grains is recognized to be substantial. In rural areas, storage structures have been developed on the basis of traditional practices and experience. Locally available materials, such as, wood, bamboo, straw and clay are used for their construction, which are not fit for long-term storage of food grains. The storage structures vary in type, size and shape, which are spherical, cylindrical or rectangular.

Most of the traditional storage structures are not completely moisture-proof or air-tight. The platform on which the structure is built, is not high enough, and rodents find easy access into the structure. So, the grains are usually damaged by insects, fungi and rodents. These structures are also not suitable for fumigation. Moreover, their life is short and frequent maintenance is required.

The traditional storage structures are *bukhari,* mud bin *(kuthla),* straw bin, *kothar* type structure, etc. (Bhole *et al.,* 1985).

Paddy grains are stored in loose form in these structures

a) *Bukhari*: A *bukhari* is a cylindrical structure and is made of mud and split bamboos. It is placed on a wooden or a masonry platform to prevent the cylinder from coming in contact with the ground. The capacity of the structure ranges from 3 to 10 tonnes.

b) *Mudbin (Kuthla)*: The mudbins are made of unburnt clay mixed with straw, having 2.5-8 cm thick wall, and they are oval or rectangular in shape. A small hole is kept at the base for taking out the grains, and a bigger hole is kept at the top for filling it with grain. Both the holes are plugged, when grain is stored.

c) *Straw bin*: Dried plants are used for making straw bins, which after being filled with grains are further reinforced from outside by winding paddy straw ropes around the whole structure. The capacity of straw bin is 2-6 quintals.

d) *Kothar* type structure: It is very much similar to a timber box placed on raised platform, which is generally supported on pillars. Both the floors and the walls are made of wooden planks, and the tiled or thatched roof is placed over it as a protection against sun and rain. The capacity of the structure ranges between 9 and 35 tonnes.

Large-scale storage: The modern structures used for large-scale storage of foodgrains include bag-storage structure and bulk-storage structure. Paddy can be stored safely for longer periods, if following three conditions are met:

i) The moisture content of grain is usually maintained at 14% or less, while 12% or less for seeds. The longer the grain needs to be stored, the lower the moisture content is to be maintained. So, the optimum moisture content of paddy grain for different storage periods are: 14-18% (2-3 weeks), 13% or less (8-12 months) and 9% or less (more than 1 year), (Chandrashekaran *et al.*, 2008).

ii) Paddy grains are protected from insects, rodents, and birds.

iii) Paddy grains are protected from re-wetting by rain or imbibing moisture from the surrounding air.

1.7. Parboiling

Parboiling of paddy was developed in ancient times, which probably originated with ageing as its primary purpose. More than half of rice consumed in India, Bangladesh, Sri Lanka and Nepal is parboiled. It is estimated that about one-fourth of the milled rice is parboiled.

Parboiling of paddy is carried out in three steps: (i) soaking of paddy, (ii) steaming of soaked paddy and (iii) drying of steamed paddy (Pandey, 1998).

i) Soaking of paddy is done at or below gelatinization temperature. The lower the temperature, the slower is the process of soaking and *vice versa*.

ii) The steaming of pre-soaked grain is done to gelatinize the starch with minimum volume expansion. Higher steam pressure and longer steaming time make paddy harder and darker in colour. Keeping steamed paddy in a heap is equivalent to prolonged steaming, which induces same effect.

iii) Parboiled paddy may be dried in shade or in the sun or with hot air. Shade drying takes longer time but gives greater milling recovery. Rapid drying in the sun or with hot air causes higher breakage during milling.

1.7.1. Method of parboiling

The method of parboiling may primarily be classified as:

A. Conventional method

a) ***Single boiling***: The traditional process consists of soaking of paddy in cold water for 2-3 days in large cement tanks, followed by steaming of soaked paddy for 5-10 minutes and drying in the sun. During prolonged soaking, fermentation sets in and an off flavour develops in grain.

b) ***Double boiling***: In this method, steam is first injected into raw paddy in the steaming kettle before soaking. Hot paddy raises the temperature of soaking water to 45-50°C, which helps to reduce the soaking time to 24 hours or a day. Then the soaked paddy is steamed. Sometimes, the soaking water itself is heated to about 50°C before the raw paddy is dumped in it and hence, first steaming is not required.

B. Modern method

a) **CFTRI method**: The hot soaking process for parboiling of paddy was developed at Central Food Technological Research Institute (CFTRI), Mysore. It is done by soaking of paddy in water at 65-70°C for 3-4 hours. Water is drained out and soaked paddy in the same vessel is steamed for 5-10 minutes, and thereafter, the paddy is dried under the sun or in mechanical drives.

b) **Jadavpur University method**: In this method, soaking of paddy is done in high-temperature water (60-70°C) for 1-3 hours, while steaming requires only 3-5 minutes. After steaming, paddy is rapidly cooled before drying. Drying takes place in a rotary steam-jacketed high-temperature dryer.

1.7.2. Effect of parboiling

Parboiling is nothing more than rice pre-cooked with the hulls intact. Parboiling of paddy is a hydro-thermal treatment and it affects the physical, chemical and organoleptic properties of rice kernel:

i) The primary change is the gelatinization of starch along with somewhat discolouration resulting in harden and more translucent grain. Therefore, the parboiled rice is easier to dehull and more resistant to breakage during milling.

ii) Parboiling results in partial heat decomposition of B-complex vitamins and facilitates their diffusion into the endosperm. Thus, parboiled milled rice is richer in B vitamins than raw milled rice at same degree of milling.

iii) During parboiling, oil globules in the aleurone and germ layers are disrupted and oil migrates outward.

iv) The protein bodies are disrupted and the solubility of the protein greatly decreases.

v) Parboiled milled rice absorbs water more slowly and is more resistant to disintegration and dissolution during cooking than raw grain. The cooked rice becomes firmer.

1.7.3. Advantages of parboiling

i) The milling recovery of parboiled paddy is greater than raw rice.

ii) The milled rice becomes translucent and shiny.

iii) The parboiled milled rice remain firmer during cooking and the cooked rice is flaky in nature and more digestible.

iv) The bran obtained from parboiled rice gives higher oil recovery.

1.7.4. Disadvantages of parboiling

i) Parboiled rice requires more time during cooking than raw rice and it may have characteristic off flavour, which may not be liked by raw rice eaten people.

ii) The heat treatment during parboiling destroys some natural antioxidants in rice and reduces aroma to some extent.

1.8. Milling

1.8.1. Principle

The utilization of rice as food requires milling of rough rice or paddy to remove husk, bran and embryo from the grain. Milling quality is the measure of rough rice (paddy) performance during milling process. It is influenced by three factors: (i) characteristics of rice lot, genetic and environmental factors; (ii) type of milling unit used; and (iii) processing conditions including the extent of milling (Barber and Barber, 1979).

The milling process generally consists of five fundamental steps (*Dela* Cruz and Khush, 2000):

Step I : Cleaning of rough rice to remove impurities

Step II : Shelling or dehulling of rough rice with removal of husk

Step III : Separation of brown rice from paddy

Step IV : Milling and polishing of brown rice with aspiration of bran

Step V: Grading or separation of whole grains from brokens and germs

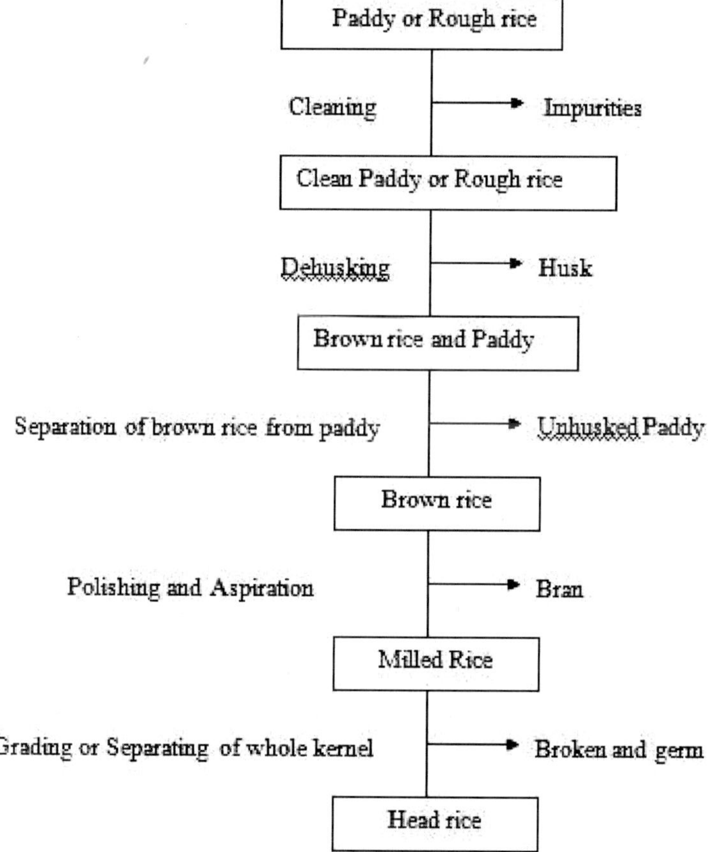

Fig. 1: Rice Milling process

1.8.2. Method

Paddy received at rice mills usually contain impurities like stones, dust, etc. which need to be removed before milling. Besides, the moisture content of paddy grain is determined to get optimum milled rice recovery. Moisture content (MC) in paddy grain is expressed as a percentage of total weight.

$$MCw.b. = \frac{Wm}{Wt} \times 100$$

where, MCw.b. = Moisture content (%) on wet basis

Wm = Weight of moisture in grain

Wt = Total weight of grain

The clean rough rice is passed through a shelling device to remove the hulls from the kernels, brown rice is obtained. Generally, the hulls form 20% of rough rice, but the variation of 18-25% is observed.

$$Hulling\,(\%) = \frac{Weight\ of\ brown\,rice}{Weight\ of\ rough\,rice} \times 100$$

The broken rice is milled to remove the bran and embryo, and then polished to obtain milled rice. Bran and embryo constitute about 8-10% of rough rice.

$$Milling\,(\%) = \frac{Weight\ of\ milled\ rice}{Weight\ of\ rough\,rice} \times 100$$

The milled rice is separated from brokens and embryo to get head rice. Although the average head rice recovery is 65 -70%, but it may be as low as 50% due to greater proportion of brokens.

$$Head\ rice\ recovery = \frac{Weight\ of\ head\ rice}{Weight\ of\ rough\,rice} \times 100$$

Head rice recovery is dependent upon grain size, shape and appearance. Varieties having long and chalky kernel usually record low head rice yield. Besides, temperature and humidity during ripening and post-harvest operations influence the the grain breakage during milling. The optimum moisture range for processing is 12-14%. If the moisture is too low(<12%), the grain become brittle and breakage is higher; while too high moisture (>16%) also reduces the head rice yield.

Milling has a profound effect on rice composition, more specifically the nutritional value. The bran contains more of essential nutrients than the endosperm. Major proportion of fat, fibre, ash, B-vitamines, iron, zinc and aromatic volatiles of rice are removed with bran during milling process. Calories, which come mostly from starch, are largely retained and protein is fairly well-retained in milled rice. Thus, brown rice is more nutritious than the usually consumed milled rice.

1.8.3. Rice Mill

a) **Engelberg type huller (sheller-*cum*-polisher):** The Engelberg type huller is the simplest and cheapest among paddy milling machines, which follows old milling process. It is basically a sheller-*cum*-polisher. The huller consists of a solid heavy fluted cylinder, which rotates at 500-600 rpm inside a hollow stationary cylinder with uniform perforations. The dehusking and polishing are affected by friction. Skill and close attention are required for efficient operation to prevent excessive breakage.

b) **Huller mill:** It removes the hull and bran in one operation in a solid fluted cylinder rotating within a hollow cylinder. The hulling process is affected by the shearing action of rotating solid cylinder ribs against a stationary blade, whose clearance can be varied according to the variety of the rice being milled. The mixture of crushed hull, bran and broken drop through the perforated lower half of the hollow cylinder. The milled rice is discharged at the outlet end. The average milling recovery is 60%.

c) **Disc sheller mill or '*Cono*' type mill:** In this type of mill, hulling and scouring are accomplished in separate operations. The hulling process is done in twin horizontal disks coated with emery stone. The upper disk remains stationary, while the lower disk rotates. Centrifugal force and abrasive action causes the shearing of hull. Scouring is done in a frustum of a cone coated with emery stone rotating inside a screen mesh of the same shape provided with three rubber brakes. The rubber brakes choke the movement of the grain and due to induced attrition, the bran is removed and expelled through the screen mesh while the pearled rice is discharged at the bottom outlet. Mean milling recovery in disc sheller mill is 65%.

d) **Rubber roll sheller mill:** The mill has separate units for hulling and scouring. The hulling process is an application of the principles of granular dynamics, which is achieved through the frictional action of two rubber rolls that differ in their circumferential speeds. Fractional force between the rubber rolls separate the hull from the grain and produces brown rice. Scouring and pearling is then accomplished in units similar to that of 'cono' type mill. The rubber roll mill may give milling recovery of 70%.

Modern rice mills are basically rubber-roll sheller mills, which are equipped with cleaner or destoner, rubber roll shellers, efficient paddy separator, bran separator, polisher, colour sorter and automated packaging system. Apart from modern milling equipments, there are also facilities for bulk storage of paddy with mechanical handling, drying and parboiling units, and sometimes, oil extraction units from rice bran. Such modern-mill complex, properly managed, can not only affect the maximum rice recovery, but also helps to utilize efficiently the by-products of milling.

1.8.4. By-products

There is a common perception in the farming community that each part of the rice plant is useful. The by-products of rice include paddy straw after threshing; while husk, bran and broken rice at rice milling industry.

a) **Paddy straw:** Grains are separated from paddy straw bundles during threshing. The ratio of grain and straw of paddy usually varies from 1:3 to 2:3, depending on types and varieties. Usually, the local or indigenous rice produce more straw than grain, while high-yielding varieties have more or less equal grain : straw ratio and hybrid varieties produce more grains than straw by weight.

Paddy straw have different end-uses from the time immemorial. The pulverized paddy straw is used for preparation of cattle feed after adding molasses and urea. Long and straight erect straw are used by the poor villagers for thatching the roof top. It has little uses in mushroom production system.

b) **Husk :** Husk or hull is the outer covering of rice, botanically known as lemma and palea. It is separated from the brown rice during dehulling operation in milling process. Husk usually constitutes about 20-22% of rough rice or paddy.

The husk is mostly used as boiler fuel, wherever parboiling is practiced. In some places, where steam engines run the mill, the husk supplies the fuel for the boiler and, if adequately managed, the mill can be independent of external sources of electric power. It has limited uses as insulating materials in walls of buildings and cold storage facilities. Husk, as a potent source of silica, is used for manufacturing of silicates. Certain portion of husk is used as cattle or poultry feed.

c) **Bran :** Bran is a brownish-red thin covering of rice between husk and kernel, which constitutes about 8-10% of rough rice. Pure rice husk is a by-product of modern rice mill, while husk and bran mix together in huller mills. Bran contains about 15-20% oil, which is presently a supplementary source of edible oil in India. Rice bran oil contains *'oryzanol'*, which forms the health related issues, compared to other vegetable oils. In addition, rice bran has limited use as animal feed.

d) **Broken rice:** Although the nutritional value of broken rice is similar to head rice, but broken has low market value due to its non-uniform size and shape and less acceptable preference of the rice eating people. The proportion of brokens to rough rice varies from 3 to 15% depending on moisture content of paddy, type of milling unit used and degree of milling. Although broken rice is not suitable for use as table rice, but very poor people use brokens as their rice even today. Clean brokens are further processed to produce rice flour for use to prepare *pitha* (home made cake) in eastern and northern India, and *idli, dosha* in South India.

1.9. Cooking

1.9.1. Washing

Milled rice is washed usually three times or more before it is cooked to remove dust, insect, husk, and other foreign materials, and this process has a direct effect on its nutrient content. The extract of loss of solids in water during washing depends on the damage to the structure of rice during polishing process. The washing loss is less in "crystal" parboiled rice than ordinary parboiled or raw rice. On an average, 50% of total solids in washing water is nitrogen-free extract, carbohydrate material. The extent of washing losses of N- free extract, protein, calcium, iron and vitamins is 1-2%, 5-7%, 18-26%, 18-47% and 7-44%, respectively depending on the degree of bran removed during milling.

1.9.10. Cooking

Cooking of rice varies in types, methods and time. The proportion of raw or parboiled rice and water is important to prepare the desired form of cooked rice. Normally, rice is cooked by boiling with water just enough to be taken up by the rice to large excess volume. Where, the excess water on completion of cooking is discarded. The solid loss in excess cooking water is more, when the rice is cooked in a vessel directly over fire. The ebullition of the boiling water makes the grains hit against each other, resulting in leaching out of starch into cooking water and fermentation of a thick sticky gruel. Less or little water results in cooked rice with hard and brittle in texture, whereas much water gives sticky and pasty form along with excess water. Well-cooked rice is fluffy, tender and firm (easy to separate). Pressure cooking is a better method than the conventional cooking, as it retains maximum nutrients. Besides the types of rice or varieties, purpose of cooking and nature of end product are also determinant factors for cooking.

Newly harvested rice, when cooked, becomes a sticky or partly mass, swells slightly and losses a fair amount of solids into cooking water, yielding a thick gruel. The rice, when cooked after storage for a few months, swells more easily, the cooked rice becomes more flaky, the grains remain integral and the gruel becomes thin.

1.10. Grain elongation

Linear elongation of rice kernel on cooking with minimum increase in girth is considered as a highly desirable trait in some high-quality rices. It mainly depends on gelatinization temperature and localization of starchy cells in the endosperm. Anatomical studies show that nearly penta or hexagonal cells are arranged in honey-comb fashion in elongation types of rice.

The elongation ratio of cooked rice is determined by the following formula:

$$Elongation\ ratio = \frac{Length\ of\ cooked\ rice\ kernel\ (mm)}{Length\ of\ raw\ rice\ kernel\ (mm)}$$

1.11. Aroma

Aroma is a special quality of rice that is emitted during cocking and eating of rice-based preparations. Scented rice is preferred by the people all over the world, which draws premium price in domestic as well as international market. There more than 100 different chemical constituents, which are responsible for aroma in rice. Among them, 2-acetyl-1-pyrroline (2-AP) with a threshold of 0.1 ppm in water is the major aroma principle in cooked rice (Buttery *et al.*, 1983).

2. Processed and Value-added Products

2.1. Processed products

About 4-5% of total rice production in India is converted into diverse rice products like parched rice, puffed rice and flattened rice:

a) **Parched rice (*mudi*):** Parboiled rice is soaked in salt water till its moisture content increases to about 20%. Parched rice is prepared by throwing rice in sand heated to a high temperature in iron or earthen pan. On stirring, rice begins to crackle and swell. Then the contents of the pan are removed and sieved to separate the parched rice from sand. Parched rice is a crisp product with a greyish to white colour and is sold either salted or unsalted. It is eaten as such or mixed with butter milk and then consumed.

b) **Puffed rice (*khai*):** Paddy is filled in earthen jars and is moistened with hot water. After 2-3 minutes, the water is decanted and the jars are kept in an inverted position for 8-10 hours. The paddy is exposed to sun for a short time and then parched at 190-210°C for 40-45 sec. During parching, the grains swell and burst into soft white product. The parched grains are sieved to remove sand and winnowed to separate the husk.

c) **Flattened rice (*chuda*):** Parboiled rice is soaked in water for 2-3 days to soften the kernel, followed by boiling in water for a few minutes and the water is drained off. The paddy is heated in a shallow earthen vessel or iron pan till the husks break open. It is pounded by an iron pestle or using iron roller, which flattens the kernel and removes the husk. The husk is separated by winnowing. Flaked rice is thin, papery and white in colour. Now-a-days, flattened rice is made in commercial flaking machines (edge runner type) and is thinner than that is traditionally prepared in households. *Chuda* is commonly used as breakfast food in rural India and is greatly used as life saving eatable item at the time of flood, etc.

2.2. Fermented products

Several breakfast food of rice are prepared after fermentation, which have diverse uses:

a) ***Idli***: It is a popular and easily digestible food in South India. *Idli* is a parboiled rice and black gram *dal* based fermented and steamed food. Parboiled rice and dehusked whole black gram are soaked separately in the ratio of 3:1. They are ground separately, where blackgram into a fine paste and rice to slightly coarsed condition. The ingredients are mixed together, salt is added. The mixture is to be allowed to ferment overnight, usually for 16 hours at 30^0 C. *Idlis* are made by steaming in perforated containers.

Parboiled rice having relatively hard texture than raw rice is more suitable for making soft and spongy *idlis*. The speciality of black gram in *idli* preparation is owing to its muciliaginous material content, which helps in retention of CO_2 evolved during fermentation. The common bacterium recognized in *idli* fermentation are*Lactobacillus delbrueckii, Streptococcus lactis*, etc. With the progress of fermentation, there is an increase in better volume, acidity and non-protein nitrogen. Fermentation increases B- vitamin content.

b) ***Dosa***: It is another important food for South Indian people. Parboiled rice and black gram are soaked together in the proportion of 4:1 to 6:1. Ingredients are ground fine and salt is added. It is allowed to ferment overnight and then it is fried in a thin batter on a pan. *Dosa* usually have soft and spongy texture and low bulk density.

3. References

1. Barber, S. and Barber , C.B.D. (1979). Outlook for rice milling quality evaluation systems. *In Proceedings of the workshop on Chemicl aspects of rice grain quality*. International Rice Research Institute Malina, Philippins. pp. 209-221.

2. Bhole, N.G.; Ojha, T.P. and Padya, A.C. (1985). Harvesting and drying machinery and storage structures. In: *Rice Research in India*, Indian Council of Agricultural Research, New Delhi, India.

3. Buttery, R.G.; Juliano, B.O. and Turubaugh, J. (1983). Cooked rice aroma and 2-acetyl-1-pyrroline. *Journal of Agriculture and Food chemistry*. 31: 823-826.

4. Chandrashekharan, B.; Annadurai, K. and Kavimani, R. (2008). Rice science, Scientific publishers (India), Jodhpur, Rajasthan, India.

5. DelaCruz, N. and Khush, G.S. (2000). Rice grain quality evaluation procedures. In: R.K. Singh, U.S. Singh and G.S. Khush (Eds.), *Aromatic Rices* , Oxford & FBH Publishing Co. Pvt. Ltd., New Delhi, India.

6. Pandey, P.H. (1998). Principles and practices of post harvest technology, Kalyani publishers, Ludhiana, Punjab, India.

❑❑❑

Chapter 2

Waxy Cereals

C. Lalmuanpuia*, Baljit Singh and Savita Sharma

Dept of Food Science and Technology, Punjab Agricultural University, Ludhiana, Punjab-141004, *e-mail: tetea.chhakchhuak@gmail.com

ABSTRACT

Mutant genotypes of certain cereals that lacks or contain less than 3 per cent of amylose are called waxy cereals. They have opaque grains and becomes sticky, when cooked. This property is due to the nature of starch which lacks amylose. They stain reddish-brown colour with iodine while normal starch stains blue colour with iodine. There is recent rise in interest towards waxy cereal starches due to their reported high digestibility, functional properties like high solubility, swelling, gelatinization and low retrogradation. This chapter explores their utilization to improve the quality of certain food products like better expansion in extruded snacks, anti-staling agent in bread for shelf life stability, for production of specialty foods like udon noodles and its potential utilization as a freeze thaw stable ingredient in foods.

1. Introduction

Cereal grains store energy in the form of starch. Starches are the major constituents of the cereal grains and generally consisting of 60-70% by weight [1][2]. Among cereal grains, rice contains the highest amount of starch followed by wheat, maize and sorghum. The functionality of cereal flour or meal in foods, therefore, significantly depends on the properties of starch.

In cereals, synthesis of starch occurs in specialized organelles known as amyloplasts. In the amyloplasts, amylose synthesis is carried out by the granule-bound starch synthase (GBSS), also called the 'waxy' protein [3]. Mutation occurs, which alters the properties of starch with high levels of amylopectin giving rise to waxy cereals. Waxy cereals from maize [4], barley [5] and other plant species are reported to lack both amylose and GBSS [3].

The term 'waxy' was first used to describe amylose-free mutants of maize because the endosperm of dried kernels appeared smooth and opaque, when cut and resembled texture of a hard wax unlike the translucent appearance of regular (non-waxy) kernels [6]. The term "waxy" became more often used in the literature on maize and the term "glutinous" more often used to describe high amylopectin varieties of rice, millet, and sorghum. Even though the term "glutinous" and "waxy" are not accurately satisfactory in a chemical sense, the latter term has been used in this chapter for lack of a better one. This chapter will focus on the recent advancements in the studies of waxy cereals, their properties and impact of the ratio of amylose/ amylopectin on the end-use product qualities.

2. Origin and distribution

2.1 Waxy rice

The glutinous nature of rice arose due to a mutation in an intron 1 splice donor site of the *Waxy* gene. The glutinous phenotype is not detectable in the archaeological records, the region of its earliest cultivation has not been documented. Therefore, its evolutionary and geographical origins have remained unclear. In a study, conducted by Olsen and Purugganan (2002) to examine the allele genealogy of the *Waxy* locus in order to trace the evolutionary and geographical origins of this phenotype. 105 glutinous and non-glutinous varieties were studied from across Asia, and they found an evidence that the splice donor mutation probably arose in South-east Asia. Glutinous rice have a very long cultural importance across a wide geographical area in East Asia [7].

2.2 Waxy maize

The first reports about the occurrence and cultivation of waxy maize (Zea mays L. var. *certainaKulesh*) in China come from 1909 and was described by Collins as a new type of corn found in China. He stated: "The appearance and physical composition of the seeds of this Chinese corn were so distinct from that of other varieties that the possibility of a difference in chemical composition naturally suggested itself, but analyses did not yield any very unusual results" [8]. However, it was later found that the changes in functional properties were not due to difference in chemical composition but rather due to changes in the ratio of amylose and amylopectin.

According to study on karyotype, isozyme and molecular markers, the geographical origin of waxy maize was considered to be from South western China, where waxy maize is consumed as an important staple crop [9]. Amylose-free (waxy) maize has been a staple food in Eastern Asia (covering regions of Cambodia, Laos, Myanmar, Thailand, Vietnam, etc.).

The name 'waxy' is derived from the waxy appearance of the endosperm when exposed in a cleanly cut cross-section. These waxy corn starch stains a reddish brown with 2 per cent potassium iodide solution, whereas normal corn starch stains blue colour. This staining reaction is also expressed in the pollen of the waxy gene, which is an aid in breeding.

Waxy maize starch(WMS) are commonly consumed as food in Asia. It has many distinguishing properties which makes it a valuable raw material that can be used in food industries, paper-making, textiles and feedstuffs[10].

The amylose-free starch show low pasting temperature, setback, and final viscosity but high peak viscosity, which means it can easily swell, gives sticky texture and barely retrograde [11]. In addition, WMS is more easily digestible than normal maize starch. They are a suitable raw material for food processing, especially in the production of puffed foods or baked foods.

However, there are some limitations from gels or pastes of WMS in native form in terms of being easy to break down either from prolonged heating, high shear or acidic conditions, producing weak-bodied, cohesive, rubber pastes and undesirable gels [12]. To resolve these problems, WMS and hydrocolloids are often used together in food systems to modify the rheological and textural properties of food products [13][14].

2.3 Waxy wheat

Waxy wheat (Triticumaestivum L.)was developed in Japan in 1995 via classical breeding and genetics by Nakamura *et al.* (1995) [15]. Waxy wheat having low amylose content was developed by eliminating all the three isoforms of the GBSS, which are responsible for the biosynthesis of amylose [3] [16].

Two valuable co-products can be obtained from waxy wheat flour, which includes vital wheat gluten and waxy wheat starch [17].There have been many suggestions for uses of waxy wheat including a source of blending flour to improve processing quality, palatability and shelf-life stability, [18] [19] [20] or as an alternate for waxy maize to produce modified starches (Reddy and Seib, 2000). There have been breeding programs for waxy wheat in a number of countries [21].

The uniqueness of properties and use ability of waxy wheats in noodles, bread, and tortillas had been broadly reviewed [22][23]. Due to its swelling and water-holding capacity, the addition of waxy wheat flour in bread can provide increased softness. Its greater retrogradation resistance is also known to provide a longer shelf life in bread. Therefore, the potential anti-staling effects of waxy wheat flour could be of interest for bakeries. In bakery products, waxy wheat flour may also be used to lower the fat content and to replace some additives commonly used in bakery [24]. WWS can be converted to maltodextrins or modified to produce thickeners with low pasting temperatures and a low rate of retrogradation [23]. Baked products incorporated with waxy wheat have characteristics that resembles extruded foods(light and improved texture). Due to its unique properties and many possible food applications, waxy wheat have started to draw special interest in the field of food engineering.

2.4 Waxy barley

Barley (Hordeum vulgare L.) is one of the oldest crops domesticated in the beginning of the agriculture revolution. Therefore, the barley family has evolved in many different regions and into a number of varieties. They have been categorized into three types with respect to their amylose content: Low or 'waxy' amylose (0-10%), Normal amylose (~25%) or High amylose (>35%).

Barley contains high levels of β-glucans, which is an important contributor to dietary fibre. Waxy barley tends to be linked with higher levels of β-glucan, which has been reported to have a number of health benefits including reduction of blood cholesterol and reducing the risks of colon cancer. Waxy barley starches tend to swell to a much greater extent than normal starches [25]. Waxy barley has high freeze-thaw stability and is useful as a food (or non-food) thickener, due to its ability to hydrate rapidly and form viscous gels [26].

Waxy barley starch appears to be more susceptible to enzyme hydrolysis than normal barley starch [27], which may be beneficial for malt quality. Formation of amylose-lipid complexes can reduce starch solubility and lower extracts, making high amylose types less suitable for malting.Limited evaluation of waxy types for malting has been undertaken.

There is increased interest in barley due to realization of its high soluble fibre, low fat and low Glycemic index. The high fibre content in barley also gives a feeling of fullness and satiety to the consumer, as a consequence, it can be an useful tool in weight control. The soluble fibres present in barley are reported to lower blood cholesterol and reduce risk of heart disease. They are, therefore, extracted as nutritional supplements. Waxy hulless barley are known to contain a high content of beta-glucan with a fast cooking grain and excellent pearling and flaking quality.When heated with water, waxy starch undergoes changes, which are responsible for the unique character of many foods. These includes viscosity and mouth-feel of gravies, texture of some confectionaries, pie fillings and to a lesser extent, bakery products.

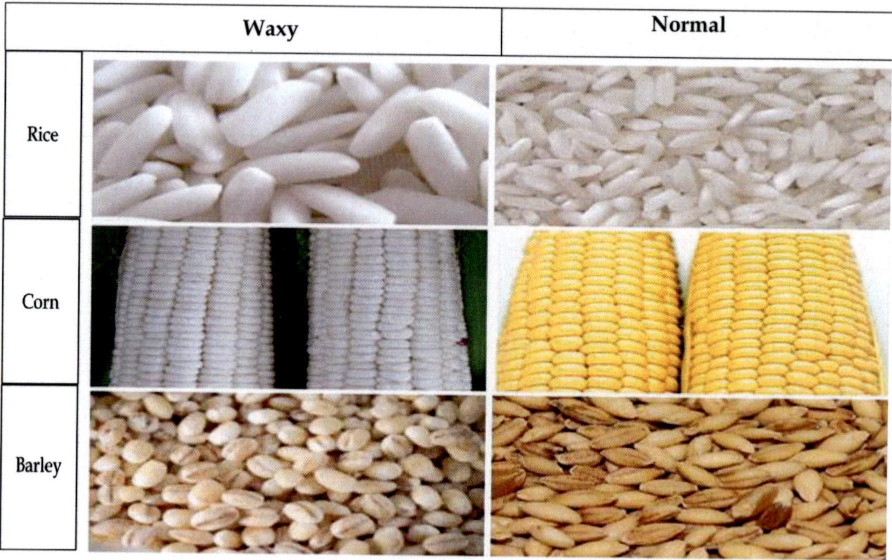

Fig. 1: Cereal grains of waxy and normal varieties.

3. Properties of Waxy Cereals

3.1 Physicochemical properties

Starch molecules have an hemiacetal group at the reducing end. Glucose monomers are linked with either α-1,4 or α-1,6 bonds to this hemiacetal group. These different linkages give different structure to the starch molecule, linear chain of α-1,4 as the amylose or branched chain of α-1,6 as the amylopectin. The linear chain of amylose have a degree of polymerization (DP) ranging from 500–6000 glucose residues. However, there are reports that certain fraction of the amylose molecules as being slightly branched by á-(1,6)-linkages. On the other hand, amylopectin is a very large, highly branched chain molecule with a DP ranging from $3X10^5$ to $3X10^6$ glucose units.The ratio of amylose/amylopectin in starch led to differences in structure of granules, physicochemical properties and product quality. The amylose/amylopectin ratio varies among different cereal starches but typically starch comprises of amylose (25–28%) and amylopectin (72–75%) [28]. Waxy starch are qualitatively determined by staining it with iodine, which gives a reddish-brown colour, and are found in varieties of rice, maize, sorghum, millet, and barley. Gris in 1860, found a "red-staining starch from glutinous rice" [29], The red-staining starch discovered in corn was mistakenly considered to be an erythrodextrin

[30]. However, Tanaka (1912) reported that the iodine colour was not due to erythrodextrin because it was not present in sufficient quantity to account for the colour development [31].

Starch are generally regarded as the most important constituents of cereals in terms of functional attributes of, such as pasting behaviour, gelatinization, retrogradation, swelling capacity, viscosity, solubility, binding and bulking capacities, which influence the product quality [32]. Due to this, various food processing industries employed cereal starches with different characteristics during food formulations.

Mutant genotypes of certain cereals like maize, barley, rice and wheat, etc. containing 100 % amylopectin are called "waxy cereals". Different varieties having high content of amylose, around 70 % known as 'amylotypes' have also been discovered. Kempton, 1919, reported that the waxy character of the endosperm behaves as a recessive character to the normal varieties, the two characters being definitely alternative, and naturally no blending or intermediate stages having been found [33].

The recent rise in interest towards waxy starches is pertained to their resistance to retrogradation. Among the waxy or glutinous cereals, maize had so far offered the best possibilities as a commercial source of waxy starch because of its high yield and economy.

3.2 Digestibility

Digestibility of the cereal starches decreased in the order of waxy > normal > high amylose. The susceptibility of different cereal starches and their modified chemical derivatives to the hydrolysis by amylolytic enzymes, such as α-amylase and glucosidase had been used to study the *in vitro* digestibility. Due to their branched structure amylopectin tends to have more sites for enzymatic action giving greater access of the enzyme to the substrate [34]. Vesterinen *et al.* (2002) also reported that digestibility of cereal starches generally decrease as the amylose content increases [35]. High amylose varieties are more resistant to digestion than waxy varieties. The linear amylose chains yield sugars more slowly than highly branched amylopectin. Thus, the waxy cereals have faster rate of digestion than normal varieties.

3.3 Swelling capacity and solubility

One important functional property of cereal grains that determine its usefulness is its ability to absorb water and swell up upon heating and with further heating, the starch disintegrates and gets solubilized. The swelling of starch granules as a consequence of water absorption causes amylose to leach in intergranular spaces and which further causes loss of crystallinity, increased granular size and solubility [36]. This property is influenced by the ratio of amylose/amylopectin, chain length, degree of polymerization and molecular weight [37] [38]. The swelling capacity and solubility of starch granules is in indirect relationship with its amylose content because amylose restricted its swelling power[39].

3.4 Gelatinization and gelation

Gelatinization behaviour of cereal starches is an important aspect to understand because in most foods gelatinization process is essential in order to give suitable rheological properties, water-holding capacity, mouthfeel, texture and other sensory attributes. The process of gelatinization is often characterized in terms of gelatinization temperature or gelatinization temperature range [40].The gelatinization behaviour of waxy starch is affected by the chain length of branched amylopectin and molar size. Swollen gelatinized starch granules solubilize faster but in normal starch, the amylose contents restricted the granular swelling. Therefore, waxy starches with low amylose to restrict swelling has higher rate of solubilization [41].

The paste viscosity is higher in waxy starches but they are highly susceptible to mechanical breakdown. The waxy starch from cereals formed a weak gel with low hardness and is often liquid like. Gels from waxy starch have good clarity, low retrogradation and high breakdown value [42].

Table 1: Gelatinization temperature for waxy and normal cereal starches

Cereal	**Gelatinization temperature**		**Reference**
	Waxy	**Normal**	
Rice	67.6	75.5	Waterschoot *et al.*, 2014 [43]
Maize	71.9	74.2	Zhou *et al.*, 2015 [44]
Wheat	63.8	65.4	Schirmer *et al.*, 2013 [41]
Barley	64	68	Schirmer *et al.*, 2013 [41]

Gelation can be described as the process by which dispersed starch molecules in an aqueous medium begin to re-associate and form three dimensional network structures. Upon gelation, waxy starches show a higher peak viscosity and a larger breakdown than regular starches, which is due to their higher swelling power. Waxy starches doesn't show hardly any setback while native starches have high setback values, which is due to the formation of an amylose network in normal starch [43].

3.5 Retrogradation

Retrogradation of starch refers to the re-association or re-crystallization of starch, when they are cooled after being cooked or gelatinized. It is the main cause for staling of bread, increasing firmness of crumb, decrease in starch digestibility, loss of crispiness and change in flavour and aroma. Retrogradation of gelatinized starch often leads to separation of liquids from the polymer network due to contraction. This is a process known as syneresis.

For determining the retrogradation, light microscope and X-ray diffraction analysis are normally used. Many studies reported that retrogradation occurs more rapidly with amylose than with amylopectin due to their smaller unbranched molecule that rapidly crystallizes a few minutes after gelatinization is completed [45]. In amylopectin, retrogradation is associated with the ability of the outer branches of the molecules to form double helices. Therefore, retrogradation in amylopectin can be avoided by enzymatic hydrolysis of the outer branches, or by making them unable to form double helix structures by complexing with lipids [46] [47].

In most food products, retrogradation is undesirable, as it brings on decline in quality and therefore, must be prevented. Waxy cereals containing high amylopectin have a much lesser tendency to retrograde; therefore, the use of waxy cereals in foods can decrease the degree of retrogradation.

4. Uses of Waxy Cereals

4.1 Extruded snacks

Starch rich foods like cereals, tubers and pulses have been used by snack food industries specially as functional ingredient to achieve specific textural attributes. Extrusion cooking is commonly used to produce puffed or expanded snacks, in which the desired texture can be obtained by manipulating the amylose/amylopectin ratio. For example, addition of waxy starch is done to increase the expansion of extruded snacks.

However, the limitation of using waxy starch in extrusion is the excessive breakdown of amylopectin molecules by the high-temperature and high-shear processing conditions.To increase the resistance of starch molecules from breaking down under extruder processing conditions, waxy starches are often chemically modified by "cross linking". A cross linked waxy corn starch have improved resistance of amylopectin to breakdown and therefore, have controlled expansion capabilities in the puffed snacks. High-amylose corn starches can be used, when increased crunchiness and strength are required.

4.2 Baked snacks

For snacks that expands by baking, where internal temperature of snacks increase much slower than in extrusion or frying, it is recommended to use pre-gelatinized waxy corn starches. Pre-gelatinized starches are made by pre-cooking in water and then dried; they require no further cooking before baking to contribute to texture development. Pre-gelatinization is an important feature because conditions in baked expanded snacks do not allow enough gelatinization of regular waxy starch. Pre-gelatinized waxy starch allows better expansion of the baked snacks.

4.3 Bread-making quality

Bread is composed of a continuous phase consisting network of cross-linked gluten molecules and leached starch polymer molecules, mainly amylose, which may be complexed or uncomplexed with polar lipid molecules, and a discontinuous phase of gelatinized, swollen, entrapped, deformed wheat starch granules [48]. Therefore, the characteristics of gluten and starch plays important role in determining the textural quality of bread.

Waxy wheat containing high percentage of amylopectin appears to be more susceptible to α-amylase during fermentation with higher gas production and lower gas retention, as compared with the non-waxy wheat flour [20]. Breads made from high amylose wheat flour

(HAWF)usually a much smaller loaf volume, while waxy wheat flour (WWF) produced bread with bigger loaf volume than the regular wheat flour. Lee *et al.* (2001) reported increased in porosity in the crumb structure of bread with respect to increase in the proportion of waxy starch [20]. The quality of bread is also affected by protein quality and other compositions of flours. Therefore, the high percentage of amylopectin is not entirely responsible for the quality of bread.

There is an increasing interest in the use of waxy wheat for bread making inview of extending shelf-life and avoiding the use of additives due to its low susceptibility to retrogradation. The use of waxy wheat would avoid the addition of the additives like enzymes, emulsifiers, etc. which are normally used in bread-making to extend the shelf-life of baked products [49].

The major cause for staling of bread is believed to be due to starch retrogradation [50]. There had been few studies on the possibilities of using waxy starch to change the amylose/amylopectin ratios of flour in order to extend the shelf-life of baked products [51] [52] [53][20] [54]. These studies reported that incorporation of waxy wheat starch in dough resulted in extended shelf-life of breads by retardation of the staling and retaining more moisture in breadcrumbs. Morita *et al.* (2002) conducted a storage study and reported that breadcrumbs made from WWF remain softer than those of HAWF and non-waxy wheat flour after storage for 7 days [55].

Biliaderis (1992) reviewed that the process of retrogradation of starch occurs in two phases,it started with complexing of amylose via helix-helix interactions succeeded by a slower re-crystallization of amylopectin[56]. Therefore, reducing the amylose content, should rule out the rapid and early phase of staling. However, staling may not be entirely due to of starch retrogradation; some other factors like protein quality, inherent water-holding capacity, water mobility, flour lipids and interactions between protein and starch also contribute to retrogradation [57].

Thus, the addition of waxy starch to reduce amylose content in bakery product improves its quality by retardation of retrogradation, elongated shelf-life and formation of a new texture of breads with the soft, viscous and glutinous breadcrumbs [20]. Morita *et al.*, 2002 reported that partial substitution of upto 40% waxy wheat flour can be used to non-waxy wheat in bakery product application [55].

4.4 Noodle making quality

Noodles are made from unleavened dough which is stretched, extruded, or rolled flat and cut into long thin strips or variety of other shapes. They are consumed as staple foods in many Asian countries. There can be various types of noodles depending upon the raw materials used and methods of manufacturing. The type of wheat starches and flours significantly effects on the final quality of noodles.

Waxy and partial waxy wheats are of great value for the production of certain Asian noodles. There are diverse types of Asian noodles and they vary with the type of raw materials and manufacturing methods. One such type of noodles produced in Japan is udon, also known as white salted noodles, generally prepared from medium-protein soft wheats.

For textural properties of udon noodles commonly consumed in Japan and Korea, consumers prefer a soft noodle with a firm surface [58]. The desirable attributes includes high paste viscosity and breakdown, high swelling power of starch and low gelatinization temperature, which are associated with reduced amylose contents [59] [60][61].

A study conducted by Oda *et al.*, 1980, reported strong correlations between amylose content and sensory scores of udon noodles. Waxy cereals with low amylose contents have higher starch swelling properties, high peak viscosity and resistance to retrogradation, which are required for producing superior quality of udon noodles [63] [63][64][65]. On the other hand, low starch paste viscosity is desirable for yellow alkaline noodles also known as Chinese ramen [66] [67].

Hence, waxy and high-amylose wheats with different amylose contents contribute to give rise to various noodle types with desired eating qualities. Waxy wheat flour are often added to normal wheat flour to reduce amylose content. The noodles made from the reduced amylose flours have increased cohesiveness, springiness and resilience, whereas firmness, gumminess and chewiness of the noodles decreased [68].

Yamaguchi *et al.* (2000) also reported that substitution of waxy wheat flour results in improved textural and sensory properties of noodles [69]. Waxy and regular starch were re-constituted with gluten protein to produce Japanese udon noodles. High amylose content of starch positively correlate with hardness and negatively with cohesiveness of cooked noodles. The protein content of flour was found to be negatively correlated with hardness of cooked noodles, indicating that flour with high protein content produce a good quality of Japanese udon noodle[70] [71]. It was reported that the optimal range for amylose content of flour to produce superior udon noodles was around 21-24%[68] [70].

4.5 Waxy Starch as a Freeze-Thaw Stable Ingredient

Freeze-thaw stability is the most important property of starch used in foods that that usually undergo freezing and thawing. When a gelatinized starch or a paste of cooked starch are subjected to freezing temperatures, starch particles specially amylose chains re-associate and release water. This phenomenon is called syneresis. Due to their slow rate of retrogradation, waxy cereals are desirable for refrigeration and frozen food products. There are number of foods like sauces, soups and frozen meals that are often kept in freezing temperatures. However, freezing often have negative effects on the food quality and to prevent this, specialty ingredients like modified starches derived from cereals are commonly used. These modified flours are not able to carry the "natural" label and industries are driven by consumers demand that favours natural products. In searching for an alternative to modified flours, Kowalski *et al.*, 2016 analyse whether other native unutilized flours like waxy wheat flour (WWF) possess this freeze-thaw stability [72].

To determine this, WWF gels and regular wheat flour gels were analysed and compared before and after being subjected to freeze-thaw cycles. Waxy wheat samples shows minimal changes in rheological behaviours compared to regular wheat samples.The changes which occurs in regular wheat flours were attributed to the retrogradation of starch formed by amylose. Both the flours were subjected to ten freeze-thaw cycles and studied for differential scanning calorimetry and syneresis before and after they are each freeze-thaw cycle. WWF showed less than 5% water loss while regular wheat showed 25–40% water loss. DSC after the freeze-thaw cycles displays negligible retrogradation enthalpy values in waxy wheat compared to 1.3–1.8 J/g in regular wheat samples. Results indicated that waxy wheat can be used as a novel and natural freeze-thaw stable ingredient in foods, such as soups, dressings, and frozen meals. The minimal changes in waxy wheat flour indicate that it can be used as a substitute to other freeze-thaw stabilizers, such as gums and modified starches commonly used in food industry.

5. Conclusion

Waxy varieties of cereals reportedly arise due to mutations, which results in amylose free starches due to absence of Granule Bound Starch Synthase (GBSS) responsible for synthesis of amylose. The waxy varieties of rice and maize were reported to have originated from China, where they have gained cultural importance as traditional cuisines. Waxy wheat was developed in Japan via classical breeding and genetics by Nakamura in 1995. They are so-called waxy due to wax-like appearance of the endosperm, when cut and exposed in a cleanly cut cross-section. These waxy cereals have distinct properties due to which there is an increased interest of its use as a blending flour or raw material in many food industries. The high amylopectin containing starches have lower gelatinization temperature, higher swelling power and solubility and low rate of retrogradation. There are many studies suggesting their potential use in many food applications. Some of the potential uses of waxy flours includes anti-staling agent in bakery products for shelf-life stability, to increase expansion and sensory quality in extruded snacks,production of specialty foods like udon noodles, freeze thaw stable ingredient in foods that undergo freezing.

6. References

1. Manelius, R. (2000). Enzymatic and acidic hydrolysis of native and modified starch granules. Diss. Åbo: ÅboAkademi University. ISBN 952-12-0748-5.

2. Hoseney, R.C. (1986). Principles of cereal science and technology. American Association of cereal chemists, St. Paul, MN.

3. Sivak, M.N. & Preiss, J. (1995). Starch Synthesis in Food. In J. Kigel, & G. Galili (Eds.), *Seed Development and Germination* (pp. 139-168). New York: Marcel Dekker.

4. Nelson, O.F., and Rines, H.W(1962). The enzymatic deficiency in the waxy mutant of maize. *Biochemistry and Biophysiology and Research Communication*, 9: 297-300.

5. Eslick, R.F. (1981). 'Mutation and Characterization of Unusual Genes Associated with the Seed' in Barley Genetics IV, Proceedings of the Fourth International Barley Genetics Symposium (Asher, MJC., Ellis, R.P., Hayter, A.M. and Whitehouse, RNH., eds.), pp. 864-867, Edinburgh University Press.

6. Boyer, C.D. and Hannah, L.C. (1994). 'Kernel Mutants of Corn' in Specialty Corns (Hallauer, A., ed.), pp. 2-28, CRC Press, Inc.

7. Kenneth M. Olsen and Michael D. Purugganan (2002). Molecular evidence on the origin and evolution of glutinous rice. Genetics 162: 941–950.

8. Collins G.N. (1909). A new type of Indian corn from China. *Bur Plant Ind* 161: 1–30.

9. Zeng M.Q., Yang T.X., Wang P. (1981). The relative analyses of maize cultivar Menghai four-row wax. *Acta Genet Sin* 8: 91–96. (in Chinese).

10. Tian M.L., Tan G.X., Liu Y.J., Rong T.Z., Huang Y.B. (2009). Origin and evolution of Chinese waxy maize: evidence from the Globulin-1 gene. *Genet Resour Crop Evol* 56: 247–255.

11. Hibi, Y. (2001). Effect of retrograded waxy corn starch on bread staling. *Starch/Staerke*, 53, 227–234.

12. Whistler, R.L., & BeMiller, J.N. (1997). Carbohydrate Chemistry for Food Scientists. St. Paul, MN: Eagan Press.

13. Chaisawang, M. & Suphantharika, M. (2005). Effects of guar gum and xanthan gum additions on physical and rheological properties of cationic tapioca starch. Carbohydrate Polymers, 61: 288–295.

14. Nagano, T., Tamaki, E. & Funami, T. (2008). Influence of guar gum on granule morphologies and rheological properties of maize starch. Carbohydrate Polymers, 72: 95–101.

15. Nakamura, T., Yamamori, M., Hirano, H., Hidaka, S., Nagamine, T. (1995). Production of waxy (amylose-free) wheats. Mol. Gen. Genet. 248: 253–259.

16. Van Hung, P., Maeda, T. & Morita, N. (2007). Dough and bread qualities of flours with whole waxy wheat flour substitution. *Food Research International*, 40: 273-279.

17. Guan, L., Seib, P.A., Graybosch, R.A., Bean, S., Shi, Y.C. (2009). Dough rheology and wet milling of hard waxy wheat flours. *Journal of Agriculture and Food Chemistry* 57: 7030-7038.

18. Hoshino, T., Yoshikawa, R., Ito, S., Hatta, K., Nakamura, T., Yamamori, M., Hayakawa, K., Tanaka, K., Akashi, H., Endo, S., Tago, S., Ishigami, S. (2000). Flour blends for breads, cakes, or noodles, and foods prepared from the flour blends. United States Patent 6,042,867.

19. Hoshino, T., Yoshikawa, R., Ito, S., Hatta, K., Nakamura, T., Yamamori, M., Miyamura, H., Murayama, Y., Kawamura, Y., Hayakawa, K., Tanaka, K., Tago, S., Ishigami, S., Mizukami, M., Tanaka, Y. (2000). Flour blends for deep fried foods, steamed Chinese manju, dough sheets, okonomi-yaki, and takoyaki. United States Patent 6,139,894.

20. Lee, M.R., Swanson, B.G. & Baik, B.K. (2001). Influence of amylose content on properties of wheat starch and breadmaking quality of starch and gluten blends. Cereal Chemistry, 78: 701–706.

21. Graybosch, R.A. (1998). Waxy wheats: origin, properties and prospects. Trends Food Sci. Technol. 9: 135–142.

22. Hung, P.V., Maeda, T., Morita, N. (2006). Waxy and high-amylose wheat starches and flours Characteristics, functionality and application. Trends in Food Science and Technology 17: 448-456.

23. Maningat, C.C., Seib, P.A., Bassi, S.D., Woo, K.S., Lasater, G.D., (2009). Wheat starch: Production, properties, modification and uses. In: BeMiller, N., Whistler, R.L. (Eds.), Starch Chemistry and Technology. Academic Press: New York, pp. 441-510.

24. Alexander, R.J. (1992). 'Carbohydrates Used as Fat Replacers'. *Developments in Carbohydrate Chemistry* (Alexander, R.J. and Zobel, H.F. eds.), American Association of Cereal Chemists. pp. 343-370.

25. Goering K.J., Eslick Rand B.W. DeHaas (1973). *Cereal Chemistry* 50: 322-328.

26. Vasanthan T. and R.S. Bhatty (1996). Cereal Chern. 73: 199-207.

27. Macgregor A.W. and R.S. Bhatty, Eds. (1993). In Barley: Chemistry and Technology. Am. Assoc. Cereal Chemists, Inc., 90-92.

28. Delcour, J.A., and Hoseney, R.C. (2010). *Principles of cereal science and technology* 3rd edition, AACC International Inc., St. Paul, Minnesota, USA. ISBN: 978-1-891127-63-2. P 25-30: 35-42.

29. Brink, R.A. and Abegg, F.C. (1926). *Genetics,* 11: 163-199.

30. Weatherwax, Paul (1922). *Genetics,* 7: 568-572.

31. Tanaka, Y.J. (1912) Ind. Eng. Chem., 4: 578-581.

32. Hagenimana A. and Ding X. (2005). Preparative study on pasting and hydration properties of starches and their mixtures. Cereal Chem., 82: 70-78.

33. Kempton, J.H. (1919). Inheritance of waxy endosperm in maize. U. S. Dept. Agric. Bull. 754. 99 PP.

34. Planchot, V., Colonna, P., Gallant, D., Bouchet, B. (1995). Extensive degradation of native starch granules by α-amylase from aspergillus fumigatus. Journal of Cereal Science, 21(2): 163-171.

35. Vesterinen, E., Myllärinen, P., Forssell, P., Söderling, E., Autio, K. (2002). Structural properties in relation to oral enzymatic digestibility of starch gels based on pure starch components and high amylose content. Food Hydrocolloids, 16: 161-167.

36. Conde-Petit, B., Nuessli, J., Arrigoni, E., Escher, F., & Amadò, R. (2001). Perspectives of starch in food science. Chimia, 55(3):201-205.

37. Hoover, R. (2010). The impact of heat-moisture treatment on molecular structures and properties of starches isolated from different botanical sources. Critical Reviews in Food Science and Nutrition, 50(9): 835-847.

38. Ratnayake, W.S., Hoover, R. & Warkentin, T. (2002). Pea starch: composition, structure and properties - a review. Starch/Staerke, 54(6): 217-234.

39. Singh, N., Singh, J., Kaur, L., Sodhi, N.S. & Gill, B.S. (2003). Morphological, thermal and rheological properties of starches from different botanical sources. Food Chemistry, 81(2): 219-231.

40. Gonera, A. & Cornillon, P. (2002). Gelatinization of starch/gum/ sugar systems studied by using DSC, NMR and CSLM. Starch/Stärke, 54: 508–516.

41. Schirmer, M., Hochstotter, A., Jekle, M., Arendt, E., Becker, T., (2013). Physicochemical and morphological characterization of different starches with variable amylose/amylopectin ratio. Food Hydrocoll. 32: 52-63.

42. Kong, X., Kasapis, S., Bao, J. (2015). Viscoelastic properties of starches and flours from two novel rice mutants induced by gamma irradiation. LWT - Food Sci. Technol. 60: 578-582.

43. Waterschoot, J., Gomand, S.V., Willebrords, J.K., Fierens, E., Delcour, J.A. (2014). Pasting properties of blends of potato, rice and maize starches. Food Hydrocoll. 41: 298-308.

44. Zhou, Z., Robards, K., Helliwell, S., Blanchard, C. (2002). Ageing of stored rice. Changes in chemical and physical attributes. J. Cereal Sci. 35: 65-78.

45. Silverio, J., Svensson, E., Eliasson, A.C. & Olofsson, G. (1996) Isothermal microcalorimetric studies on starch retrogradation. J. Thermal Anal., 47: 1179–1200.

46. Lundqvist, H. & Eliasson, A.C. (2005). Influence of whey proteins β-lactoglobulin and α-lactalbumin on retrogradation of starch. Prog. Food Biopolym. Res., 1: 87–99.

47. Wursch, P. & Gumy, D. (1994). Inhibition of amylopectin retrogradation by partial beta-amylolysis. Carbohydr. Res., 256: 129–137.

48. Gray, J.A. & BeMiller, J.N. (2003). Bread staling: Molecular basis and control. Comprehensive Reviews in Food Science and Food Safety, 2: 1–21.

49. Šárka, E. & Dvoøáèek, V. (2017). Waxy starch as a perspective raw material: A review. *Food Hydrocolloids,* 69: 402-409.

50. Hug-Iten, S., Handschin, S., Conde-Petit, B. & Escher, F. (1999). Changes in starch microstructure on baking and staling of wheat bread. Food Science and Technology, 32: 255–260.

51. Bhattacharya, M., Erazo-Castrejon, S., Doehlert, D.C. & McMullen, M.S. (2002). Staling of bread as affected by waxy wheat flour blends. Cereal Chemistry, 79: 178–182.

52. Ghiasi, K., Hoseney, R.C., Zeleznak, K. & Rogers, D.E. (1984). Effect of waxy barley starch and reheating on firmness of bread crumb. Cereal Chemistry, 61: 281–285.

53. Hayakawa, K., Tanaka, K., Nakamura, T., Endo, S., & Hoshino, T. (2004). End use quality of waxy wheat flour in various grain-based foods. *Cereal Chemistry,* 81: 666-672.

54. Morita, N., Maeda, T., Miyazaki, M., Yamamori, M., Miura, H. & Ohtsuka, I. (2002). Effect of substitution of waxy-wheat flour for common flour on dough and baking properties. Food Science and Technology Research, 8: 119–124.

55. Morita, N., Maeda, T., Miyazaki, M., Yamamori, M., Miura, H. & Ohtsuka, I. (2002). Dough and baking properties of high amylose and waxy wheat flours. Cereal Chemistry, 79: 491–495.

56. Biliaderis, C.G. (1992). 'Structures and Phase Transitions of Starch. *Food Systems' in Food Technology* 46: 98-109.

57. Stampfli, L. and Nersten, B. (1995). 'Emulsifiers in Bread Making'. *Food Chemistry*. 52: 353-360.

58. Nagao, S. (1996). Processing technology of noodle products in Japan. In J.E. Kruger, R.B. Matsuo, & J.W. Dick (Eds.), Pasta and noodle technology (pp. 169–194). St. Pual, MN: American Association of Cereal Chemists.

59. Endo, S., Karobe, S. & Nagao, S. (1988). Factors affecting gelatinization properties of starch. Nippon Shokuhin Kogyo Gakkaishi, 35: 7–14.

60. Crosbie, G.B. (1991). 'The Relationship between Starch Swelling Properties, Paste Viscosity and Boiled Noodle Quality in Wheat Flours' in *J. Cer. Sci.* 13: 145-150.

61. Panozzo, J.F. & McCormick, K.M. (1993). The rapid viscoanalyser as a method of testing for noodle quality in a wheat breeding programme. Journal of Cereal Science, 17: 25–32.

62. Jane, J., Chen, Y.Y., Lee, L.F., McPherson, A.E., Wong, K.S., Radosavljevic, M., *et al.* (1999). Effects of amylopectin branch chain length and amylose content on the gelatinization and pasting properties of starch. Cereal Chemistry, 76: 629–637.

63. Miura, H. and Tanii, S. (1994). 'Endosperm Starch Properties in Several Wheat Cultivars Preferred for Japanese Noodles'. *Euphytica* 72: 171-175.

64. Noda, T., Tohnooka, T., Taya, S. & Suda, I. (2001). Relationship between physicochemical properties of starches and white salted noodle quality in Japanese wheat flour. Cereal Chemistry, 78: 395–399.

65. Wang, L. & Seib, P.A. (1996). Australian salt-noodle flours and their starches compared to US wheat flours and their starches. Cereal Chemistry, 73: 167–175.

66. Akashi, H., Takahashi, M. & Endo, S. (1999). Evaluation of starch properties of wheats used for Chinese yellow-alkaline noodles in Japan. Cereal Chemistry, 76: 50–55.

67. Miskelly, D.M., & Moss, H.J. (1985). Flour quality requirements for Chinese noodle manufacture. Journal of Cereal Science, 3: 379–387.

68. Guo, G., Jackson, D., Graybosch, R. & Parkhurst, A. (2003). Asian salted noodle quality: Impact of amylose content adjustments using waxy wheat flour. Cereal Chemistry, 80: 437–445.

69. Yamaguchi, I., Otobe, C., Ushiyama, T., Yanagisawa, T., Seko, H., Amano, Y., *et al.* (2000). Wheat starch with novel characteristics. US patent 6: 165-535.

70. Baik, B.Y., & Lee, M.R. (2003). Effects of starch amylose content of wheat on textural properties of white salted noodles. Cereal Chemistry, 80: 304–309.

71. Chibbar, R.N. & Chakraborty, M. (2005). Characteristics and uses of waxy wheat. Cereal Foods World, 50: 121–126.

72. Ryan J. Kowalski, Alexander Meldrum, Siyuan Wang, Helen Joyner (Melito) and Girish M. Ganjyal (2017). Waxy wheat flour as a freeze-thaw stable ingredient through rheological studies. Food Bioprocess Technol. DOI 10.1007/s11947-017-1899-y.

❑❑❑

Chapter 3

Advances in Post Harvest Management of Lentil (*Lens culinaris* Medik)

Anirban Roy[1], Arghya Mani[2], Paramita Roy[3] and Tanmoy Sarkar[4]

[1]Research Scholar, Department of Genetics and Plant Breeding, BCKV, Mohanpur, [2]Research Scholar, Department of Post Harvest Technology, Mohanpur, [3]Master Degree student - Department of Vegetable Science, Mohanpur, [4]Research Scholar, Department of Fruit Science, Mohanpur, Corresponding author – manifruitpostharvest@gmail.com

ABSTRACT

Lentil is one of the popular legume crop and is one of the most preferred pulse crops in India. As compared to the other pulse crops, lentil is a considerably new crop. There is a void in the standardization of agro technology related to this crop. Post Harvest aspects of lentil are not under that spotlight. Mismanagement during harvesting time and improper storage reduces its viability and other property. Sometimes weather phenomenon causes a challenge for lentil harvesting. Most challenging element during harvesting is proper harvesting without any delay. Insect damage during storage of lentil is a key target for managing lentil during post harvest that can be controlled from field level itself. Breeding for insect resistance, harvesting at physiological maturity stages are most important to minimize lentil on farm loss after production.

Keywords: Lentil, Post Harvest.

1. Introduction

Around 2.6 billion tonnes of grains are produced but 10-30% post-harvest losses occur, which could be used to feed additional 1.35 billion people.(Singh, 2017). Lentil is one of the important pulses grown in India. Preferences for this pulse are quite high after chickpea. As a highest producer of pulse after Canada and highest consumer of pulse, in Indian agricultural economy Lentil is important. But post harvest losses ruins that profit. It is loss that occurs due to post harvest mismanagement. In that context post harvest management should minimize losses by introducing several measure and genetic improvement as well quality of lentil reduces drastically, if not harvested in time, even during storage lentil seed coat over time turn brown due to oxidation of tannins oxidation occurs quickly in conditions of high temperature, high humidity and exposure to sunlight. Sometimes, harvesting after sudden rain causes seed infected with various pathogen and seed loses its keeping quality.

1.1. About Lentil

Lentil is a very well known Mediterranean crop of cool dry climate thrives well in harsh condition and is the oldest cultivated legume plant (Rehman *et al.*, 1994). Lentil is cultivated in Canada, USA, Mexico, India, Chile, Peru. Lentil is a bushy herbaceous annual crops having haploid chromosome number 7. Usually lentil pod contain generally two seed. There are two subgroup in lentil are Macrosperma and Microsperma mainly divide on the basis of their origin and seed size. India is second in production after Canada.(Singh and Singh, 2014). The most detailed and complete study of the cultivated lentil was made by Barulina (1930).

1.2. Importance of Lentil

With about 30% of their calories from protein, lentils have the third-highest level of protein, by weight, of any legume or nut, after soybeans and hemp. Proteins include the essential amino acids isoleucine and lysine, and lentils are an essential source of inexpensive protein in many parts of the world, especially in West Asia and the Indian subcontinent, which have large vegetarian populations. As a common feature of being a pulse lentils are deficient in two essential amino acids, methionine and cysteine. However, sprouted lentils contain sufficient levels of all essential amino acids, including methionine and cysteine. Lentils also contain dietary fiber, folate, vitamin B1, and minerals. Red (or pink) lentils contain a lower concentration of fiber than green lentils (11% rather than 31%). Health magazine has selected lentils as one of the five healthiest foods. The low levels of Readily Digestible Starch (RDS) 5%, and high levels of Slowly Digested Starch (SDS) 30%, make lentils of great interest to people with diabetes. The remaining 65% of the starch is a resistant starch that is classified RS1, being a high quality resistant starch, which is 32% amylose. Lentils also have some anti-nutritional factors, such as trypsin inhibitors and relatively high phytate content. Trypsin is an enzyme involved in digestion, and phytates reduce the bioavailability of dietary minerals.

1.3. Proper Harvesting Time of Lentil

In general, the harvest takes place 10 to 15 days after the grain has reached physiological maturity. At the time of maturity, the grain has specific moisture content and special physical characteristics. The most appropriate time of harvest is determined, based upon the length of the growing cycles (which differ according to the crop and varieties) and also, the degree of maturity of the grain. Lentil can harvest when upper leaves dry and pods become pale yellow.

1.4. Losses due to Improper Harvesting Practices

Yield losses in lentil increase significantly due to delayed harvesting. Lentil is very prone to pod splitting and pod drop with delays in harvest. This especially occurs, when weather events (rain and strong winds) occur after the plant has dried. Lentil plants that have been affected by rain are also more difficult to thresh out at harvest, and often contain much higher levels of unthreshed pods and pod material. Early harvested lentil grain is much more resilient against breakage during harvesting and subsequent handling, even at low moisture contents. Grain quality deteriorates the longer that mature lentils are left exposed to weathering in the paddock. The seed coat of lentil is very prone to wrinkling, if it has been exposed to wetting and drying events due to rain or heavy dew during the harvest months. Expansion of the seed as it absorbs moisture, and then contraction as it dries, weakens the seed coat. This renders it much more susceptible to mechanical damage during harvest and handling operations. Levels of cracked and damaged grain can be as high as 50% in extreme cases of delayed harvest combined with rain and/or wind events. Yield of split & pulses in traditional mills are only 65 to 75% due to the above losses as compared to 82 to 85% potential yield.

1.5. Maturity Indices

1.5.1. Harvesting details of lentil

It is impact forces that easily damage lentil. It is important to minimize any mechanical damage of lentil grain because the appearance of the grain is a very important quality parameter. Ease of threshing and the amount of seed splitting differs between varieties, so some testing is necessary. If the crop has lodged, it is usually best to harvest in the opposite direction, or at right angles, to the direction the crop has fallen. Crop lifters may help. From another aspect that present practices of harvesting lentil is a highly labour-intensive operation requiring 20 workers to harvest 1 acre in one day (80-100 person hours per hectare) (FAO, 2016), where labour force is limited and to enter in this particular area using combine harvesters for rice, adapted for pulse crops, which could harvest 3-4 acres per day can be utilized. It is to be noted that for mechanized harvesting cultivating determinate varieties need to cultivated. Also, as the maturity stage of the crop for harvesting is determined by subjective methods, it is best to introduce moisture meters so that the best moisture content for harvesting can be determined.

1.5.2. Storage of lentil

Monitoring the quality of grain before and during harvest is extremely important. The lentil seed coat and kernel (cotyledon) can be discoloured by crop-topping or premature desiccation, if sections of the paddock have uneven maturity. Staining of seed caused by green plants in the crop or admixture of splits, weeds, stones, etc., will reduce the value of the grain and result in deductions to the final price received. Visual appearance of lentil is critical. Human food markets demand a quality sample without cracking, staining, de-hulled grain or insect damage. Grain samples indicating no mechanical damage from harvesting will always be more acceptable to a buyer. The smaller seed of lentil makes it less prone to mechanical damage than larger seeded crops like faba bean.

1.5.3. Seed drying

The purpose of drying grain is to remove excess moisture to prevent spoilage during storage. Continuous-flow or batch dryers provide reliable drying, although they can reduce quality, if run at too high a temperature. Temperature should not exceed 45°C, when using heat to dry lentil. Check the specifications or talk to the manufacturer about safe conditions for drying lentil.8 High capacity aeration systems can also be used to dry grain, and are ideally suited for drying grain harvested at 15–16% moisture content. Aeration drying has a lower risk of cracking and damaging lentil, which can occur with hot-air dryers. Aeration drying requires a larger capacity fan to move high volumes of air through the grain at a faster rate than that required for cooling only.

1.5.4. Post harvest losses of lentil

a) Loses due to post harvest disease

b) Loses due to pests

c) Insects are not considered a major problem in stored lentil. While seed beetles or Bruchids (Bruchinae) are considered primary pests of pulse crops, very few *Bruchid* species (present in Australia) attack lentil. It is important to note that there are exotic Bruchids (not present in Australia) that do attack lentil. One exception appears to be in cases, where lentil is loaded into storages containing residues of cereal grain already infested with cereal insect pests, such as flour beetles (*Tribolium* spp.) and grain borers (*Rhyzoptherha* spp.) These prior infestations can develop and spread in lentil. Removal of these residues prior to loading is essential. In some cases, grain insects can cross-contaminate lentil storage from cereal storage.

d) Loses due to viability reduction

Viability and vigour is the most vulnerable in case of lentil storage after storage pest. Even if the storage condition is good to avoid insect contamination, reduction in vigour and viability is seen on next year crop germination and initial crop establishment. Maintenance of good viability and vigour is totally genetically controlled.

1.6. Other Post Harvest Practices for Lentil

It is a common practices after lentil maturity that after uprooting keep harvested plant for 2-3 days, if there is no probability of rain, as it reduces the moisture which is very much important for good threshing.

1.7. Value Added Product of Lentil

However, lentil is mainly consumed as 'Daal' in India. Lentil is mixed with cereal flour used as baby food and sometimes, used in other preparation. Now-a-days a product from lentil pest, in local language, it is known as 'Daler Bori' is popular in India, which is increasing the income source of women and a way-out in value addition of lentil.

1.8. Improvement of Post Harvest Traits

The traits which determine quality post harvest attributes of lentil are quality parameters, which include seed colour, keeping quality, and insect resistance, which is important for seed purpose. Besides post harvest quality traits related to different products of lentil also important from commercial point of view. But some of the traits that are important mostly are resistance to strorage pest mainly bruchid, storage disease resistance, maintaining high viability and vigour. Some of these problems can be managed by good storage and keeping practices but exploitation of genotypic potential is best approach. Incorporation of these traits in popular varieties now-a-days is reasonable as well as economic in long-term.

A1. Conventional

Conventional breeding approaches for post harvest trait improvement is the basic approach. But as these traits are not like yield improvement, so strategy is different. Conventional cross-breeding approach through genotype selection, based on response to post harvest challenges are common.

B1. Biotechnological

As most of the intervention related to reducing post harvest losses are fruit and vegetable crops oriented. Where for grain crop, post harvest management lean towards minimizing storage pest.

C1. Mutation

Besides conventional cross-breeding, generation of variability is possible through mutation. Though probable outcome from mutation breeding is very much low in frequency. Still if mutation occurs in exact chromosomal position controlling, the desired traits can be selected from M2 onwards.

2. References

1. Abbas F., Riaz A., Zafar T., Hassan M., Muhammad H. Umar I., Hassan J., Alam W., Muhammad 5, Mahmood M., Sohail H., Tiwana U.A., Iqbal M.T., Bilal M. Gene Action Studies in the Inheritance of Economic Traits in Diallel Cross of Lentil (*Lens culinaris* Medik) *Amercan Journal of Plant Science*, (2016). 7: 756-768, Working Paper-13.
2. Singh C.B. (2017). Post harvest Management and Storage of Pulses in India: Challenges and Opportunities Spandan-India.org Article.
3. Formulation and Operationalization of National Action Plan for Poverty Alleviation and Rural Development through Agriculture (NAPA) FAO, (2016). Ministry of Agriculture, Livestock and Irrigation.
4. Singh, K.M. and Singh, A.K. (2014). Lentil in India: An overview ICAR Research complex for Eastern Region, Patna.
5. Rehman, S. and Altaf, C.H.M. (1994). Karyotipic studies in *Lens culinaris* Medic, Ssp. Macrosperma cv. Laird X Precoz. *Pak. J. Bot.*, 26(2): 347-352.
6. Barulina, H. (1930). Lentils of the USSR and other countries. *Bulletin of Applied Botany: Genetics and Plant Breeding*, 40: 265–304.

□□□

Chapter 4

Value Addition to Pulses by Post Harvest Technology and Processing

Brati Acharya[1], Bishal Mukherjee[2], B. Lalramhlimi[1], Subhramalya Dutta[1], Ivi Chakraborty[3] and Rajib Nath[2]

[1]Department of Vegetable Science, Bidhan Chandra Krishi Vishwavidyalaya, Mohanpur, Nadia, [2]Department of Agronomy, Bidhan Chandra Krishi Vishwavidyalaya, Mohanpur, Nadia, West Bengal, India, [3]Department of Post-Harvest Technology of Horticultural Crops, Bidhan Chandra Krishi Vishwavidyalaya, Mohanpur, Nadia, *Corresponding author Email: ivcpht@gmail.com

1.1 Introduction

Ensuring food security has become a major problem for the burgeoning billions in the present 21st century. Grain legumes and cereals have been a part of balanced diet in different continents across the globe since early civilization. India is one of the developing countries of the world, where malnutrition and poverty has become one of the major concerns till date. To meet their dietary requirements, the people who are mostly vegetarian in India depend on pulses and cereals. The Indian Council of Medical Research has recommended an intake of 80g/day of pulses in diet for active male adult[29]. **A sustainable diet for vegetarians may not be possible without the protein-rich grain legumes**[63]. Grain legumes popularly known as pulses are dietary protein in the daily diet of human beings as well as animal feed also. The daily consumption of at least three servings of fruits or vegetables is advised for healthy diets and the variation of food items should be derived from different plants and those plants should belong to different botanical families[58].

Pulses are the common name of legumes given to the seeds of the dried vegetables, such as beans, chickpeas, lentils, pods, peas and cowpea[43]. Legumes are the mature seeds of plants belonging to the family Fabaceae (earlier known as Leguminosae), which is the largest family of 700 families and 18,000 species in flowering plants[51]. Legumes can have a herbaceous structure as well as a woody structure. Seeds have hard-shelled, usually kidney-shaped or rounded structure. Food legumes utilized as dry seeds are often referred to as pulses or grain legumes. Pulses and legumes terms are often used interchangeably. However, Codex Alimentarius Commission of FAO/WHO Food Standard programme has given clear distinction for pulses and legumes, on the basis of their fat content. FAO calls pulses only legumes with dry, edible seeds, with low fat content. Pulses are dry seeds of leguminous plants and legumes include oil seeds, such as soybean and peanut.

Pulses is an important food ingredient for more than 2 billion people in the world and about 69 million tonnes of beans, lentils and chickpeas are produced in about 78 million hectares of land in the world. Our country makes 1.5% of pulses production in the world, especially 3rd in chickpea production and 4th in lentil production[53, 61]. Nutritional arena of pulses can make the marked difference in their utilization and reduction in rampant malnutrition around the globe. Food and Agriculture Organization (FAO) and United Nations have declared 2016 as "International Year of Pulses" (IYP) and seeds as **"nutritious seeds for sustainable future"**[44] . Indian Pulse and Grain Association has also portrayed pulses the future ingredient to commensurate the nutritional and health benefits and Food Security Portal, as they possess rich source of plant protein, energy, dietary fibre, a wide variety of micronutrients and bioactive compounds.

Pulses contain edible protein (18-32%), which provides a chief source of essential amino acids and bioactive peptides that may act as safeguard for the people with meager income groups thriving to attain food self-sufficiency in rural areas of our country. For these groups of people, it has been unachievable. Pulse protein possesses functional properties, such as water holding, fat binding, foaming and gelation which could expand their potential use in the development of a number of food products[10]. Besides these, pulses are enriched with several health benefitting properties and help in preventing illnesses, such as cancer[62] and diabetes[51] and heart disease.

India has to depend on foreign countries to some extent for import of pulses to meet demand in the country. This is not always only because of inefficiency of the local producers but also postharvest losses in pulses, which is the major causes for unavailability of healthy pulse products in the market. Post harvest losses in pulses are estimated to the extent of about 25-30%. These losses decrease availability of good quality products, forcing national policy-makers to import such commodities and this may have serious impacts on the earnings of local pulse growers of different states. By reducing the high post harvest losses of crops particularly protein rich legumes, nutritional security in India and in other developing countries can only be achieved[36]. In fact, there is lack of knowledge and information among field staffs, extension workers, farmers and rural entrepreneurs regarding the post harvest management, processing of pulses and derivation of value added products. So, improving the different post harvest aspects of pulses is the most important way forward.

1.2 Briefing Pulses: Their importance

Pulses are no doubt the richest and cheapest source of high quality vegetable protein. They are considered as "Poor man's meat" as well as "Protein tablets". Pulses are rich in carbohydrates, mainly starches (55-65% of the total weight); proteins, including essential amino acids (18-25%, and much higher than cereals); fat (1-4%), energy, dietary fibre, a wide variety of micronutrients and bioactive compounds. Pulses are considerably richer in calcium than most cereals and contain about 100 to 200 mg of calcium per 100 g of grain. They are also considerably rich in iron, thiamine, riboflavin and nicotinic acid, as compared to cereals. The remainder consists of water and inedible substances.

Pulses are annual leguminous crops yielding from one to twelve grains or seeds of variable size, shape and colour within a pod. They are cultivated mainly for their seeds. The term "pulses" is limited to crops harvested solely for dry grain, thereby excluding crops harvested green for food (e.g. green peas, green beans, etc.) which are classified as vegetable crops. Pulses also excluded those crops used mainly for oil extraction (e.g. soybean and groundnut) and leguminous crops (e.g. seeds of clover and alfalfa) that are used exclusively for sowing purposes. They are grown agriculturally, primarily for human consumption, for livestock forage and silage and as a soil enriching green manure also. Legumes are very rich in an enormous variety of non-protein amino acids that can be beneficial or toxic.

The ancient civilizations in the Middle East and in America included grain legumes and cereals in well-balanced diets. The Egyptian pyramids presented various legume seeds including other things in the funeral offerings. Apparently, legumes were a food of special consideration to be offered to kings at that time, in contrast to the present day reputation of being the meat of the poor. Pulse crops were initially domesticated at the start of agriculture over 8000 years ago in an exceedingly fertile band of land stretching from Turkey to Iran[64]. Milling of pulse seeds conjointly dates back to earlier period[7].

India is the largest producer, consumer and also the largest importer of pulses in the world. In India, 24-26 million hectares of area are occupied by pulses with a production of 17-19 million tonnes of pulses annually. India accounts for more than one-third of the total world area and over 20% of total world production. In India, pulses are mainly cultivated on marginal lands under rainfed conditions. Only 15% of the area under pulses has assured irrigation. Farmers are getting attracted towards cash crops like cotton, maize and oilseeds (mainly soybeans) instead of pulse cultivation despite high wholesale pulse prices in recent years because of better return and lower risk of cash crops. The high level of fluctuations in pulse production (due to biotic and abiotic stress) and prices (in the absence of an effective government price support mechanism) make the farmers lose confidence on taking up pulse cultivation.

Besides their food value, pulses are able to fix atmospheric nitrogen through root nodules enriching the soil thereby play an important role in cropping systems. Pulses have immense potential for alleviating human malnutrition in developing countries by virtue of many inherent advantages, which are:

1. They can adapt and grow vigorously under a wider range of environments and on relatively poor soils without supplemental nitrogen, which is particularly advantageous for consistent agricultural practices in remote areas.
2. They require low water supply, which offer opportunities to make use of India's dry lands, which constitute more than 70% of total cropped area, thereby, increasing food production, farmers' income and nutrient contribution in the diet.
3. Pulses have a high supplementation value for cereal protein. Combination of pulses with cereal rich diet is promising in updating the dietary protein quality. Pulses provide sufficient quantity of lysine, an amino acid, which is deficient in cereals.

Legumes are consumed by humans in many forms. The nutrient availability from legumes depends upon nutrient content and factors, such as post harvest handling, processing methods, presence or absence of anti-nutritional and/or toxic factors and possible interaction of nutrients with other food components i.e. Polyphenolic compounds, certain minerals, etc. Some important pulse crops[18] are:

1. **BEANS, DRY** *Phaseolus spp.*: kidney, haricot bean *(Phaseolus vulgaris);* lima, butter bean *(Phaseolus lunatus);*adzuki bean *(Phaseolus angularis);* mungo bean, golden, green gram *(Phaseolus aureus);* black gram, urad *(Phaseolus mungo);* scarlet runner bean *(Phaseolus coccineus);* rice bean *(Phaseolus calcaratus);* moth bean *(Phaseolus aconitifolius);* tepary bean *(Phaseolus acutifolius).*
2. **BROAD BEANS, DRY** *Vicia faba:* horse-bean (var. *equina*); broad bean (var. *major*); field bean (var. *minor*).
3. **PEAS, DRY** garden pea (*Pisum sativum*); field pea (*P. arvense*).
4. **CHICK-PEAS** chickpea, Bengal gram, garbanzos (*Cicer arietinum*).
5. **COW PEAS, DRY** cowpea, black eye pea/bean (*Vigna sinensis; Dolichos sinensis*).
6. **PIGEON PEAS** pigeon pea, cajan pea, Congo bean (*CajanusCajan*).
7. **LENTILS** (*Lens esculenta*; *Ervum lens*).
8. **VETCHES** spring/common vetch *(Vicia sativa).*
9. **LUPINS** (*Lupinus spp.*).

Fig. 1: Different Pulse Crops

1.3 Nutritional Values and Uses of Pulses

In India, from time immemorial, many legumes and pulses have been consumed as part of a primarily cereal-based diet. They are consumed as whole gram or as split pulses (dhals). Traditional consumption for pulses is as whole grain in meals, salads and soups. It is also used as snack (roasted chickpea) and canned (beans, chickpeas, green lentils, etc.) food. Apart from this, it is possible to use pulses in various kinds of product (bread, ready-made soup, baby foods, candies, breakfast cereals, snacks, meat products, diet products, gluten-free products) in the form of culture flour (chickpea flour, pea flour, lentil flour) as raw materials or additives, as well as in the form of fractions in the health sector as re-inforcing agents (leguminous proteins, fibers, oil)[17].

Pulses are source of constant supply of nutrition round the year and nutritional value of pulses per 100 gram is much higher than any other vegetarian food. Pulses have health benefitting properties and help in preventing illnesses, such as cancer[62], diabetes[51] and heart disease. Pulses are a source of carbohydrates (60-65%) of which starch is the primary carbohydrate. Slowly digesting nature of starch in pulses make pulses diabetic friendly [low glycemic index (GI)][52].

Pulses are the main source of protein in the primarily vegetarian Indian diet. Protein values are about twice as many as grain grains and are 18-31.6%. Their digestibility is between 69-90%. Besides proteins, pulses are also good sources of vitamins, minerals, omega-3 fatty acids and dietary fibre or non-starch polysaccharides (NSP). Several legumes contain non-nutrient bioactive phytochemicals that have health-promoting and disease-preventing properties. The important health promoting 'non-nutrient' compounds present in pulses and legumes include non-starch polysaccharides (NSP), phytosterols, saponins, isoflavones, a class of phytoestrogens, phenolic compounds and antioxidants, such as tocopherols and flavonoids. Epidemiological observations and a number of human and animal studies have shown that the NSP in pulses have a role in protecting against degenerative diseases, such as diabetes, cardiovascular disease and cancer. NSP protects against chronic diseases, such as diabetes by controlling blood glucose levels. Their hypocholesterolemic property protects against cardiovascular disease. They also help prevent bowel cancer by binding carcinogens and toxins.

The fat percentage in pulses is generally very low (app. 0.8-1.5%) and the fat is as polyunsaturated structure. Moreover, it does not contain cholesterol. Pulses have low glycemic index, 16-32, and are rich in starch and dietary fiber. Besides, it contains folic acid, vitamins of group B, mineral substances (potassium, phosphorus, calcium and iron) and phenolic substances. It is also effective on heart and blood vessels, diabetes, obesity, cancer diseases when consumed frequently, regularly and properly cooked[43,53]. Besides human nutrition, it is important to contribute to the soil as a rare plant that can bind free nitrogen of the air, and to animal husbandry with high protein content[61].

1.4 Antinutritional Factors (ANFS) of Pulses

Legume seeds are recognized as a chief source of proteins just after milk in India. Pulses are known to have many plausible health benefits not to mention reduced risk of cardiovascular disease, cancer, diabetes[9], osteoporosis, gastrointestinal disorders, hypertension, adrenal disease and reduction of LDL cholesterol[26, 49, 57]. Moreover, the nutritional benefits obtained are gluten-free[18]. But keeping in mind, the legume seeds are also proverbial to possess a few nutritionally combative elements (antagonistic factors) that impede the digestibility of consumers. As a consequence, these may limit the availability of otherwise good proteins in the diet and may cause diseases resulting from malnutrition[8].

A substance which hinders with metabolism of certain nutrients causing toxic or unwanted physiological effects, such as flatulence in human or animal is defined as an antinutritional factor. Regardless of these, ANFs provide a defensive role against insects[48] predators and pathogens attacks[12]. Presence of anti-physiological or toxic substances, such as trypsin and chemotrypsin inhibitors, phytates, lectins, polyphenols, flatulence causing agents, cyanogenic compounds, lathyrogens, esterogens, goiterogens, saponins, anti-vitamins and allergens may badly affect the nutritional value of pulses. The nutritive value of foods is lowered by these substances, as they inhibit the digestibility and utilization of proteins. Thus, it is a necessity to get rid of these substances either by processing or by genetic manipulation. It is crucial to undergo a critical quantitative evaluation of such anti-nutritional compounds[55]. These anti-nutritional compounds may either be heat-labile or heat stable and in pulse crops, they are categorized as: protein ANCs and non-protein ANCs[15]. The effect of these compounds ranges from relatively inoffensive polyphenols to the relatively harmful protease inhibitors. Alkaloids[38], phytic acid, phenolic compounds, such as saponins and tannins[14, 27] are some non-protein ANCs.

The colour of the flower of a plant is controlled by phenol compounds and they are bitter in taste. They are responsible for reducing the bioavailability of the mineral (Zinc). They also form complex with protein molecules and lower their solubility, therefore affecting digestibility. On the other hand, phenol compounds are helpful as anti-cancer and anti-atherosclerotic. Vicine and convicine present in faba beans are responsible for acute hemolytic disease (favism). They are also useful in lowering cholesterol. A neurotoxin known as B-N-oxalyl di-aminopropionic acid (BOAA) is found in *Lathyrus sativus* (kesari dal). Prolonged consumption of this compound for more than six months in sizable amount can lead to paralysis in lower limbs in men because this substance interact with production of normal collagen in connective tissues[3]. Processing methods used traditionally include dehulling, milling, soaking, cooking, pressure cooking, roasting, germination and fermentation; and in disguise, these methods lower ANFs to varying degrees. On the other hand, modernized technological methods like fractionation, ultrafiltration extraction and extrusion can pacify them as functional food or nutraceutical ingredients[3].

Boiling, autoclaving and microwave cooking affect the composition, anti-nutritional factors, flatulence factors and nutritional quality of chickpeas. However, microwave cooking caused slight losses in minerals, while boiling and autoclaving caused significant losses. All cooking treatments improved the *in-vitro* protein digestibility and protein efficiency ratio of lentils, chick pea. It is quite clear that cooking by microwave not only saves time but also retains the most nutritive value[25].

Surprisingly, some anti-nutrients may give beneficial health effects at low concentrations. For instance, phytic acid, lectins, tannins, saponins, amylase inhibitors and protease inhibitors reduce the availability of nutrients and when it is used at low levels, it can inhibit growth and are proven to lower the blood glucose and insulin responses to starchy fods and/or the plasma cholesterol and triglycerides. Moreover, phytates, tannins, saponins, protease inhibitors, goetrogens and oxalates have been associated to lessen cancer risks. This means that anti-nutrients might not always have adverse effect even if they lack nutritive value[23]. Regarding the balance between beneficial and hazardous effects of plant bioactives and anti-nutrients depending on their concentration, chemical structure, time of exposure and interaction with other dietary components, they can be treated as anti-nutritional factors with negative effects or non-nutritive compounds having positive effects on health.

Soluble and insoluble dietary fibre can be both found abundantly in pulses[60, 11]. Prebiotic galacto oligosaccharides, counting raffinose, stachyose and verbascose made up between 3.5 and 6.9% of the cooked pulses (dry weight basis). Pulse fibre is rich in pectin and soluble fibre fraction made up of diverse mixture of polysaccharides shows a range of viscosities depending on the type of pulse. Having a large cellulose component the sugars constitute the insoluble fibre fraction. According to microscopic examination, cooked whole pulses showed that a large fraction of the starch was partially gelatinized and contained within intact cell walls, where boiled flour pastes had comparatively few ungelatinized granules. Some starch remained unfeasible to digestive enzymes after simulated upper-gut digestion. Cell walls of pulses help modulating starch gelatinization and play a major role in reducing enzymatic hydrolysis, which leads to low glycemic response attributed to pulses[11].

Milling, dehulling, soaking, germination, fermentation and cooking are the different processing methods subjected to pulses[33]. These processing techniques are quicker, energy saving and showed low fuel consumption. They also provide more nutritional leverage and produce edible products, which have a higher nutritional value and comparatively lower toxic compound as well. The type of pulses and the processing technique control and regulate the degrees of elimination of toxic compound[30]. For instance, cooking generally lowers saponin content of seed/grain in chickpea, mungbean and uradbean[54], pressure cooking curtails saponin, as compared to ordinary cooking[32]. Baked cooked lima beans (BCLB) tops the list in the highest protein content (14.97%), the least fat (1.02%) and the highest fibre content (5.88%), as compared to steamed cooked lima bean (SCLB). Phenolics, flavonoids and lycopene content were found to be higher in BCLB (18.5 GAE/g, 104.5 mg/100g and 12.7 mg/100g), respectively, when compared to SCLB which was found to be significantly higher ($p < 0.05$) in tannin, phytate and trypsin inhibitor compared to BCLB. Nonetheless, comparing with SCLB samples, BCLB samples were the most favoured in all the tested sensory attributes[21].

1.5 Importance of Post Harvest Technology and Processing in Pulses

Prevention of losses and value addition to the raw food commodities through preservation and processing are the two major objectives of post harvest technology. Cleaning, sorting and grading are done for raw food materials, followed by conditioning either for storage or processing. Importance of post harvest technology lies within the indisputable fact that it has the potential to fulfill food demands of growing population by eliminating losses creating more nutritive food products from raw commodities by proper processing and value addition. 70% of pulses produced in India are processed; therefore, post harvest technology plays an important role in per capita availability. Value addition to pulses after harvest is gaining importance in our country due to diversity in socio-economic conditions, industrial growth, urbanization, etc. It is not only to satisfy the producers or processors by getting high monetary return but also with better taste and nutrition.

Milling is a process by which materials are reduced from a larger size to a smaller size. The tools for milling seeds (e.g. quern stones) date back to the Neolithic period, c. 5600 BC[16] and were used to produce either dhal or flour. *Dhal* (also spelled *dahl, dal* or *daal*) or *splits* refer to the two cotyledons that remain after dehulling and splitting a pulse seed. The word *dhal* is derived from the Sanskrit word *dal,* which means "to split"[45]. The Sanskrit language dates back to about 1500 BC. Chickpea (*Cicer arietinum*) and pigeon pea (*Cajanus cajan*) *dhal* were mentioned specifically by Charaka, c. 700 BC[1]. Charaka was sometimes referred to as the "Father of Anatomy", as he was one of the founders of medicine and lifestyle advice promoting health and well-being in ancient India. Chickpea flour was also mentioned by Susruta in 400 BC[47]. Sushruta was recognized as the "Father of Indian Surgery" and wrote texts on general medicine and surgical procedures. Pulses were thought to be an important part of a healthy diet.

Considering the dates of pulse cultivation, Sanskrit manuscripts and the age of early milling equipment, the origins of pulse milling would date somewhere between 3500 and 7600 years ago. Pulses consumed in the form of *dhal* or flour account for around 80% of all pulse consumption in India[38]. Hence, milling is the most significant process for human consumption of pulses, followed by various methods of cooking and food preparation. Traditional milling in India yields a low 65–70% *dhal*[50] as compared to a theoretical yield of 80–90% (deduced from the seed coat comprising 10–20% of the seed). These two facts justify the importance of scientific research in pulse milling related areas. In the case of pulses, there are several processes that can be defined in this way:

1. Dehulling or decortication - loosening and removal of the seed coat to supply *polished* seed, *"footballs"* or *"gota"*.
2. Splitting - loosening and cleavage of the two cotyledons to produce *splits* or *"dhal"*. Dehulling and splitting often occur simultaneously in the same "milling" process for many pulses.
3. Flour milling or grinding - reducing whole seed or cotyledons to flour.

Post harvest loss reduction technology in pulses is confined to use of optimum harvest factors, reduction of losses in handling, packaging, transportation and storage with modern infrastructure machinery, processing into a wide diverse products, home scale preservation with low cost technology. Use of thermal processing, low temperature, drying, chemical reactions coupled with other preservation techniques are applied to enhance the storability. Containers and packaging materials confer portability besides extending the shelf-life. Adoption of those techniques could create obtainable outsized amount of food by avoiding losses and supply higher quality food and nutrition, more raw materials for processing, thus ensuring profitable returns to the farmers.

Post harvest technology has potential to create rural industries. India, where 80% people live in the villages and 70% of them rely on agriculture has experienced that the process of industrialization has shifted the food, feed and fibre industries to urban areas. This has resulted in capital drain from rural to urban areas, decreased employment opportunities in the rural areas balance trade in favour of urban sector and mismatched growth in economy and standard of living between rural and urban people. It is attainable to evolve applicable technologies, which can establish agricultural based rural industries. The farmer whose role has been reduced to producer can be re-modelled into producer cum processor and thus getting more dividends for hard labour, input, kind of risk taken and generating resource for socio-economic advancement keeping pace with the modern times.

1.6 Post Harvest Losses and Prevention in Pulses

"Post harvest losses" in pulses is the amount of quantitative and measurable qualitative losses in pulses throughout numerous phases after harvesting operation. Besides economic losses, post harvest losses may also result from poor post-harvest systems in pulses. From the harvest onwards, legume grains undergo a series of operations during which either quantitative or qualitative any kind of losses may happen in any time. According to various estimates[35, 56], total post harvest losses accounted for 25-30%. One estimate puts the milling losses at 1.5 million tonnes valued at Rs. 1500 crores[4].

Table 1: Estimated post harvest loss of pulses[36]

Stages	Losses (%)
Harvesting	1.0-3.0
Handling	1.0-7.0
Threshing	0.5-5.0
Drying	1.0-5.0
Transport	0.5
Primary processing	1.0
Storage	5.0-10.0
Milling	15.0-20.0
Total	25.0-50.0

During post harvest operations, such as threshing, winnowing, transportation, processing and storage, sizeable quantitative and qualitative loss of pulses occurred. Hence, due emphasis should be given to reduce qualitative as well as quantitative losses of pulses during various post harvest operations.

Table 2: Type of loss at different post harvest stages in pulses[36]

Stage of operation	Type of loss
Late harvest	Shattering losses, losses due to birds and pests attack
Insufficient drying of grains	Losses due to development of moulds and insects
Improper threshing	Broken grains and threat of insects
Poor storage	Losses happen due to combined attacks of insects, moulds, rodents, etc.
Improper milling	Broken grains and powdering loss
Transport	Quantitative loss
Defective packaging	Quantitative and qualitative loss

The extent of losses takes place at different phases varying between grains to grain. Fluctuation in post harvest losses of pulses depends on the climatic conditions, varieties, locations as well as the processing techniques applied in pulses. These losses can be minimized in the process of threshing, winnowing, storage, processing, handling and transportation[20].

i) Threshing and Winnowing

About 0.5-5.0% losses and 0.60 % losses have been reported in threshing and winnowing, respectively. To reduce the losses, threshing and winnowing operations are required to be completed within a short period through improved equipments on the pucca platform in the farm.

ii) Transport Losses

About 0.5 % losses has been reported during transport of produce from field to threshing floor and 0.19 % losses for transporting the produce from threshing floor to storage. Efficient and quick transportation supported by good packaging material can reduce the losses.

iii) Processing

Losses upto 1.0% has been reported due to use of outdated methods in processing. To reduce the milling losses and to increase the output, improved dal milling method developed by C.F.T.R.I,

Mysore should be adopted at the farm level and skilled personnel are required for giving proper trainings to the poor and illiterate farmers.

iv) Storage

About 5.0-10.0% loss is reported due to improper and inefficient methods of storage. Improved scientific storage facilities should be adopted to reduce the losses considerably.

Quantitative losses mainly results from spoilage, dryness of portion of produce, infestation by insects, rodents or birds. The storage losses caused by rodents, insects and micro-organisms have been reported to be maximum. Lack of storage facilities forces the farmers to sell their produce at lowest selling price. During storage, it must be checked that the pulses should remain in good condition and not undergo any deterioration due to fungal and insect infection or attack by rodents. In villages, grains are stored in mud-bins, paddy straw mud plastered bin, bamboo mud plastered bin, reed, mud bricks, baked mud plastered pitches, etc. The Indian Grain Storage Institute has also developed modern farm storage capacity of 14.5 million tonnes to meet farmers' requirements.

To avoid the post harvest losses following preventive measures should be adopted by the pulse producers, which can be much helpful:

1. Right stage of harvest at proper maturity will reduce losses.
2. Use proper method of harvesting.
3. Adopt modern scientific mechanical methods, to avoid the losses in threshing and winnowing.
4. Use improved techniques of processing.
5. Adopt cleaning and grading for remunerative prices to avoid financial loss.
6. Use good packaging materials for storage and in transport i.e. B-T will Jute bags or HDPE bags.
7. Adopt proper technique in storage.
8. Apply pest control measures during storage to avoid pests like bruchid beetle.
9. Proper handling i.e. (loading and unloading) of pulses with good transportation facilities at farm and market level reduces losses.

10. The quality of the stored grains should be examined by quality testers based on moisture content in grains, colour, odour, cleanliness, etc. before sending it to the market.

1.7 Post Harvest Systems of Pulses

In the food chain of pulses, harvesting is the 1st stage between the actual and theoretical agricultural production, which is done either manually or by the help of machines. But harvesting is not done until the crop attains its optimal maturity. The maturity of the crop can be characterized by shattering of leaves or browning or yellowing of leaves and pods. After the harvest, it is necessary to dry the crop to reduce the moisture content of the grains before the threshing operations. After threshing, the grains must be cleaned and stored properly for further processing operations like milling of grains, quality testing, etc. Finally, the grains are to be sent in the market for sale to the consumers, to commercial dal mills or to other agro-food industries. The sequence and interactions of these operations constitute a system i.e. post harvest system in pulses. This complex system can be described by the following flowchart easily:

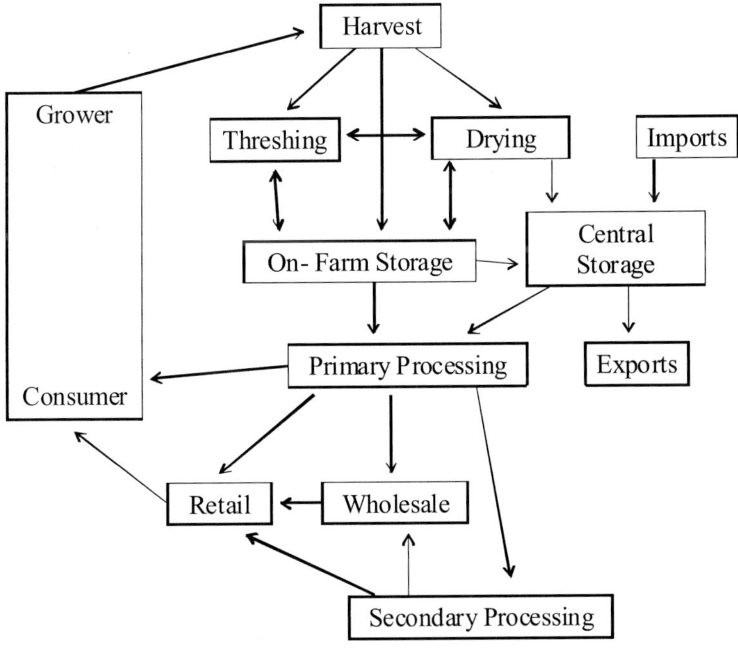

Fig. 2: Post harvest system of Pulses

Table 3: Traditional and improved technologies for post harvest operations

Post harvest operations	Traditional technologies	Improved technologies
Harvest	Manual	Manual and mechanized
Pre-drying	Standing or in shocks	Standing or in shocks
Storage of harvested crops	In fields or in threshing floor	In fields or in threshing floor
Threshing	Manual	Mechanized
Pre- cleaning	Hand winnowing	Mechanized
Drying	Natural	Artificial
Cleaning and sorting	Winnowing in the wind	Mechanized
Storage of grains	In traditional granaries	In bags or in bulk
Processing/Milling	Manual	Mechanized

1.8 Pulses Processing Technologies

A series of processes are involved before the legumes are ready to be used either as a plated item or as food preparation ingredient. Processing is the most important marketing function in the marketing of pulses for the present date. Through processing harvested produce are brought closer to human consumption in the form of value added products. Pulses are converted into dal by splitting the whole seed. Over 75% of the total legumes produced in the country is split into dal. Pulse processing is a small scale industry comprising of thousands of dal mills distributed throughout the country. Mills are mostly concentrated in producing areas, such as Indore (Madhya Pradesh), Jalgaon and Akola (Maharashtra), and in other major cities, such as Mumbai, Kolkata, Chennai, Hyderabad and Delhi. Several steps are involved in pulse processing. These processes include cleaning, drying, sorting, splitting, milling, and fractionating. Depending on the particular pulse and its use, other steps like de-hulling (decorticating), puffing, roasting and grinding may be included as well [24].

Colour inspection and quality tests are carried out by experts on receiving the field-dried pulse grains at the processing facility and then transferred to storage bins. Proper storage of products reduces the insect pest attack and thereby ensures quality and optimum preservation of the crop. Temperature and relative humidity are controlled to maintain optimum conditions for preservation and eradication of pest and fungal attack as different crops require different storage conditions.

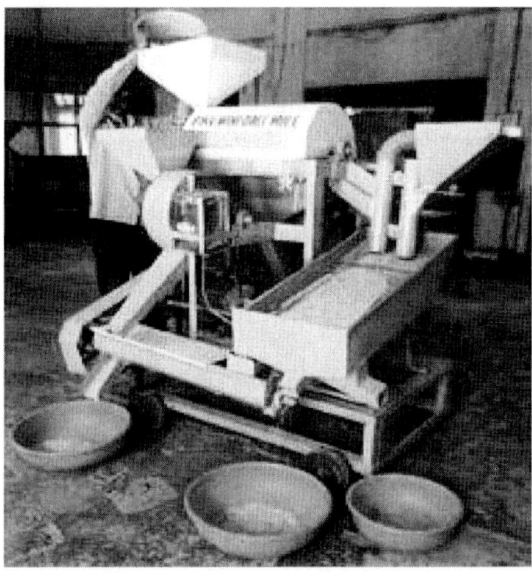

Fig. 3: Dal Mill Mechinery

When processing is completed, the product is bagged and sent to the market for selling. Pulses may be packed and shipped after cleaning or it may go under further processing. De-hulling and splitting are common practice, where the seed coat is removed and makes the pulses easier to digest by the human beings while improving other qualities.

For improving dehulling and splitting process, soaking of pulses in cold water overnight for at least 5 to 12 hours is a common practice. Use of warm water enhances hydration, which helps to prevent seed hardness. Thick and tough seed coats of pulses are often mechanically cracked before soaking to help facilitate moisture uptake. This also promotes faster cooking and more digestible cotyledons. Heating is more costly for processors and should be carried out in controlled process in order to maximize the yield while preventing unwanted microbial issues. In Asia, producers often hold their food legumes at ambient temperatures for several days after soaking to promote germination, which activates certain enzymes that then partially hydrolyze the proteins, starch, and oligosaccharides. The sprouted grains are then consumed directly or dehulled, roasted and ground for use in other foods.

Pulses could be consumed either as whole grain or in the form of ingredients like flour. Flour production involves grinding whole or split grain legumes into dry flour or wet batter for other food uses. Pulse composition, the fineness of the grinding, the ratio of particle size grades, and the cooking conditions affect the properties of the product, such as mouth feel, texture, and others. Techniques followed for processing are given in details below:

1. Cleaning Process

Cleaning criteria and cleaning methods are crucial to get high quality product and contribute to food safety. Cleaning of pulses is done by removing the dust and foreign substances to improve the physical properties of the products. Clean, healthy and high quality products have a high level of nutritional value, encourage sustainable agriculture and contribute to the reduction of the effects of climate changing scenario.

Cleaning is the first step in processing of legume. Screen cleaner and aspirator take out coarse and fine impurities as well as dust in the first process of cleaning. Screen cleaners are available in different types and sizes but they usually consist of flat sieves with slotted or round holes to separate fine and coarse impurities as well as dust particles based on size.

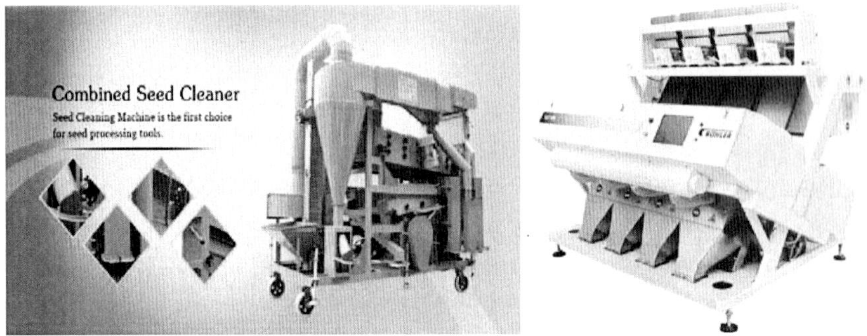

Fig. 4: Combined Pulse Seed Cleaner and Buhler Sorting Machine

Most of the pulse processors use a de-stoner to take out high density impurities, such as stones, pieces of metal and glass. In a de-stoner, the product will be spread over a perforated deck. The air passing through the perforated surface will create an air cushion on which the pulses will float slowly towards the machine outlet by gravity. Processors run the pulse grains over gravity tables, which act as a filter to rid them of light density foreign material like pebbles, dirt and any undeveloped, broken or damaged pieces. Sorting of legume grains could be based on size, density and colour, etc. Size sorting of pulses could be done on separators. For chickpeas, the separators usually use 5 sieves varying in dimension from 6 mm to 10 mm.

2. Calibration Process

The cleaned product is subjected to calibration process. Calibration process is done by screening machines. The purpose of calibration is to obtain standard groups by lowering the foreign materials to specified level, and to improve and classify the quality for commercial purposes. Homogeneous structure and standardized product in terms of cooking time for the final consumer are obtained through calibration[43].

Fig. 5: Calibrating Apparatus in Pulse Processing

Pulses like dry beans, chickpeas, green lentils are subjected on the sieves for calibration. Red lentils do not require calibration. While the caliber values of the pulses presented to the inner market and the outer market are different, the price of the pulses plays a determining role in the price of the product[43].

3. Dehulling or Decortication Process

Two steps are involved in dehulling of legumes: (1) loosening the hull (by the dry or wet method) and (2) removing the hull and cleaning. Methods involved in loosening of hull are:

1. Hull is loosened by prolonged sun-drying.
2. Applying tiny amount of edible oil, followed by sun-drying and tempering.
3. Soaking the legumes in water for several hours, followed by coating them with red-earth slurry and sun-drying.

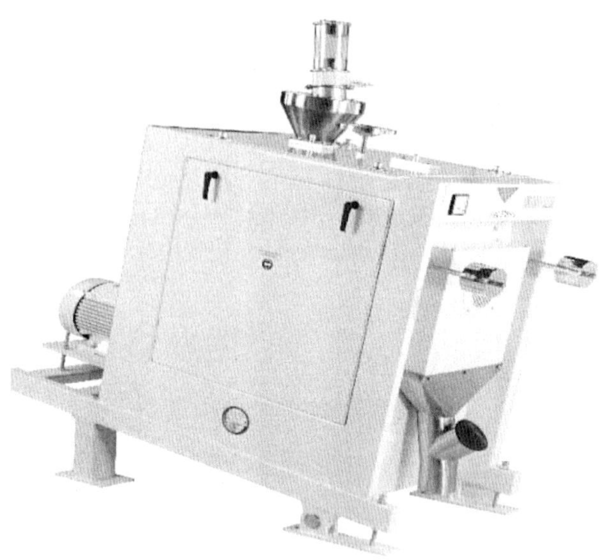

Fig. 6: Dehulling Machine of Pulse Processing

4. Soaking in water for many hours to loosen the hull before manufacture of food product.

Dehulling or decortication produces refined cotyledons with good appearance of the grains, texture and cooking qualities, which are more easily digestible and efficiently utilized by the body. Various dehulling methods are used, however, the oldest and most common technique involves spreading out the seeds to dry in the sun or mixing them with a bit of water before pounding them in a mortar with a pestle, followed by winnowing the hulls off to get the clean cotyledons. Commercial mills use similar methods, though being much larger in scale, they are adapted for greater yield and operational efficiency of grains.

4. Splitting Process

Various operations particularly decorticating and splitting are mechanized. Though both splitting and dehulling are effective as independent operations, both are often done in parallel. Splitting is enhanced by adding water before dehulling. After the hulls are discarded by winnowing, the split cotyledons are separated by sieving. The whole grains leftover that have not been split are similarly processed until as much of the grain as possible is dehulled.

5. Milling Process

Flour can be produced by further processing pulses. The type of the flour and therefore the milling method will be determined by the final product, which will be made of the flour. For instance, the particle size range and distribution have a huge effect on the functioning of flour like water absorption and viscosity, which will ultimately affect the properties of the final product like texture and porosity. The 4 principal techniques used to bring about the size reduction necessary for processing are impact milling, attrition milling, knife milling, and direct-pressure milling, etc.

Fig. 7: "Dal" Mill Machine

6. Packaging Process

Pulses are packaged in a proper way to protect from moisture loss and insects like bruchid beetles using cellophane, polyethylene (PE) pieced cardboard boxes, polyethylene bags. New more popular packaging options include paper/foil/PE spectacle packaging material. In general, volumetric automatic filling system and vertical type packaging machines are used in the packaging of pulses. The shelf-life of pulses that have been cleaned, calibrated, sorted, packaged and served, is as long as two years[42].

Fig. 8: Automated Packaging Machine

1.9 Value Addition to Pulses Products

Where there is a big requisite for promotion of short cooking time, microbial safety and high quality foods, value addition in pulses can be game changer in the market. Pulses are a good and cheaper source of plant proteins, dime a dozen than nuts, milk, cheese, meat and fish. They can also be used in bakery products like pasta, bread, snacks, etc. Talking about processed foods like bread, pasta, designer foods for the snacks, baby foods and sport foods, pulses are no less than mainstream ingredients. They have a high potential to step up breakfast cereals, microwaveable or partly prepared pulse based meal in fulfilling people's demands for convenient meal solution.

Value is added to the product at every stage of processing. Estimated value additions to the raw food materials through primary and secondary/tertiary processing in India are 75% and 25%, respectively. Primary processing having the greatest share clearly shows that it has a gigantic role to play in aiding the economic conditions of the farmers.

It is being depicted in Purans and Mahabharata that pulses have been an essential part of Indian diet as dal-chawal, dal-roti and in popular snacks like sattu, besan ke laddoo, etc. They are also by and large consumed in wide varieties of cuisines thoughout the country. Going back to the ancient times, Sattu is considered an 'Instant' food which is made by preparing flour from roasted chickpea and barley or wheat since the time of the "Rigveda" (8000 BC) and was commonly used by the poor mediocre people living in harsh conditions. One of the most popular traditional snack foods is papad, which is primarily made from pulses flour like black gram, green gram with salt, oil and spices. Papads are now an easy target for value addition by experimenting various processes, such as adding different cereal flours, green leafy vegetables, etc. Another widely popular product is the Hummus prepared from chickpea and has been found to reduce the post prandial glucose response 4 times less than white bread. In recent past, Indian Institute of Pulses Research (IIPR), Kanpur, India commended that pulses have a huge potential to be promoted as "Health Food" or "Nutri-Rich Food".

Industrialization has brought a new face in our country by revolutionizing the rural economy constantly. Farmers can lead a better lifestyle considering their economic and social status, if we put enough time and effort upgrading the agricultural produce, processing and value addition and its trade to urban areas. Thus, it's a high time to bring a revolution through value added products.

1.10 Values Added Products of Pulses

Value addition to pulses happens both through primary as well as the secondary processing. The opportunities for value addition extends from development of gluten free in baked food products, snacks and breakfast cereals; improvement of texture and colour of gluten free breads and batter; texture and colour improvement of fresh and dry gluten-free pasta; development of gluten-free soups, sauces, gravies; protein enrichment in fruit smoothies and vegan protein drinks; egg-replacement products, etc.

The most basic form of value addition to pulses occurs in the form of converting the grain legumes into "Dal"[36], which is obtained by splitting of the whole seed of the respective food legumes. Processing of grain legumes into dal involves the use of dal mills, which mostly operate in the small and medium scale sector. Traditional dal mills have lower recoveries of dehusked and splits (65-70%) as compared to the newly introduced improved dal mills in the region[46]. The improved dal mills are highly versatile and energy efficient with dehusking efficiency of about 95% and the yield of split pulses of about 80-85% depending on the variety of pulse and conditioning of the pulse grain. The steps involved in the processing of pulses are detailed in below (Fig. 9)[13].

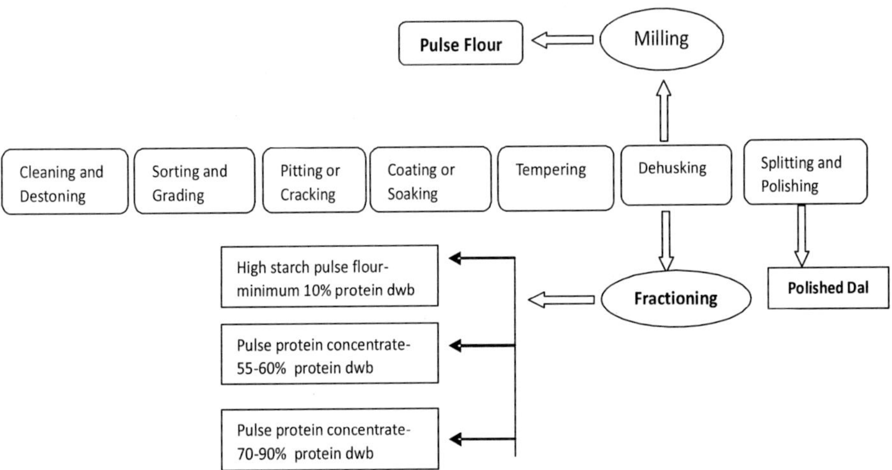

Fig. 9: Processing of pulses to dal and other value added products

1.11 Value Added Produts from Fractionating

In the search for new food protein and fiber resources, commercial facilities have begun focusing on extracting protein concentrates from pulses via a process called fractionation that allows researchers to separate out component ingredients to obtain the desired concentrates and isolates. Fractionation typically takes the form of a dry or wet method: air classification or wet milling. Dry and wet separation processes have been used for some time to fractionate grain legumes for both experimental purposes and industrial applications.

Wet Method

The traditional wet process is intended for food applications. Using this method, the hulls are removed from the seeds and then milled or ground into flour. The legume flour is pulped using a decomposing agent like an alkaline solution to pull out the protein, which is then dried. The solid matter left after the protein has been separated out is screened through a series of sieves to recover the starch.

For chickpeas and dehulled split yellow peas, starch fractionation involves steeping the seeds in warm water with *toluene* (i.e., a colourless, water-insoluble, flammable liquid often used as a solvent) to prevent fermentation. This is followed by wet grinding and repeated screening. Lentils are better served by a similar method that also includes re-suspension in a 0.2 per cent sodium hydroxide (NaOH) solution, which dissolves most of the protein.

Dry Method

The dry method uses a mill and air classification process to break down the dehulled seeds and separate out the starch and protein fractions. Dry processes have been employed more successfully with grain legumes than with other legume varieties because in legumes starch is the principal storage compound rather than oil. Removal of the loosened hulls from the grain in the dry-milling technique is commonly done in small machines. These usually take the form of under-runner discshellers or grinders with emery or stone contact surfaces. A plate mill is sometimes used to both hull and split the soaked and dried grains.

Wet Versus Dry

In a comparative study of dry and wet milling, dry air classification of pea flour containing 22 per cent protein and 55 per cent starch yielded fractions containing 53 per cent protein and 83 per cent starch. The protein fraction also contained some broken starch granules in addition to most of the lipid, ash, sugars, flavor, and colour compounds in the flour. The protein isolated from wet milling contained 88 per cent protein and refined starch contained less than 1.0 per cent protein. The refined fiber was light coloured and relatively free of other constituents. The main drawback of the wet milling method is the resulting loss of protein and starch in the whey and washes, as well as the expensive effluent recovery requirements.

Parching of pulses

Legumes, such as Bengal gram and peas are parched to give highly acceptable products. Bengal gram is tied in a moist cloth and kept overnight before it is parched. Peas are soaked in water for 5 minutes dried partially in the sun for 15 minutes and then parched. Salt and turmeric powder are sometimes added to the steeping water or the grains smeared with the paste of the same before they are parched. Parching is done in a hot iron vessel, containing sand at 190-200°C for 60-80 seconds. Parching Bengal gram has been used in the treatment of protein calorie malnutrition in children[59].

1.12 Pulses Based Product Concepts

Dal (split chickpea/pigeon pea) and food items prepared from besan (dal flour) and other pulses are regularly consumed in the countries of the region. 'Phutna'-roasted grains, 'Pakora'- oil fried, 'Kadi'- boiled in buttermilk, 'Roti'- chickpea flour in combination with wheat flour, 'Dhokla'- fermented pulse based product and 'Satu'- roasted, chickpea flour with cereal flours[31] are few such examples. Soyabeans among pulses have been most successfully exploited for development of a number of value added innovative products. From soyabeans, the following meat replacement products are made: tempeh, tofu (also known as soya curds or tahoe) and soya nuggets.

Consumers, today, are striving to have greater variety and healthier foods in the diet with the convenience of ready-to-eat (RTE) and fast cooking meals. Pulse protein flours, concentrates and isolates, therefore, offer opportunities for novel product development. It is reported that the typical flavour intensity of pulses decreases in low moisture applications and higher the heat treatment, the lower the flavour intensity. Colour of different pulses also vary from light yellow to green or red and can be used to obtain natural colour diversity in snack products and to differentiate these products from non "free-from" goods[41].

In addition to the above mentioned products a wide variety of pulse-derived fermented foods can also be developed. The micro-organisms involved in legume fermentation hydrolyze and metabolize pulses constituents resulting in the production of valuable products. Fermentation of pulses causes desirable changes including the improvement of nutritional value and sensorial properties and increased nutrient and phytochemical bioavailability (due to reduction/ elimination of anti nutritional factors), and also favours the formation of new bioactive compounds[22].

Table 4: Innovative Pulse Based Products

S. No	Pulse based products	Ingredient	Benefits of the product
1.	Pulse Pasta Flour	Made from chickpea flour and whole wheat flour	High protein and high fibre pasta.
2.	Pulse power bar	Made with a blend of chick pea and pumpkin flour mixed with bananas, hemp seed and egg.	Energy dense and high protein product.
3.	Muffins	Made from coarsely ground dried peas and pea fibre	Gluten free, high fibre baked delight.
4.	Candies	Pea protein concentrate	Fruit flavoured gummies fortified with pea protein and essential fatty acid.
5.	Twisted Roll	Made with yellow pea and heat flour	Higher fibre than traditional steamed bun.
6.	Pea tarts	Made with sweet potato and chick pea	Vegetarian and gluten free.
7.	Zuki bar	Made from whole small red bean flour and yellow pea hull	Healthy and all natural snack bar covered with a thin layer of chocolate.
8.	Pulsations	Black beans, dates, coconuts, dried cranberries and oats	High fibre and protein snack.
9.	Smoothie	Pea protein curd	Provide an allergen source of protein
10.	Pulse noodles	Blend of wheat and pulse flour including lentil, chick pea and split pea flours	Having a neutral flavour and can be accompanied with seasoned soups or sauces, also provide a balanced amino acid profile leading to enhanced protein quality.

(*Source*: Datta Majumdar *et al.*, 2017)

1.12 By-Product Utilization in Milling Pulses

Milling of pulses draws several by-products which comprise of different fractions or broken parts of pulses, which needs to be separated from the whole/split pulse. The by-products of dal milling like husk, dust powdery grain parts and small brokens are usually sold as cattle feed. The pulses husk which is low in bulk density (B.D.) has been conventionally used as the cattle feed forming about 10-12% of the raw material and is sold at a lower price. While the dal powder and small brokens, as cattle feed too, found to be richer in nutrients are sold at a higher price.

1.13 Value Addition in Storage

Storage of pulses has been considered as the most crucial factor, among all post harvest operations, due to losses caused by insect infestation. Storage alone contributes 7.5% of total post harvest losses, which are in total to the tune of 25-30%. Traditionally, storage is done in jute bags, *Thekka,* mud and metal bins with application of insecticides, fumigation and aeration. Since long back, controlled or modified storage structures were being used in the form of airtight structures, but their scientific studies have taken momentum in recent past.

The following requirements should be fulfilled for safe storage of pulses

1. Selection of site (location)

The storage structure should be located on a raised well-drained place. It should be easily accessible. The storage structure should be protected from humidity, excessive heat, direct sunrays, insect and rodents.

2. Selection of storage structure

The storage structure should be selected according to the quantity of pulses to be stored.

3. Cleaning of storage structures

The storage structures should be properly cleaned before storing pulses. There should be no left over grains, cracks, holes and crevices in structure, which may be harbour of insects. Before storage, the storage structure should be fumigated.

4. Cleaning and Drying

Before storage, the pulses should be properly cleaned and dried. Grains should be free from foreign matters and excessive moisture to avoid quality deterioration and pest attack.

5. Cleaning of bags

As far as possible, new gunny bags should be used. The old gunny bags should be properly cleaned, dried and fumigated before use.

6. Separate storage of new and old stock

To check infestation and to maintain hygienic condition of godown, the new and old stock should be stored separately.

7. Cleaning of vehicles

The vehicles used for transporting pulses should be properly cleaned with phenyl.

8. Use of dunnage

Dunnage should be used before stacking bags to avoid absorption of moisture from floor. Bags should be kept on wooden crates or bamboo mats preferably along with a cover of polythene sheet.

9. Proper aeration

There should be proper aeration during clear weather condition but care should be taken to avoid aeration during rainy season.

10. Regular inspection

Regular inspection of stored pulses should be carried out to check infestation. It is necessary to maintain proper health and hygiene of the stock.

1.14 Value Addition Effects in Pulses Marketing

The world today is in great need for high quality pulses that have both superior taste and nutrition. Initiations have already started for the development of new and improved post harvest storage, processing and value addition methods in marketing pulses crops. The liberalized trade regimes have brought a huge competition in the market, thus requiring an advanced and effective post harvest handling, processing, storage and distribution to access markets further[2]. Slow but steady developments are making in fresh quick packing items and minimally processed products of pulses. The marketing concept for minimally processed pulses is based on the consumer's desire for natural, less processed and high quality homemade style preparations. Chilled ready-to-eat foods are becoming a popular in the market. The nutrition, colouration, dehulling, cooking properties and moisture content have to be taken care of in achieving quality standards in the production.

Pulses demand is on the rise as people now-a-days have high appeal for value added products, such as bread and snack foods and initiations have started to increase their nutritional value and starch from pulses for use in the proper industries. Taking an example, the pulses in Australia have generally been used by the stock feed industry and any human consumption has been in a pre-cooked form, such as baked bean, thus leading to quite a lamentable belief, quite wrongly, by some as poor quality food. Moreover, supermarket snack rarely use a pulse component. On the contrary, South-east Asian countries have top selling snacks made from legumes. "Breakfast cereals" made from pulses with added nuts or fruits can be a fresh new flavour to try. Notably, one of the fastest growing segments of the retail market are pre-prepared meals. Manufacturers quest for flavour, appearance and an ability to retain texture and colour when reheated from a frozen or cold state. These properties will attract the consumers want these healthy nutritious product. Many pulses like lentil, peas, etc. can satisfy these requirements.

Regional, seasonal and particularly varietal differences must be taken into consideration while marketing pulses. Consumers in Tamil Nadu and Bihar states have a huge appeal for deep yellow colour in dal. Thus, to trigger the demand, dal is mixed with yellow colour additives (even non-permitted dyes) in water solution, dried and sold. The same method of coloring is also being used by millers to conceal the small patches of husk remaining on the dal, due to incomplete milling. While in western and north-western India, consumers are attracted to an oily looking dal, so, extra oil coating is given to the finished dal. This gives an oily shine to the product. In fact, the pulse industries compete in the market with various alternatives and therefore needs to differentiate itself. Here, the most important factor is to be creative and feature the varieties in pulses, rather than to mask them into a similar product and have the same effect.

1.15 Recommendations

It's a high time, where pulse consumption and production should be boosted in our country for many reasons including nutritional value, benefitting from health, promoting sustainable agriculture, soil health improvement, reducing climate change effects and pulse processing technologies should also be promoted accordingly. Pulse processing units need to be more advanced and skilled so that it can ensure human health and digestibility along with profitable income to the pulse producers, millers as well as traders also. Awareness is the key to achieve such feat by beating the challenge of food supply in world today, where population explosion is a colossal threat.

1.16 Conclusion

The post harvest technology in pulses in recent times has attracted attention as a powerful tool for rural industrialization. There is a sizeable quantitative and qualitative loss of pulses during different post harvest stages. Hence, it is appropriate to give due emphasis to reduce qualitative as well as quantitative losses of pulses during post harvest operations. With the onset of industrialization most of the processing activities have been shifted to urban areas, where infrastructural facilities are available. There is a strong need to reverse this trend and create means to increase income of rural people and generate employment opportunities in rural areas. Processing and value addition can be very powerful tools to achieve this goal. It provides an important opportunity for marketers that they should come up with strong brands in pulses to make consumers loyal to them by emphasizing on image building of the brands. With the UN General Assembly having declared the year 2016 as the "International Year of Pulses", it is the right opportunity to identify and launch appropriate interventions supported by appropriate policies in order to promote value addition through product development and marketing of pulses in the region. In addition to identifying new market opportunities, there is also the need to progress towards adoption of industry-wide grading standards and classifications of individual pulses, in order to address to demands of different segments of the trade.

One of the important policy interventions that need to be considered is support and advocacy towards encouragement of value addition and use of pulses in the governmental nutritional programs.

This could be in the form of directives and awareness campaigns on the benefits of incorporation of new pulse based, protein and micronutrient rich foods in the national programs of nutritional support to children, adolescent girls and pregnant women and school mid-day meal programs.

The onus solely lies on the marketers, to orchestrate effective campaigns and deliver quality products to create brand loyalty among the consumers. Marketers also need to revise prices to bring it in consonance with the consumers' perceived level. This step may boost the sales of these products. Marketers need to come up with different strategies for different class of cities to meet consumers' preferences and win their favour and retain their loyalties.

2. References

1. Acharya, K.T. (1998). *A historical dictionary of Indian food.* Oxford University Press, Delhi.

2. Agarwal, A. (2016a). Value Addition of Pulse Products in India. *Journal of Nutritional Health and Food Engineering,* 5(2): 590-592.

3. Agarwal, A. (2016b).Duality of anti-nutritional factors in pulses. *Journal of Nutritional Disorders and Therapy,* 6(1): 1-2.

4. Ali, N. and Srivastava, P. K. (1993). Conditioning, storage and milling of food grains and oilseeds. *AET.* 17(3-4): 1-18.

5. Anonymous (2010). Processing methods for dry Peas, lentils and chickpeas. *USA Pulses Technical Manual,* Chapter 4, USA Dry Peas and Lentil Council.

6. Bahwere, P., Balaluka, B., Wells, J.C.K., Mbiribindi, C.N., Sadler, K., Akomo, P., Dramaix-Wilmet, M. and Collins, S. (2016). Cereals and pulse based ready-to-use therapeutic food as an alternative to the standard milk and peanut paste based formulation for treating severe acute malnutrition: Non-inferiority individually randomized controlled efficacy clinical trial. *The American Journal of Clinical Nutrition,* 103: 1145-1161.

7. Bar-Yosef, O. (1998). The Natufian culture in the Levant, threshold to the origins of agriculture. *Evolutionary Anthropology: Issues, News, and Reviews: Issues, News, and Reviews,* 6(5): 159-177.

8. Bhagyawant S. S. (2013). Antinutritional factors in pulses and their diversity. *Innovare Journal of Food Sciences,* 1(1): 27-29.

9. Bharathi, R.R. and Shunmugam, A.S.K. (2014). Anti-Dietetic Factors in Legumes - Local Methods to reduce them. *International Journal of Food and Nutritional Sciences*, 3(3): 84-89.

10. Boye, J., Zare F. and Pletch, A. (2010). Pulse proteins: Processing, characterization, functional properties and applications in food and feed. *Food Research International*, 43: 414-431.

11. Brummer, Y., Kaviani, M. and Tosh, S.M. (2015). Structural and functional characteristics of dietary fibre in beans, lentils, peas and chickpeas. *Food Research International*, 67: 117–125.

12. Chrispeels, M.J. and Raikhel, N.V. (1991). Lectins, lectin genes, and their role in plant defense. *Plant Cell*, 3(1): 1-9.

13. Mazumdar, S.D., Durgalla, P., and M. Gaur, P. (2017). Chapter 5-Utilization of Pulses - Value Addition and Product Development. pp. 65-89. https://www.researchgate.net/publication/312937783.

14. Davis, K.R. (1981). Effect of processing on composition of tetrahymena relative nutritive value on green and yellow peas, lentils and white pea beans. *Journal Cereal Chemistry*, 58: 454–460.

15. Duranti, M. and Gius, C. (1997). Legume seeds: Protein content and nutritional value. Journal *of Field Crop Research*, 53: 31–45.

16. Encyclopædia Britannica (2010). Quern. Encyclopaedia Britannica Online. Retrieved from http://www.britannica.com/EBchecked/topic/487081/quern.

17. Erdil, D.N. (2016). Application studies of pulses in bakery products. *Healthy Nutrition and Healthy Life with Pulses Platform Project.* Ýstanbul, Turkey.

18. FAO, (2016). *The United Nations, declared 2016 as "International Year of Pulses" (IYP).* Retrieved from http://www.fao.org/fileadmin/user_upload/pulses2016/docs/factsheets/Nutrition_EN_PRINT.pdf.

19. FAO/FSNFORUM, (2016). Pulses are praised for their health, environmental and economic benefits. How can their full potential be tapped?. Global Forum on Food Security and Nutrition. Retrieved from www.fao-org/fsnforum 2016.

20. FAO/WHO Food Standards Programme, (2007).

21. Farinde, E.O., Obatolu, V.A. and Fasoyiro, S.B. (2017). Microbial, Nutritional and Sensory Qualities of Baked Cooked and Steamed Cooked Lima Beans. *American Journal of Food Science and Technology*, 5(4): 156-161.

22. Frias, J.E., Penas, C. and Villaluenga, M. (2016). Fermented Pulses in Nutrition and Health Promotion. In: *Fermented Foods in Health and Disease Prevention*. Academic Press, pp. 385-416.

23. Gemede, H.F. and Ratta, N. (2014). Anti-nutritional factors in plant foods: Potential health benefits and adverse effects. *International Journal of Nutrition and Food Sciences*, 3(4): 284-289.

24. Hatice P. (2017). Pulse Processing Technology, *International Conference on Technology, Engineering and Science (ICONTES)*, 1: 336-338.

25. Hefnawy, T.H. (2011). Effect of processing methods on nutritional composition and anti-nutritional factors in lentils (*Lens culinaris*). *Annals of Agricultural Science*, 56(2): 57–61.

26. Hu, F.B. (2003). Plant based foods and prevention of cardiovascular disease: An overview. American Journal of Clinical Nutrition, 78: 544-551.

27. Hudson, B.J.F. and El-Difrawi, E.A. (1979). The Sapogenins of the seeds of four lupin species. *Journal of Plant Foods*, 3: 181–186.

28. India Pulses and Grains Association (IPGA) (2014). Handbook on minor and imported pulses of India. Foretell Business Solutions Private Limited and Commodityindia.com (www.Commodity India.com), ISBN 9788175110502, South Africa.

29. Indian Council of Medical Research (2010). Nutrient Requirement and Recommended Dietary Allowances for Indians-A Report of the Expert Group of the Indian council of Medical Research. *National Institute of Nutrition*, Hyderabad.

30. Jain, A.K., Kumar, S. and Panwar, J.D.S. (2009). Antinutritional Factors and Their Detoxification in Pulses - A Review. *Agricultural Review*, 30(1): 64–70.

31. Jambunathan, R. and Singh, U. (1989). Present status and prospects for utilization of chickpea. In *Proceedings of Second International Workshop on Chickpea Improvement*. ICRISAT, Hyderabad, India, pp. 41-46.

32. Kataria, A., Chauhan, B.M. and Punia, D. (1989). Anti-nutrients and protein digestibility (in vitro) of mung bean as affected by domestic processing. *Food Chemistry*, 32: 9.

33. Khokhar, S. and Apenten, R.K.O. (2003). Anti-nutritional factors in food legumes and effects of processing. In: Squires, Victor R. (Eds), *"The Role of Food, Agriculture, Forestry and Fisheries in Human Nutrition"*,

Encyclopedia of Life Support Systems (EOLSS), Oxford, U.K. pp. 82 - 116. www.http:// www.eolss.net/ebooklib/sc_cart.aspx?File=E5-01A-06-05.

34. Kumar, A. and Gurung, T.R. (2015). SAARC Outlook on Water-Energy Food Nexus in SAARC Region, retrieved from http://www.sac.org.bd/archives/publications/.

35. Kurien, P.P., Desikachar, H.S.R. and Parpia, H.A.B. (1973). Processing and utilization of grain legumes in India. *Tropical agriculture research series.*

36. Lal, R.R. and Verma, P. (2007). Post Harvest Management of Pulses, Technical bulletin, Indian Institute of Pulses Research, Kanpur. pp. 46.

37. Mangaraj, S. and Singh, K.P. (2011). Optimization of machine parameters for milling of pigeon pea using RSM. *Food and Bioprocess Technology,* 4(5): 762-769.

38. Mangaraj, S., Agrawal, S., Kulkarni, S.D., Kapur, T. (2005). Studies on physical properties and effect of pre-milling treatment on cooking quality of pulses. *The Journal of Food Science and Technology,* 42: 258–262.

39. Marero, L.M., Payumo, E.M., Aguinaldo, A.R., Matsumoto, I. and Homma, S. (1991). Anti-nutritional factors in weaning foods prepared from germinated cereals and legumes. *LWT - Food Science and Technology,* 24: 177–181.

40. Markievicz, M., Kolanowka, A. and Gulewics, K. (1988). The chemical composition of debittered lupine seeds and their extract. *Bulletin of the Polish Academy of Sciences-Biological Sciences,* 36: 1-3.

41. Meissner, K. (2016). Improving nutrition and choice in the free-from sector Innovation opportunities using pulse-based ingredients in free-from products. *Agro Food Industry Hi-Tech,* 27(3): 22-24.

42. Ministry of Education (MEB) (2013). Module of packing cereals and products. *Food Technology,* 1-47. Ankara, Turkey.

43. Ministry of Education, MEB (2011). Module of calibration in dry pulses and cereal grains. *Food Technology,* 1-72. Ankara, Turkey.

44. Ministry of Women and Child Development (MoWCD) (2011). Report on the Working Group on Nutrition for The 12th Five year plan (2012-2017). Government of India. Retrieved from http://planningcommission.gov.in/aboutus/committee/wrkgrp12/wc d/wgrep_nutition.pdf.

45. Monier-Williams Sanskrit-English Dictionary (2010). 21 April, 2010. Retrieved from http://webapps.uni-koeln.de/tamil/.

46. National Bank for Agriculture and Rural Development (NABARD) (2016). Dal Mill retrieved from https://www.nabard.org/english/dal4.aspx)

47. Nene, Y.L. (2006). Indian pulses through the millennia. *Asian Agri - History*, 10: 179-202.

48. Peumans, W.J. and Van Damme, E.J. (1995). Lectins as plant defense proteins. Plant *Physiology*; 109(2): 347–352.

49. Philanto, A. and Korhonen, H. (2003). Bioactive peptides and proteins. *Advances in Food and Nutrition Research*, 47: 175-778.

50. Planning Department of the Government of Uttar Pradesh (2000). Model project on improved dal mill. Government of Uttar Pradesh. 21 April 2010. Retrieved from http://planning.up.nic.in/innovations /inno3/ae/dal.htm.

51. Ramdath, D., Renwick, S. and Duncan, A.M. (2016). The Role of Pulses in the Dietary Management of Diabetes. *Canadian Journal of Diabetes*, 40(4): 355–363.

52. Rizkalla, S.W., Bellisle, F. and Salma, G. (2002). Health benefits of low glycemic index foods, such as pulses, in diabetic patients and healthy individuals. *British Journal of Nutrition,* 88(S3): 255-262.

53. ªahin, N. (2016). Importance and usage area of legumes for food industry. *Symposium of Healthy Nutrition and Healthy Life with Pulses.* Ýstanbul, Turkey.

54. Sharma, A. and Sehgal, S. (1992). Effect of processing and cooking on the antinutritional factors of faba bean (*Vicia faba*). *Food Chemistry,* 43: 383.

55. Shi, L. (2015). A Quantitative Assessment of the Anti-nutritional Properties of Canadian Pulses (Master's Dissertation). University of Manitoba, Winnipeg, Manitoba.

56. Srivastava, H.C., Bhaskaran, S., Menon, K.K.G., Ramanujam, S. and M.V. Rao. (1984). Pulse production, constraints and opportunities In: Birewar, B.R. Edn. (1982), "Post-harvest technology of pulses". *Proceedings of Symposium on Increasing Pulse Production in India – Constraints and Opportunities,* Oxford and IBH Publishing Company, New Delhi, 167-176.

57. Tharanathan, R.N. and Mahadevamma, S. (2003). Grain legume-A boon to human nutrition. *Trends in Food Science and Technology,* 14: 507-518.

58. Thompson, H.J., Heimendinger, J., Haegele, A., Sedlacek, S.M., Gillette, C., O'neill, C. and Conry, C. (1999). Effect of increased vegetable and fruit consumption on markers of oxidative cellular damage. *Carcinogenesis*, 20(12): 2261-2266.

59. TNAU Agritech Portal. Post Harvest Technology, Value Addition in Pulses, http://www.21food.com/userImages/jyglobe/jyglobe$38171957.jpg.

60. Tosh, S.M. and Yada, S. (2010). Dietary fibres in pulse seeds and fractions: Characterization, functional attributes, and applications. *Food Research International*, 43(2): 450–460.

61. Uysal, O. and Subaþý, O.S. (2014). A study on determination of strategy for pulse industry through SWOT and SOR analyses in Mersin Province. *Turkish Journal of Agriculture- Food Science and Technology,* 2(6): 256-259.

62. Vohra, K., Dureja, H. and Garg, V. (2015). An Insight of Pulses: From Food to Cancer Treatment. *Journal of Pharmacognosy and Natural Products,* 1: 108.

63. Wood, J.A. and Malcolmson, L. (2011). Pulse Foods: Processing, Quality and Nutraceutical Applications. In: Tiwari, B.K., Gowen, A. and Mckenna, B. (Eds.): Pulse Foods, Oxford Academic Press, pp. 193-221.

64. Zohary, D. and Hopf, M. (1973). Domestication of pulses in the Old World: legumes were companions of wheat and barley when agriculture began in the Near East. *Science*, 182(4115): 887-894.

❑❑❑

Chapter 5

Processing Quality and Value Addition in Potato

Hirak Banerjee[1], Sukamal Sarkar[1], Krishnendu Ray[2], Lalita Rana[1] and Sourav Garai[1]

[1] *Bidhan Chandra Krishi Vishwavidyalaya, West Bengal, India,* [2] *Shasya Shyamala Krishi Vigyan Kendra, West Bengal, India*

ABSTRACT

A steady increase in production of potato over last five decades has brought about the potato revolution in India. However, the increasing trends in potato production cannot be sustained without diversification in its utilization. Therefore, potato processing industry will be essential to sustain the steady production rate. Fast growth in processing sector is expected in India due to increased urbanization, rise in per capita income and expansion of tourism trade. However, potato processing sector has been handicapped for want of good quality and sufficient quantities of raw materials throughout the year. Potato can be processed into mainly chips and others like french fries, canned products, starch, alcohol, custard powder, papad, etc. Costly storage and transportation, limited availability with considerable quality of raw materials are the main drawback faced by processing sector. Some of the recently released potato cultivars are Kufri Chipsona-1, Kufri Chipsona-2, Kufri Chipsona-3 and Kufri Chipsona-4. The primary considerations of good quality potato tubers are tuber shape (round, round-oval, long-oval), size (35-55 mm), depth of eyes (shallow) and chemical constituents (reducing sugar of 250 mg $100g^{-1}$ fresh weight, dry matter 20% or more, etc.). Crop management practices like irrigation (at 40 mm cumulative pan), boron (0.1% foliar spray of B at 40, 50 and 60 days after planting) and zinc nutrition (4.5 kg Zn ha^{-1}) and harvesting time (at full maturity) determine the quality of potatoes. Hence, potato processing is important to solve the problem of expensive refrigerated storage, to avoid market gluts and to save precious food against spoilage. It is essential to create awareness among farmers about several uses of potatoes, to encourage village level processing of potatoes and to generate additional income for rural people.

1. Introduction

The potato (*Solanum tuberosum* L.) is an important high yielding, nutritive food crop after rice, wheat and maize. Countries like India and China have witnessed potato revolutions and have become major potato producers in the world. Currently, India is the second largest producer of potato next to China (45.3 million tonnes) in the world with an average yield of 22.8 t ha^{-1} (Sarkar *et al.*, 2018). A steady increase in area, production and productivity of potatoes in the country over last five decades has brought about the potato revolution (Mondal *et al.*, 2015). However, the increasing trends in potato production can not be sustained without diversification in its utilization. Therefore, potato processing industry will be essential to sustain the steady production rate. Fast growth in processing sector is expected in India due to increased urbanization, rise in per capita income and expansion of tourism trade. However, potato processing sector has been handicapped for want of good quality and sufficient quantities of raw materials throughout the year.

1.1. Potato Processing

In India, the first attempt at processing of potatoes on a commercial scale was made by Col. Rennick in 1911, who established a factory near Narkanda, Himachal Pradesh to produce potato meal. He utilized hand-operated machines for peeling and slicing and cauldrons and furnaces for boiling and drying. The yield of dried product was between 16-20%. Dehydrated potato chips are being made in Indian homes for a long time, but the commercial large scale production of fresh fried potato chips has started only about 25 years back.

Potato can be processed into different products like chips, French fries, canned products, starch, alcohol, custard powder, papad, etc. In India, potato chips is the most common processed potato product. About 90% of the potato chips are produced by the unorganized sector and less than 10% produced by organized sector (Ezekiel and Shekhawat, 1999). In organized sector, potato processing has been made in a traditional way. Hence, their quality of product is very poor and difficult to sell in the established market. Besides potato chips, there is a demand of French fries for home consumption as well as for export now-a-day. It is very essential to increase the organized sector for potato processing, because there exists very limited potato processing industries particularly in North-western plains around Delhi region. Potatoes for processing are available only during January to March in north-western plains of India. Very limited quantities are available during August to November either from high temperature (10-12°C) with CIPC application/pit stored potatoes from Malwa region of Madhya Pradesh or from *kharif* crop in plateau regions involving costly storage and transportation (Singh *et al.*, 2002). So, it is needed to increase the processing industries mainly in higher potato production area of different states of the country. At the same time, quality of potato for processing should be available in nearby processing industries throughout the year.

1.2. Present Scenario of Potato Processing

Potato processing in India has become important due to rising urbanization and demand for processed food. Processing also helps in reducing the storage space, which is the main problem in India. Due to reducing in storage space, it adds value to potatoes and gives better returns (Srivastava and Kumar, 2002). Traditionally, processing potato varieties can be stored in on-farm storage for three months and increased the availability of potato to processing industries. In India, potato is grown in almost all the states but only the certain areas are suitable for producing processing varieties because of favourable environmental conditions during crop growth.

1.3. Potato-based Processed Products

Potato can be processed into number of products, such as chips, dices, potato flour, granules, flakes, waris, papads, canned potatoes, etc. by means of dehydration for preserving products for long time. Potato chips are the most popular processed product both in organized and unorganized sectors. Good quality potato chips should have low oil content and no brown colouration. French fries are also an important product, which is found in restaurants and bars of big cities. Because of higher prices, it can not be consumed at village level. There is a little demand of French fries in India because of high cost of French fries. Therefore, in India, it will take long time for French fries to become popular like chips. Besides chips and French fries, several value added products, such as canned products, custard powder, papad, soup thickener, biscuits, weaning foods, etc. have been developed from potato starch, which can be prepared even at low dry matter content and high reducing sugar (Marwaha and Sandhu, 1999). In the Indian plains, there is an abundance of solar energy after harvest of potato (March onwards), which can be utilized for dehydration to process such products. The dehydrated potato products are commonly prepared by organized sector as well as at village level, because dehydration is the simplest and the lowest cost-effective method of processing. This simple type of processing should be encouraged so that it is acceptable by all class of people in the country and upgrade the value of low grade tubers (small or damaged tubers).

1.4. Morphological and Bio-chemical Characters of Potato Tuber

The primary considerations are morphological characters of the potato tuber *viz.* tuber shape, size, depth of eyes and chemical constituents (tuber dry matter and reducing sugar content). Quality requirements of various potato products are given in Table 1.

Table 1: Quality requirements of potatoes for processing

Characteristics	Dehydrated	French fries	Chips	Canned
Tuber size (mm)	30	50	40-60	35
Specific gravity	1.080	1.080	1.085	1.080
Dry matter (%)	22-25	20-24	22-25	18-20
Starch (%)	15-19	14-16	15-18	12-24
Reducing sugars (%)	0.5	0.5	0.25	0.5
ACD	Slight	Slight	-	-
ED	Slight	-	-	-

ACD, After cooking discolouration; ED, Enzymatic discolouration
Adapted from Ezekiel *et al.* (1999)

I. Shape and size of tubers

The shape of the potatoes is considered to be the most impotent trait for processing. Round to round-oval potatoes are preferred for the preparation of the chips. Long-oval tubers ensure that larger number of French fry sticks of the desired size can be obtained. Large tubers (40-60 mm in diameter) are preferred for preparation of chips and French fries, while small sized tubers are used for canning. Use of large size potatoes result in lower peeling losses, while smaller tubers result in higher peeling losses. Similarly, peeling losses in potatoes with deep eyes are much greater than in potatoes in shallow eyes (Ezekiel *et al.*, 1999).

II. Chemical composition of potato tubers

Parameters	Quantity (%) on fresh weight basis
Water	80
Dry matter	20 (Starch 14%, Sugar 2%, Crude protein 2%, Fat 0.1 %); small quantities of fibres, vitamins and glyco-alkaloids

Adapted from Ezekiel *et al.* (1999)

II. Dry matter content

It is one of the most important characters determining the end use of potatoes in processing industries. Potatoes, containing high dry matter (20% or more), are preferred for fried and dehydrated products, while those containing low dry matter (18-20%) are preferred for canning. Potatoes with high dry matter are preferred for fried and dehydrated products because (a) Lower energy requirement for removing water during frying and dehydration, (b) lower uptake of cooking medium during frying and (c) higher yields of the products (Ezekiel *et al.*, 1999).

III. Specific gravity

Specific gravity of potatoes is directly proportional to their dry matter content, i.e. the dry matter content of potatoes increases with an increase in their specific gravity.

IV. Reducing sugar content

Sugar content of potatoes plays an important role in the development of desirable colour of chips and French fries during frying. Processed products prepared from potatoes with high sugars turn brown and black, and finally become unacceptable to the consumers. At the time of harvest, the sugar content of mature potatoes is generally low (0.5-1.0% by weight). Half of this quantity is accounted for sucrose, and other half by glucose and fructose. A reducing sugar content of 250 mg $100g^{-1}$ fresh weight is considered acceptable for producing chips. While for French fries, canning and dehydrated products the reducing sugar content upto 500 mg $100g^{-1}$ fresh weight is acceptable (Ezekiel *et al.*, 1999).

V. Discolouration

Discolouration of fried products is primarily due to high reducing sugar content. But potatoes are likely to show two other types of discolouration – (a) *enzymatic discolouration:* It occurs when the potatoes are peeled, cut or injured. Some of the constituents react in the presence of oxygen (air) and starts browning. This can be overcome by keeping the cut potatoes immersed in water, (b) *after-cooking discolouration:* It develops after cooking on exposure to air.

1.5. Factors Influencing the Quality Parameters of Potato

i. Irrigation and mulching

Quality characteristics like dry-matter content, crude protein, ash content, starch and vitamin C content increased with the decrease in number of irrigation (Banerjee *et al.*, 2016a). Plants irrigated with 20 mm CPE produced poor quality tubers compared to other plants grown with other tested irrigation schedules (Table 2). Similar results were obtained by Bosnjak *et al.* (2004) and Carli *et al.* (2014), who reported that dry-matter, starch content and vitamin C content of potato tubers increased with water restriction. While plants irrigated only at critical growth stages produced tubers having better-quality parameters. Mulching with paddy straw also significantly influenced tuber quality over no-mulch condition, except specific gravity.

Table 2: Quality of produce as influenced by irrigation levels and mulching on potato cv. Kufri Jyoti (pooled value of 3 years)

Treatment	Dry matter content (%)	Crude protein (%)	Ash content (%)	Starch content (%)	Vit. C content (mg/100g)
Levels of irrigation					
I_1 - 20 mm CPE	16.0	6.3	8.06	11.9	20.1
I_2 - 25 mm CPE	16.8	7.0	8.47	12.5	22.5
I_3 - 30 mm CPE	18.2	7.3	8.89	13.2	24.1
I_4 - 35 mm CPE	18.8	8.6	9.12	14.1	26.9
I_5 - 40 mm CPE	20.0	9.0	9.82	14.5	27.3
I_6 - Critical stages*	20.6	10.0	10.03	15.1	29.0
SEm±	0.80	0.75	0.52	0.67	1.17
CD (P=0.05)	2.41	2.26	1.55	2.03	3.51
Mulching					
M_0 - No mulch	16.5	7.0	8.02	12.1	22.2
M_1 - With mulch	20.4	9.0	10.11	14.9	27.7
SEm±	1.20	0.67	0.69	0.93	1.84
CD (P=0.05)	3.56	2.01	2.07	2.78	5.49

CPE = Cumulative pan evaporation, *Control treatment

Adapted from Banerjee *et al.* (2016a)

ii. Boron nutrition

Boron application in potato cultivation significantly influences the quality parameters, namely specific gravity, total soluble solids (TSS), tuber hardness, total acidity and vitamin C content of potato tubers (Sarkar *et al.*, 2018). The potato tubers obtained from the plants treated with foliar B had higher quality attributes than those harvested from the plants treated with soil applied B (Table 3). Tubers harvested from the plants treated with RDF of NPK + 0.1% foliar spray at 40, 50 and 60 DAP exhibited significantly higher specific gravity, TSS, tuber hardness, total acidity and vitamin C content, accounting 1.87, 51.4, 33.9, 21.9 and 20.2% more than the potatoes harvested from control plots (Only NPK, without B). Boron plays an important role in the translocation of carbohydrates from leaves to other parts of the plant. Therefore, greater concentration of ascorbic acid may have been translocated to the tuber as a result of B application (Mondy and Munshi, 1993), which makes the tubers more nutritious. Their study did not show any significant variations of total dry matter content in tubers with boron fertilization. The same treatment (RDF of NPK + 0.1% boric acid spray thrice) proved to the best by producing tubers with significantly higher protein content and total starch content, accounting 70.8 and 23.2% more than control (without B). According to Sathya *et al.* (2009), B improves photosynthetic activity, enhances activity of enzymes and plays a significant role in protein and nucleic acid metabolism. Bandana *et al.*, (2016) holds the view that crisp texture of potato tuber depends mainly upon the starch content. High starch content forms more gelatinization during processing of tubers, and thereby contributes to desirable texture. The study of Sarkar *et al.* (2018) further indicate reduction of total soluble sugar content in potatoes with B application and more acceptable tubers are produced with foliar application of B (thrice). Low sugar content in tubers is considered acceptable for producing chips, while processed products from potatoes with high sugar (as obtained from control plots) turn brown or black, and finally becomes unacceptable (Mondal *et al.*, 2015). They added, B application was also effective in reducing the phenolic concentration in tubers and significantly lowest total phenol content was recorded in potatoes harvested from the plants fertilized with RDF of NPK + 0.1% boric acid spray thrice (60% less than control). This result is in agreement with the findings of Mondy and Munshi (1993), who suggested that boron could have restricted the influx of substrate into the pentose-phosphate pathway and the synthesis of phenols, thereby reducing the total phenolic concentration in the tuber. While Goldbach (1997) suggested that under B-deficiency the use of carbohydrates for the deposition of cell wall material is likely to be inhibited. This close temporal association between B-deficiency and increased phenol concentration has also been observed in the study of Sarkar *et al.* (2018). In addition, B application significantly improved the potato chip colour. Potatoes harvested from plants fertilized with RDF of NPK + 0.1% boric acid spray at 40, 50 and 60 DAP recorded lighter chip colour than those harvested from plants fertilized with any other B application (73.2% lighter than control). However, reduction of B foliar sprays upto two or one resulted in non-significant increase in chip colour of potato. Soil application of B was less effective in producing lighter chip colour than foliar sprays, and it was statistically at par with B omission treatment (RDF of NPK only). According to Bandana *et al.* (2016), the phenolic concentration in tubers influence colour development in processed potato products and the desirable chip colour can only be obtained with low phenol content in tubers.

Table 3: Quality parameters of potato cv. Kufri Chipsona-3 as influenced by varying methods and time of boron fertilization (mean of 2014-15 and 2015-16)

Fertilizer treatment	Specific gravity	Tuber dry matter content (%)	TSS (°Brix)	Tuber hardness (kg/cm)	Total acidity (%)	Vitamin C (mg/100 g of dry weight)	Protein (%)	Total starch (mg 100 g^{-1} of dry weight)	Total sugar (mg 100 g^{-1} of dry weight)	Total phenols (mg 100^{-1} g of fresh weight)	Chip colour
RDF (control)	1.070 c	21.51 a	3.50 d	7.09 d	7.34 c	12.84 b	1.54 b	53.09 d	0.81 a	25.93 a	3.36 a
RDF + 2.0 kg B ha^{-1}	1.074 bc	21.83 a	4.08 cd	7.61cd	7.42 c	13.31 b	2.08 a	55.14 c	0.77 ab	21.58 b	3.09 ab
RDF + 0.1% boric acid at 40 DAP	1.078 b	21.06 a	4.40 bc	8.46 bc	7.99 b	13.23 b	2.01 a	55.98 c	0.78 ab	20.42 bc	2.74 b
RDF + 0.1% boric acid at 40 & 60 DAP	1.080 b	22.37 a	4.84 ab	9.16 ab	8.64 a	13.70 b	2.17 a	61.10 b	0.73 b	17.14 dc	2.25 c
RDF + 0.1% boric acid at 40, 50 & 60 DAP	1.090 a	23.59 a	5.30 a	9.49 a	8.95 a	15.43 a	2.63 a	65.43 a	0.72 b	16.21 d	1.94 c

DAP, days after planting; TSS, Total soluble solids; Means, followed by a different letter, are significantly different at $p \leq 0.05$, Chip colour 1-5 score is a acceptable colour, on 1-10 scale of increasing dark colour
Adapted from Sarkar *et al.* (2018b)

iii. Zinc nutrition

In the study of Banerjee *et al.* (2016b), biochemical properties of potato tubers, estimated as TSS, total acidity, and ascorbic acid content were not significantly influenced with Zn fertilization, but the total starch content (dry weight basis) of tuber increased with increasing Zn doses (Table 4). Potato fertilized with 6.0 kg Zn ha^{-1} recorded the highest starch content (58.0 mg 100 g^{-1} of dry weight). These results are in agreement with the findings of Mousavi *et al.* (2007), who specifically observed that application of Zn at 2, 4 and 8 ppm increased starch content in potato tubers to the tune of 2.1, 5.2 and 7.1 per cent over control, respectively. Dwivedi and Dwivedi (1992) also reported that starch content was greatly affected not only by Zn but also by its application method. In their study, starch content of tuber was significantly affected by soil application (10 kg $ZnSO_4$ ha^{-1}) as well as seed soaking with Zn solution. The amount of starch accumulated is determined by the rate of photosynthesis, translocation of photosynthates from leaves to tubers and subsequent conversion to starch (Kumar *et al.,* 2008a). Therefore, the increased production of starch in tubers might be due to increased photosynthetic rates in the presence of Zn.

Zinc fertilization was effective in reducing the total phenol content in tubers as evident in Table 4. The lowest total phenol content (18.95 mg 100 g^{-1} fresh weight) was recorded in potatoes harvested from the plants fertilized with RDF + 4.5 kg Zn ha^{-1}, and it was statistically at par with total phenol content in potato tubers those harvested from plants treated with RDF + 3.0 kg Zn ha^{-1} (20.15 mg 100 g^{-1} of dry weight) and RDF + 6.0 kg Zn ha^{-1} (19.85 mg 100 g^{-1} of dry weight). This result supports previous research finding of Mondy and Chandra (1981), who also found significant reduction of total phenol content in potato with Zn fertilization.

Table 4: Biochemical properties of potato cv. Kufri Jyoti as affected by different Zn levels (values are pooled data of 2013-14 and 2014-15)

Treatments	Total soluble solids (°Brix)	Total acidity (%)	Ascorbic acid (mg 100 g^{-1})	Total starch (mg 100 g^{-1} of dry weight)	Total phenol (mg 100 g^{-1} of dry weight)
RDF + Zn_0	5.90a	8.25a	19.7a	56.7c	23.33a
RDF + $Zn_{1.5}$	5.90a	8.25a	20.2a	56.6c	21.33b
RDF + $Zn_{3.0}$	5.92a	8.37a	19.2a	57.4b	20.15bc
RDF + $Zn_{4.5}$	6.10a	8.12a	19.5a	57.7ab	18.95c
RDF + $Zn_{6.0}$	6.42a	8.97a	19.2a	58.0a	19.85bc
LSD (P=0.05)	NS	NS	NS	0.20	1.51

RDF: Recommended $N:P_2O_5:K_2O$ dose of 200:150:150 kg ha^{-1}; Means, followed by a different letter, are significantly different at $p \leq 0.05$; NS: Non-significant
Adapted from Banerjee *et al.* (2016b)

Banerjee *et al.* (2017) in their study demonstrated that the specific gravity was significantly ($p \leq 0.05$) affected with Zn fertilization (Table 5). The specific gravity was more in potato tubers harvested from plants treated with RDF + 4.5 kg Zn ha^{-1}. There was reduction in the specific gravity due to reduction of Zn level, and it was poor with RDF of NPK only (without Zn). Mousavi *et al.* (2007) also found 1.9 % increment in potato tuber specific weight with 8 ppm $ZnSO_4$ by foliar application. Zinc fertilization exerted positive and significant

($p \leq 0.05$) influence on the production of total dry matter in tubers. In addition, RDF + 4.5 kg Zn ha^{-1} produced tubers with highest total dry matter content accounting 10.2 % more than control ($^{-}$Zn). As evident from Table 5, Zn fertilization significantly ($p \leq 0.05$) improved the potato chip colour, and potatoes harvested from plants fertilized with RDF of NPK + 6.0 kg Zn ha^{-1} recorded 67.6 % lighter chip colour than those harvested from plants fertilizer with RDF of NPK only (Zn omission). This may be due to less production of phenol substances in tuber influenced by Zn^{++} ion, thus less chances of enzymatic discoloration of potato chips during frying. Reduction of Zn level upto 1.5 kg Zn ha^{-1} resulted significant increase in chip colour of potatoes. Kumar *et al.* (2008a) also find similar results with 'Kufri Chipsona-1' cultivar, where three times foliar sprays with $ZnSO_4$ (0.2%) significantly improved chip colour over control. In the study of Banerjee *et al.* (2016b), total sugar content in potato tubers was more or less constant and recorded no significant changes with varied levels of Zn fertilization.

Table 5: Effect of zinc fertilization on processing quality of tuber of potato (cv. Kufri Jyoti)

Fertilizer treatments	Specific gravity (g cm^{-3})	Tuber dry matter (%)	Chip colour score
RDF + Zn_0	1.069^{c}	19.85^{b}	7.56^{a}
RDF + $Zn_{1.5}$	1.074^{bc}	19.84^{b}	6.11^{b}
RDF + $Zn_{3.0}$	1.077^{b}	20.19^{b}	3.75^{c}
RDF + $Zn_{4.5}$	1.088^{a}	21.88^{a}	2.45^{d}
RDF + $Zn_{6.0}$	1.071^{bc}	20.32^{ab}	4.08^{c}
$LSD_{0.05}$	0.01	0.67	0.72

Values are pooled data of 2013-14 and 2014-15; RDF: Recommended $N:P_2O_5:K_2O$ dose of 200:150:150 kg ha^{-1}; Chip colour 1-5 score is a acceptable colour, on 1-10 scale of increasing dark colour; Within columns means followed by the same superscripts are not significantly different at $p \leq 0.05$; *NS* Non-significant

Adapted from Banerjee *et al.* (2017)

iv. Date of harvest

Growing of potatoes suitable for processing mainly depends on temperature during crop growth and harvest (Ezekiel *et al.*, 1999). Generally, the desirable quality required for processing is found, when harvesting is done 100-120 days after planting (DAP). Earlier harvesting of processing varieties deteriorates its quality. The processing of the produce should be done as early as possible (80 – 90 days after harvesting) to avoid deterioration in quality, when stored at ambient temperature, when processing industries are in dire need of quality potatoes (Kumar *et al.*, 2001). The processing quality of different potato cultivars after three dates of harvest has been shown in Table 6.

Table 6: Processing quality of different potato cultivars as influenced by date of harvest

Cultivar	80 DAH			90 DAH			100 DAH		
	CC	DM	SR	CC	DM	SR	CC	DM	SR
Kufri Chipsona-1	3.23	20.64	3.33	2.75	22.82	3.33	1.71	22.76	3.04
Kufri Chipsona-2	3.35	19.45	3.36	1.64	20.48	3.05	2.71	22.66	2.48
Kufri Jyoti	4.84	18.28	2.73	4.16	18.01	2.67	4.14	20.12	2.73
Kufri Lauvkar	3.43	17.90	3.15	3.15	18.98	3.15	3.62	19.62	3.63
Atlantic	1.94	21.47	3.35	1.68	21.69	2.97	2.07	21.41	2.91
FL - 1533	1.42	22.43	3.03	1.60	23.75	3.00	1.62	23.39	2.75

DAH = Days after harvest, CC = Chip colour (chip colour 1-5 score is a acceptable colour, on 1-10 scale of increasing dark colour), DM = Dry matter (%), SR = Sucrose rating (mg/g)

Adapted from Kumar *et al.* (2001)

1.6. Development of Varieties for Processing

The potato varieties released in the country till 1997 were developed mainly for table purposes for fresh consumption. These varieties generally have low dry matter (17-19%) and high reducing sugars (above 250 mg $100g^{-1}$ fresh weight) and were, therefore, not suitable for processing. Till 1997, three varieties namely Kufri Jyoti, Kufri Lauvkar and Kufri Chandramukhi were identified by the processing industries to be better than others. In the year 1990, Central Potato Research Institute (CPRI), Shimla launched a breeding programme for developing indigenous processing varieties, which has resulted in development of country's first processing varieties, namely Kufri Chipsona-1 and Kufri Chipsona-2 (Khurana, 2002). Recently, another processing variety Kufri Chipsona-3 and Kufri Chipsona-4 have been released from CPRI during the year 2005 and 2010, respectively. The crop duration, suitable area of cultivation, tuber characters and suitability of processing varieties for different processed products are given in Table 7.

Table 7: Tuber characters and processing quality of some Indian potato varieties

Varieties	Duration and adaptability	Shape / size	Eyes / texture	Dry matter (%)	Reducing sugars (mg/100g fresh weight)	Suitability for products
Kufri Jyoti	90 - 100 days; grown in the hills, the plains and the plateau region	Oval / Large	Fleet / Mealy	18 - 21	106 - 275	Chips, flakes, flour, other dehydrated products
Kufri Lauvkar	80 - 85 days; grown in the plateau region	Round / Large	Fleet / Mealy	18 - 20	200 - 250	Chips, flakes, flour, other dehydrated products
Kufri Chandramukhi	80 - 85 days; grown in the plains and the plateau region	Oval / Large	Fleet / Mealy	18 - 20	250 - 324	Chips, flakes, flour, other dehydrated products
Kufri Chipsona-1	100 days; suitable for the Indo-Gangetic and central plains	Oval to oblong / Large	Fleet / Mealy	21 - 24	45 - 100	Chips, French fries, flakes, flour, other dehydrated products
Kufri Chipsona-2	100 days; suitable for the Indo-Gangetic and central plains	Round to oval / Large	Fleet / Mealy	21 - 25	44 - 93	Chips, flakes, flour, other dehydrated products
Kufri Chipsona-3	110 days; suitable for North-western and North-central Indian Plains	Round to oval / Large	Fleet / Mealy	> 21%	<50	Chips, flakes, flour, other dehydrated products
Kufri Chipsona-4	100-110 days; suitable for West Bengal, Madhya Pradesh and Karnataka	Round to oval / Large	Fleet / Mealy	> 21%	<50	Chips, flakes, flour, other dehydrated products
Kufri Frysona	100-110 days; suitable for North Indian Plains	Round to oval / Large	Fleet / Mealy	> 21%	<50	Chips, French fries, flakes, flour, other dehydrated products

Adapted from Marwaha *et al.* (2003) and Mondal *et al.* (2015)

Evaluation of 11 varieties for French fries showed that several varieties like Kufri Chipsona-1, Kufri Chipsona-2, Kufri Lauvkar, Kufri Chandramukhi, Kufri Jyoti, Kufri Dewa, Kufri Lalima and Kufri Sindhuri precede acceptable colour (Marwaha *et al.*, 2003). However, Kufri Chipsona-1 and Kufri Chipsona-2 suited better as compared to other varieties on the basis of quality attributes like colour, texture and yield of fries (Table 8) and chips (Table 9). In last decade Indore and Ujjain districts of Madhya Pradesh have emerged as an indispensable area for production of suitable processing varieties of potato (Gaur *et al.*, 1999). Because, the region is characterized with black cotton soil and warmer climate, as compared to northern India and thus, suitable for producing good quality chips and French fries. Processing varieties like Kufri Chipsona-1 and Kufri Chipsona-2 are almost performing well in every potato growing zones of India. Present performance and validation by the industry 'online' and cultivation of Kufri Chipsona-3 and Kufri Chipsona-4 varieties in North-Western plains of India would reduce transportation cost up to the potato processing industries in North-Western region of India. It is also expected to bring down the cost of chip production with higher recovery and lesser oil consumption (Pandey *et al.*, 2001).

Table 8: Yield and quality of French fries prepared from Indian potato varieties

Varieties	Peeling loss (%)	Trimming and cutting loss (%)	Total losses during processing (%)	Colour	Texture	Yield of fries (%)
Kufri Badshah	10.8	1.8	12.6	MB	Soggy	49.5
Kufri Bahar	13.4	2.6	16.0	LB	Soggy	46.6
Kufri Chandramukhi	12.2	2.3	14.4	LB	Mealy	46.6
Kufri Dewa	14.7	4.2	18.9	L	Mealy	47.6
Kufri Jawahar	13.0	3.8	16.8	GY	Pasty	47.8
Kufri Jyoti	10.4	3.4	13.8	L	Mealy	47.5
Kufri Lalima	14.4	2.9	17.3	MB	Mealy	48.3
Kufri Lauvkar	12.6	2.3	14.9	L	Mealy	50.7
Kufri Sindhuri	16.5	4.7	21.2	L	Mealy	44.8
Kufri Chipsona-1	9.0	2.3	11.3	L	Mealy	59.3
Kufri Chipsona-2	8.4	1.7	10.3	L	Mealy	58.3

L = Light, LB = Light brown, MB = Medium brown, GY = Golden yellow
Adapted from Marwaha *et al.* (2003)

Table 9: Dry matter, chips colour and biochemical characteristics for processing cultivars of potato

Cultivars	DM (%)	CC	Reducing sugar	Sucrose	Total sugar	Total phenol	Enzymatic discolouration (OD)
Kufri Ashoka	21.4	4.3	124	216	340	72	0.230
Kufri Anand	20.6	3.0	250	275	529	55	0.196
Kufri Badshah	23.7	2.5	219	264	483	58	0.418
Kufri Chipsona-1	24.1	1.0	140	222	362	54	0.120
Kufri Chipsona-2	25.5	1.0	69	298	367	53	0.160
Kufri Jawahar	22.0	4.3	154	350	504	55	0.173
Kufri Jyoti	20.1	4.5	105	204	309	63	0.471
Kufri Sutlej	20.6	4.7	181	272	453	42	0.338
Kufri Pukhraj	18.5	4.0	192	312	504	44	0.246
Atlantic	27.2	2.7	82	236	318	44	0.149
Frito Lay – 1533	22.6	2.5	203	360	563	55	0.186
Mean	**22.2**	**3.2**	**156**	**274**	**431**	**51**	**0.242**

DM = Dry matter, CC = Chip colour 1-5 score is an acceptable colour, on 1-10 scale of increasing dark colour
Adapted from Uppal and Khurana (2003)

1.7. Methods of Preparing Different Potato-based Processed Products

Among the different processed products, potato chips are the very common or popular product. The method for the preparation of chips involves cutting of 2-3 mm thick slices directly in cold water. This is followed by blanching in hot water for 1-5 minutes and then drying in sun. The dried product can be stored in air tight containers or sealed polythene bags for 6 months. These slices are consumed after shallow frying, as and when required. These potato chips can also be used for sale in different parts of the country. The market price of chips depends on the visual appearance of dehydrated chips. Lighter the colour, more the price they fetch. During processing of dehydrated products, nutritional losses occur and the extent of losses depends on variety, sun-drying and method of processing. Research is needed to minimize nutritional losses especially during the processing of potatoes into solar dehydrated products, which are common at the village level (Rani and Ezekiel, 2003). Methods of preparing different potato-based processed products have given in flow chart as suggested by Pinky and Asha (2014).

a) **Potato chips:** Light golden yellow coloured chips are preferred, which are of 40-60 mm in size.

Potato chips

Take potatoes and wash

↓

Peel the potatoes

↓

Make thin slices

↓

Blanch these slices in hot water (80°C)

↓

Cool in cold water

↓

Surface dry the slices

↓

Frying (at 180°C)

b) **Potato flour:** It can be used in combination with other cereals and pulses flour for preparation of biscuits, cake, bhujia parantha, etc. One kg of potato will give approximately 20 g flour.

Potato flour

Take 1.0 kg of peeled potatoes in 2 litres of water in a cracker

↓

Add 5.0 g of potassium metabisulphite and boil

↓

Mash the boiled potatoes and spread on tray

↓

Dry it in oven at 60-80°C till fully dry, turn upside down in between for proper drying

↓

Grind the dried mash and sieve it to get flour

↓

Store in air-tight containers and use, when required

c) **Potato starch:** It may be used in several industries for preparation of soups, gravy, puddings. As softening agents and starch is used in cakes, breads, biscuits and cookies. In paper industry, it is useful in coatings smooth white paper because of strong binding power. One kg of potatoes will give approximatly 100 g of starch.

Preparation of starch

Take 1.0 kg of potatoes and peel it

↓

Cut into small pieces

↓

Add 1 litre of water containing 2.5 g of potassium metabisulphite to the pieces

↓

Grind the pieces in a mixer-grinder

↓

With the help of water, sieve it in a tub through muslin cloth

↓

Let the starch settle on the surface of tub

↓

After the starch settle down, rinse it with fresh water for at least 2-3 times till the water is clear

↓

Grind the starch to make fine powder and seive

↓

Store the starch in airtight container and use, when required

d) **Potato flakes:** These are generally used in ready-to-use products in making of aaloo parantha, aaloo tikki and samosa, etc.

Potato flakes

Wash and peel the potatoes

↓

Make slices

↓

Wash the slices

↓

Half fry the slice and cool it

↓

Complete drying

↓

Mash the slices

↓

Drying in drum drier

↓

Remove the dry layer from drum

↓

Pack and store

e) **Potato shreds:**

Potato shreds

Take large sized potatoes and wash

↓

Peel the potatoes

↓

Shred the potatoes

↓

Treat these sticks with a solution containing salt and potassium metabisulphite for 2 minutes

↓

Dry the sticks in solar drier for 3-4 days till they become brittle

↓

Store in airtight container

↓

Shallow-fry these dried sticks and mix spices before serving

1.8 Conclusion

Processing of potatoes is important to maximize their utilization, to solve the problem of expensive refrigerated storage, to avoid market gluts and to save precious food against spoilage. It is essential to create awareness among farmers about several uses of potatoes. There is a need to encourage village level processing of potatoes, which can be easily adopted by the rural people and will generate additional income.

References

1. Bandana, Sharma, V.B., Singh, B., Raigond, P. and Kaushik, S.K. (2016). Role of invertase activity in processing quality of potatoes: Effect of storage temperature and duration. *Journal of Environmental Biology* 37: 239-245.
2. Banerjee, H., Puste, A.M., Ray, K., Sarkar, S, Chakraborty, A. and Rana, L. (2016a). Influence of irrigation levels and mulching on growth, water use, yield, economics and quality of potato (*Solanum tuberosum*) under new alluvial soil of West Bengal. *Indian Journal of Agronomy* 61(3): 377-383.
3. Banerjee, H., Sarkar, S., Deb, P., Chakraborty, I., Sau, S. and Ray, K. (2017). Zinc fertilization in potato: A physiological and bio-chemical study. *International Journal of Plant & Soil Science* 16(2): 1-13.
4. Banerjee, H., Sarkar, S., Deb, P., Dutta, S.K., Ray, K., Rana, L. and Mazumdar, K. (2016b). Impact of Zinc Fertilization on Potato (*Solanum tuberosum* L.) Yield, Zinc Use Efficiency, Quality and Economics in Entisol of West Bengal. *Journal of the Indian Society of Soil Science* 64(2): 176-182.
5. Bosnjak, D., Ilin, Z. and Vracar, L. (2004). Potato yield and quality depending on pre-irrigation moisture level in chernozem soil. In *VII International Symposium on Protected Cultivation in Mild Winter Climates: Production, Pest Management and Global Competition* 659. pp. 447-52.
6. Carli, C., Yuldashev, F., Khalikov, D., Condori, B., Mares, V. and Monneveux, P. (2014). Effect of different irrigation regimes on yield, water use efficiency and quality of potato (*Solanum tuberosum* L.) in the lowlands of Tashkent, Uzbekistan: A field and modelling perspective. *Field Crops Research* 163: 90-99.
7. Dwivedi, G.K. and Dwivedi, M. (1992). Efficacy of different modes of application of copper, zinc and boron to potato. *Annals of Agricultural Research* 13: 1-6.

8. Ezekiel, R. and Shekhawat, G.S. (1999). Potato processing in developing countries with special reference to India. In: Abstracts, Global Conference of Potato, December 6-11, 1999, New Delhi, p. 63.

9. Ezekiel, R., Verma, S.C., Sukumaran, N.P. and Shekhawat, G.S. (1999). A guide to potato processors in India. Technical Bulletin. No. 48. CPRI, Shimla, p. 39.

10. Gaur, P.C., Pandey, S.K., Singh, S.V. and Kumar, D. (1999). Indian potato varieties for processing. Technical Bulletin No. 50. Central Potato Research Institute, Shimla, p. 25.

11. Goldbach, H.E. (1997). A critical review on current hypotheses concerning the role of boron in higher plants: suggestions for further research and methodological requirements. *J. Trace Microprobe T.*, 15: 51–91.

12. Kumar, Dinesh, Singh, S.V. and Pandey, S.K. 2001. Potato processing quality as influenced by date of haulm cutting. *Journal of Indian Potato Association,* 28(1): 167–168.

13. Kumar, P., Pandey, S.K., Singh, S.V., Singh, B.P., Rawal, S. and Kumar, D. (2008a). Evaluation of nutrient management options for potato processing cultivars. *Potato Journal* 35: 1-2.

14. Marwaha, R.S. and Sandhu, S.K. (1999). Value added products from potato. *Journal of Indian Potato Association,* 26: 89-92.

15. Marwaha, R.S., Uppal, D.S., Dinesh Kumar and Sandhu, S.K. (2003). Potato processing. In: The Potato Production and Utilization in Subtropics (S.M. Paul Khurana, J.S. Minhas and S.K. Pandey Eds.), pp. 336-346.

16. Mondal, S.S., Patra, B.C. and Banerjee, H. (2015). Micronutrient management. In: Advances in Potato Cultivation Technology, Kalyani Publishers, India, pp. 115–121.

17. Mondy, N.I. and Munshi, C.B. (1993). Effect of boron on enzymic discoloration and phenolic and ascorbic acid contents of potatoes. *J. Agr. Food Chem.,* 41: 554–556.

18. Mondy, N.I. and Chandra, S. (1981). Effect of zinc fertilization on yield, enzymatic discoloration, phenolic and nitrogenous constituents of potato tubers. *Journal of Food Science* 46(6): 1878-1880.

19. Mousavi, S.R., Mohammad, G. and Goudarz, A. (2007). Effect of zinc and manganese foliar application on yield, quality and enrichment on potato (*Solanum tuberosum* L.). *Asian Journal of Plant Sciences* 6: 1256-1260.

20. Pandey, S.K., Kumar, D., Singh, S.V., Ezekiel, R. and Tomar, T.P.S. (2001). On line performance of potato varieties Kufri Chipsona-1 and 2 for processing as chips. *Journal of Indian Potato Association,* 28(1): 156-158.

21. Paul Khurana, S.M. (2002). Research and development support by CPRI for potato export, storage and processing. *Journal of Indian Potato Association,* 29(1-2): 67-75.

22. Paul Khurana, S.M., Singh B.P., Kumar N.R., Rawal S. and Kumar, D. (2003). Production, processing and export potential of Indian potato an overview. In: Souvenir and Abstracts of Conf. Processing and Export Potential of Potatoes within Asia, March-10, 2003, Modipuram. pp. 1-9.

23. Pinky, R. and Asha (2014). Potato processing. In: *Proceedings of Summer School on "Current Trends in Quality Potato Production, Processing and Marketing"* CPRI, Shimla held during 8th to 28th July: pp. 258-261.

24. Rani Manju and Ezekiel, R. (2003). Nutritional losses during preparation of solar dehydrated potato chips and cubes. *Journal of Indian Potato Association,* 30(1-2): 179-180.

25. Sarkar, S., Banerjee, H., Chakraborty, I., Sau, S., Ray, K., Ghosh, D. and Deb, P. (2018). Assessment of growth, yield and tuber quality and profitability of potato upon boron fertilization. *Journal of Environmental Biology* 39(3): 365-372.

26. Sarkar, S., Banerjee, H., Ray, K. and Ghosh, D. (2018). Boron fertilization effects in processing grade potato on an *inceptisol* of West Bengal, India. *Journal of Plant Nutrition* 41(11): 1456-1470.

27. Sathya, S., Pitchai, G.J. and Indirani, R. (2009). Boron nutrition of crops in relation to yield and quality – A review. *Agric. Reviews,* 30: 139–144.

28. Singh, S.V., Pandey, S.K., Kumar, D., Kumar, P. and Paul Khurana, S.M. (2002). Production of potatoes for processing. Extension Bulletin No. 33. Central Potato Research Institute, Shimla, p. 13.

29. Srivastava, R.P. and Kumar, S. (2002). Potato processing. In: Fruit and vegetable preservation: Principle and practices. International Book Distributing Co., Lucknow, pp. 275-296.

30. Uppal, D.S. and Paul Khurana S.M. (2003). Processing performance of Indian and exotic potato cultivars grown in Rajasthan. *Journal of Indian Potato Association,* 30(1-2): 181-182.

❏❏❏

Chapter 6

Judging Maturity Indices of Banana

Venkata Satish Kuchi[1], Arghya Mani[2] and Ch. Sai Ratna Sharavani[3]

[1*]Department of Post harvest Technology, College of Horticulture, Dr. Y.S.R. Horticultural University, Anantharajupeta, Kadapa District, Andhra Pradesh-516105, India., [2]Department of Post harvest Technology, BCKV, Mohanpur-741252, India., [3]Department of Floriculture and Landscaping, College of Horticulture, Dr. Y.S.R. Horticultural University, Anantharajupeta, Kadapa District, Andhra Pradesh-516105, India., *Corresponding author email: newmoon_9@yahoo.com

ABSTRACT

Banana is one of the most important fruit crop in the world. It's very much demanding all over the world and there is always a race for the improvement of the fruit quality. One of the biggest parameter that determines the overall post harvest quality of any particular crop is the maturity duration at which it is harvested. In banana, there are several techniques that give us the optimum maturity indice, which includes the angularity, pulp to peel ratio, diameter and length of the fruit, colour, brittleness of the floral end, etc. But all are not the perfect indice to judge the maturity. They can determine the quality upto a certain level. But there are varietal variations as well. Different genomic groups of banana respond differently to all these parameters. So, there is a need of new sophisticated techniques with the help of which the perfect matured fruit can be identified. This includes Surface gloss meter, electronic nose, delayed light emission techniques, etc. Such hi-tech maturity determination techniques would surely ease out the process of deciding the ideal maturity.

Keywords: Surface gloss meter, electronic nose, delayed light emission techniques, maturity indice.

1. Introduction

Banana is one of the most popular fruit crop in the entire world. It's very prevalent in several tropical and subtropical regions of India with great socio-economic significance, intertwined in the cultural legacy of the country. Banana belongs to family Musaceae. Most of the present day edible bananas are triploids resulting from natural crosses of *Musa acuminate* and *Musa balbisiana.* In India, more than 50 cultivars of banana are grown, each having its own importance in the area where it is grown, as well as among the consumers and the sellers.

Banana is one of the most common desert fruit that we often find in the table of any common household. Banana is also a staple food owing to its rich and easily digestible carbohydrates with a calorific value of 67-137/100 g fruit. Being a rich source of vitamin C and minerals, it makes healthy and salt-free diet (5). It is the cheapest sources of wholesome nutrition to many in the world. It contains nearly all the essential nutrients including minerals and vitamins and has several medicinal properties. Besides, as a fresh fruit, it is also used in beverages, ice creams, confectionary and baby foods. Culinary bananas are also used for making various kinds of traditional foods in various parts of the world. Banana is rich in carbohydrates, minerals like potassium, phosphorus, magnesium and calcium. It has certain phytochemicals like serotonin and dopamine, which are neurotrans-mitter, responsible for proper functioning of the brain. It is rich in dietary fibres and helpful for easy bowel movement (28).

It is cultivated in India in an area of 830.5 thousand ha and total production is about 29,779.91 thousand tonnes. Major states for banana cultivation are Tamil Nadu, Maharashtra, Gujarat, Andhra Pradesh and Karnataka. The worldwide production of banana is around 102028.17 thousand tons of which India contributes 29.19%. Besides India, other major banana producing countries are China, Philippines, Ecuador, Brazil and Indonesia.

Fruit maturation is an imperative postharvest criterion for banana. Green-life or storage longevity and eating quality of banana was greatly influenced by the stage of maturation at which the fruit is harvested. Every fruit attains its maximum characteristics like flavour, taste and colour during storage, if it is picked at optimum time. Fruits, harvested at an advanced stage of maturity, are not suitable for long distance transportation due to their shorter storage life.

The majority of banana plants produce the flower buds within 10 to 15 months of appearance of new sucker, which is variety dependent and extent of cool weather. Flower development starts from the true stem portion of the plant 9-12 months after planting and inflorescence grows up through the center of the pseudostem. Depending on season and cultivar, bananas become matured at 90-120 days after appearing of first flower (18).

Banana bunches are harvested, when the fruits are fully developed, that is, 75% matured, the angles are becoming less prominent and the fruits on the upper hands are changing to light green; and the remnants of flower (styles) are easily rubbed off the tips. Generally, this stage reached 75 to 80 days after the opening of the first hand.

Banana being climacteric fruit, are harvested when still green, and then ripened using appropriate ripening aid. Post harvest management of banana is a tricky issue, as its quality and storage life is dependent on several factors. Maturity, harvesting method, packaging, transportation and storage conditions play vital role in presenting it in an acceptable condition in the market. By keeping the importance of maturity, this chapter is aimed to provide detailed information, regarding judging maturity indices of banana.

1.1. Maturity

Maturity of the banana bunches indicates the quality of fruit during ripening and overall marketability. Most of the post harvest losses could be managed by harvesting the fruit at proper maturity stage (6). Early harvest of banana may lack flavour and may not ripe properly, while harvesting it late may over ripe the fruit and cause splitting (29). Banana is harvested at over mature stage for local market distribution than those which are meant for export. Hence, it is important to harvest fruit at the right maturity stage to suit the purpose.

Bananas are climacteric fruits, initiation of ripening in these fruits are marked by a rapid increase in respiration rate of the fruit to a peak, after which the rate gradually falls as ripening progresses. During ripening period, the fruit softens, whereas, starch is converted to sugars, the skin colour changes from green to yellow, lose their astringency and develop characteristic flavour.

As bananas have a short storage life after ripening, those which have to be transported over long distances are harvested in the green condition and transported under refrigeration to their destination. They are then ripened under controlled conditions by the importer or seller.

The maturity stage for harvesting the fruit depends on the market, for which it is intended and is determined in terms of the marketable life required. Bananas to be marketed locally can be harvested at a more mature stage than those which are to be exported. They are harvested 75-80% maturity for long distance transport taking some three weeks, at 90% for regional and inter-island transport time of two weeks and fully mature for local domestic markets. If the fruit are too mature at harvest, they may well-split during handling, especially if they have recently been irrigated, or it has rained. Mature fruit may also ripen prematurely during transport.

1.2. Maturity indices of banana

Maturity of the fruit is assessed by characters, such as change of peel color from dark green to pale green, disappearance of angularity, finger length and diameter. Based on these maturity indices, the banana bunches can be classified into three categories viz., un-mature, mature and over-mature fruit (27). The mature fruit with pale green color with disappeared angularity having 34–35 mm finger diameter and 21–22 cm finger length are the primary requirements for exporting banana. Conventional method of assessing the maturity was done by trained personnel through manual vision for assessing colour/ angularity.

Some indices, which have been used are (11)

- Age of the bunch after emergence from the pseudostem
- The angularity of the fruit fingers in cross-section
- Ratio of the fruit peel to pulp
- Measurement of the fingers, length or diameter
- Firmness of the fruit
- Brittleness of the flower end of the bunch

1.3. The Age of the Bunch

This is measured from the time when the fruit bunch is first visible on its emergence from the psuedostem and is called 'shooting'. For given combination of cultivar and growing conditions, the approximate time between the emergence of the bunch and harvest can be estimated. In this case the bunch can be marked at the time of shooting and, subsequently, harvested at the appropriate time. In southern Ecuador, and some production areas, this is done by having a coloured stripe down the side of the polythene film bunch cover. This cover is placed on the bunch at the time of shooting. The colour of the stripe is changed weekly so that the age of the bunch can be readily known by the colour shown. Bunches are harvested between 98 and 112 days for Grand Nain or Valery cultivars in southern Ecuador (11).

Depending mostly on variety and extent of cool weather, most of the banana plants produce the flower buds within 10 to 15 months of emergence as a new sucker. Production systems of Banana are perennial. Flower development is initiated from the true stem portion of the plant within 9-12 months after planting. The inflorescence grows up through the centre of the pseudostem. Depending on season and cultivar, Bananas attains maturity at 90-120 days after appearing of first flower (18).

In dwarf and tall varieties, harvesting of Banana is done 12 to 15 months in dwarf and 15 to 18 months after planting, respectively. Ara *et al.* (2) found that crop durations of Sabri and BARI Kola 1 was 412 days and 435 days, respectively.

In BARI Kola 1 and Sabri Kola, the optimum maturity stages of were found to 120 and 100 days after emergence of flowering (DAEF) in summer and 130 and 110 DAEF in winter seasons, respectively (1).

1.4. Angularity of the Fingers

Determination of the maturity stage of Bananas by visual inspection of the fullness of the fingers is one of the most generally used methods. In the initial stages of development, the fingers are clearly angular in cross-section, but as growth progresses, this angularity decreases and the fruit becomes more rounded. The final degree of roundness varies between cultivars, but is a reasonably precise method and, with experience, is practical to use, especially for small-scale growers. Although farmers generally know and commonly use this method, it is not always applied in practice (11).

1.5. Pulp to Peel Ratio

This ratio is used as an indication of fruit maturity, but since it is a destructive method which can only be applied by taking representative samples from the bunch, it is of limited practical application for framers. In India, a minimum pulp to peel ratio of 1:1 has been accepted as being the acceptable level of maturity for ripening Dwarf Cavendish bananas (11).

1.6. Diameter and Length of Bananas

Fruit maturity is correlated to the diameter of the fingers and by measuring pre-determined fingers on each bunch with a pair of callipers an estimate of the bunch's maturity can be made. This parameter also varies between seasons; therefore, it is often difficult to establish standards even for one banana plantation. Finger length has also been use. This technique sometimes used to assess the maturity of banana hands on the bunch before harvest, but the practice is not widespread. It is more usually employed to grade hands of Bananas during packing station operations. There is no standard method of measuring the length of Banana fingers. Some measure of the outside curve from the junction of pulp and fruit stalk to the tip of the fruit, others take this measurement on the inner curve of the fruit, and yet others measure in a straight line from the fruit stalk to the tip. Where it is used, it is necessary to standardize the measurement to particular fingers as for angularity (11).

1.7. Firmness of the Fruit

A method of evaluating the maturity or ripeness of fruit is by determining its firmness, using a penetrometer to measure the pressure required to break the skin. The firmness varies during development and can be used to determine the maturity of a bunch, using representative samples (11).

1.8. Heat Units

In French Antilles, Ganry and Meyer during 1979 found the following relationship between grade (G) and temperature (T). G=15.7+0.202 ST, where 'ST' is the sum of average daily temperature above 14°C. It required about 900 degree days for Cv. Williams to arrive at a grade of 34 mm from the time of last hand emergence (28).

Studies carried out from 1996 to 1998 revealed that the best maturity index for determining the harvest time of Banana is a combination of bunch age and finger diameter. However, the bunch age and finger diameter ranges for indicating optimum harvest maturity stage have to be worked out for each of the economically important Banana varieties (31).

1.9. Surface Gloss Meter

Gloss is a perceptual trait of surfaces related to the degree to which it simulates a perfect mirror in its capacity to reflect incident light (14), and its commercial importance in the food industry is well known. Gloss properties are linked to surface composition and morphology, changes in them would most probably result in changes in gloss. Gloss measurements were made indirectly, it would first be necessary to improve the measurement procedure before gloss can be considered as a reliable criterion for the determination of ripeness. Peel gloss is a property of Bananas that is known to change. In particular, Bananas which have been exposed to below recommended temperatures enroute from harvesting to the ripening room develop very dull surface colours (15). Peel colour is known to correlate well with the conversion of starch into sugar, development of flavour and preferred textural characteristics of the pulp.

Ripening of the Bananas was accompanied by a gradual decrease in gloss, which correlated well with other chemical and physical changes in the Banana, including pH, °Brix and starch content of the pulp and peel roughness. Most importantly, the decrease in gloss correlated well with changes in peel colour, the current indicator used to assess Banana ripeness, which appropriately, correlated well with chemical and physical changes. It is believed that the decrease in gloss during ripening may be a result of a decrease in epicuticular wax content that was observed. Results suggest that peel gloss measurements have potential as criteria for determining Banana ripeness (37).

1.10. Neural Network-based Electronic Nose

There are natural factors that could be ascribed to fruit ripening, such as change in aroma, firmness, colour, shape, composition, etc. An electronic nose or e-nose makes use of one of the above characters. It resembles the human's sense of smell through an array of gas sensors that are mainly composed of electro-chemical metal oxide semi-conductors (MOS), and is capable of analyzing the volatiles emitted by a ripening fruit. The flowchart was given in Figure 1.

An e-nose is chiefly an array of electro-chemical gas sensors, however, the arrays stability deteriorates as the time progresses due to the effect of what is called a "sensor drift." Drifts are gradual change in any quantitative characteristic that is supposed to remain constant [artoson]. Sensor drifts impair there liability of gas sensing systems which in time causes the statistical models from sensor responses useless.

1.11. Hardware Components

- Thee-nose consisted of an array of 8 MQ and TGS chemo sensors with various degree of sensitivity to various atmospheric gases; 8 electro-chemical sensors mounted on this e-nose are–MQ-2, MQ-3, MQ-4, MQ-5, MQ-7, MQ-9, MQ-135 and MQ-138.
- ADHT 11 humidity and temperature sensor is to be installed within the containment to allow the user to have continuous monitoring of the internal temperature.
- Microcontroller/Microcomputer.

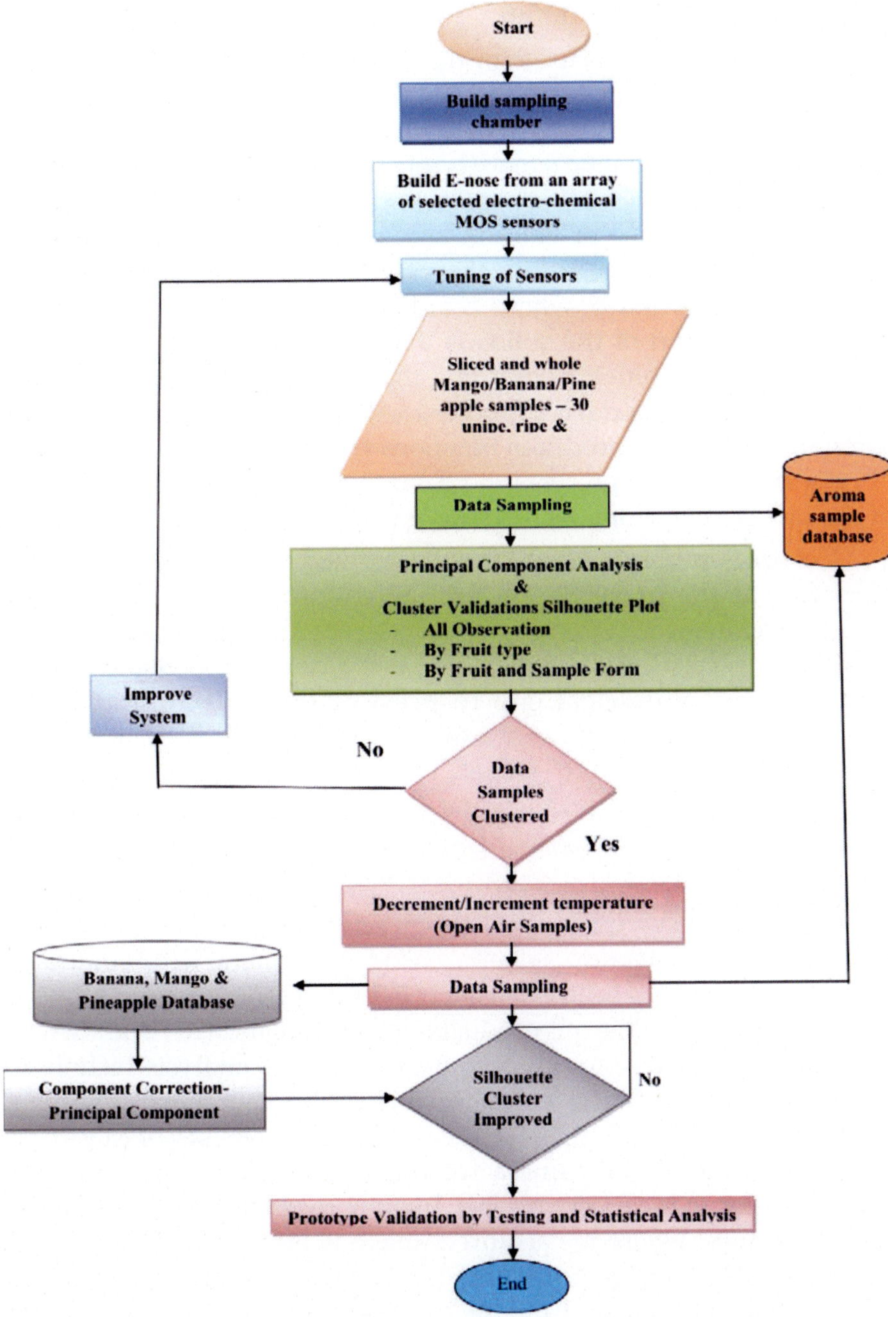

Fig. 1: Flowchart for working of nueral network-based electronic nose

1.12. Colour Measurements

Colour flex meter is used for the measurement of colour. It works on the principle of collecting the light and measures energy from the sample reflected across the entire visible spectrum. The meter utilizes filters and the mathematical models rely on "standard observer curves" that define the amount of red, green, and blue primary lights required to match a sequence of colours across the visible spectrum. The mathematical model used is called as Hunter model. It provides reading in terms of L, a, and b, where L indicates whiteness and darkness. Chromatic portion of the solids is defined by: a (red), -a (green), b (yellow), and -b (blue). These data may be sensed from the sensor (colour flex 45°/0°) and it is supported with Universal Software (V4.10) software package. It produces functions that define hue and chroma. "L, a and b" value represents the deviations of the individual values of the raw sample from the respective values of the fully ripe sample.

Colour is calibrated by fixing the defined colours like white and black tiles on the colour flex meter. Calibration is carried out through necessary changes in the sample. The skin of the fruit was peeled off and it was placed on the colour flex meter and it was then covered with a black cylindrical cup to prevent the reflection of light. Measurement of colour was done as per the above principle by determining 'L', 'a' and 'b' values.

Change in colour of fruits from deep green to light green is one of the sign of maturity of banana fruits (26).

1.13. Image Processing Technique

Automatic detection system, based on camera with computer based technology has been extensively explored for the quality analysis and grading of agricultural products in recent years. This is known as computer vision system or computerized image analysis technique and it has proven to be successful for objective measurement of various fruit crops (4, 31, 33). This system basically consists of standard illumination, a digital camera for image capturing and computer software for image processing (24). Image processing is an innovative field of science, where the acquired image is transformed into useful information. The image processing has been now widely suggested to assess the quality of fruit from the images (30). This has been possible because the image texture reflects the changes in intensity value, which might contain information about the colour and the physical structure of the objects (Figure 2a and Figure 2b).

Application of image processing in Banana started with Mendoza and Aguilera (24), who implemented a computer vision system to identify the seven ripening stages of Banana based on colour, development of brown spots and image texture information with an accuracy of 98 %. Ripeness of Banana, especially the dark spotting in full ripped Banana, could also be assessed from images by fractal texture Fourier analysis (30). These studies have illustrated the potential of using image processing in Banana fruit quality analysis after harvest at trader and consumer level. However, the physiological maturity stages of Banana using 3 point scale before harvest is also an important criteria of Banana fruit quality assessment. Flowchart for image processing technique is given in Figure 3.

1.14. Benefits

i) Detection of under-mature fresh fruit bunches can prevent early harvesting, thereby allowing the bunches to mature in the plant.

ii) Rapid detection to enable saving time and money.

iii) Avoiding unwanted transportation of under-mature or over-mature fruit to the ripening chambers.

Fig. 2: (i) (a) under-mature, (b) mature, and (c) over-mature Banana fruits (*Source:* Prabha and Kumar, 2015)

Fig. 2: (ii) (a) Original and (b) Binary image
(*Source:* Prabha and Kumar, 2015)

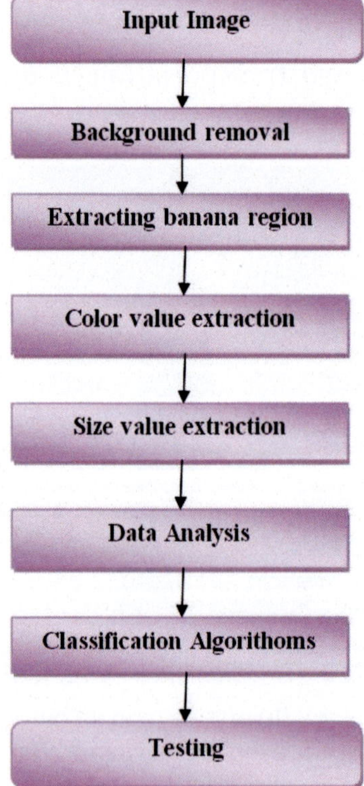

Fig. 3: Flowchart for image processing technique
(*Source:* Prabha and Kumar, 2015)

1.14. Delayed Light Emission

Biochemical degradation of chlorophyll results in exposure of carotenoids, giving fruits and vegetables a characteristic maturity symptom of yellowing (10). Since our eyes perceive colours, subjective visual evaluation has traditionally been used to judge maturity and other quality attributes of fruits and vegetables. Recent progress in the development of new methods utilizing optical properties has provided several objective techniques for rapid, accurate, more uniform quality assessment. The discovery of delayed light emission (DLE) by Strehler and Arnold (35) has offered an alternative for evaluating quality of chlorophyll-containing plant materials. The strong dependence of chlorophyll concentration and the duration and intensity of DLE have been applied in developing indices for different quality attributes of various fruits and vegetables (9). Recent investigations show that if the variables that are known to affect the duration and intensity of DLE are carefully controlled, the efficiency of sorting fruits and vegetables could be increased.

1.15. The Phenomenon of DLE

Strehler and Arnold (35) accidentally discovered the phenomenon of DLE while studying adenosine triphosphate (ATP) formation during photosynthesis using a firefly luminescent system as an indicator. They observed that when green plants are irradiated, they give off light for a considerable period after illumination. This phenomenon of DLE was explained as a reflection of certain early reactions in photosynthesis which, by virtue of their reversibility, are capable of releasing a portion of their stored chemical energy through a chemiluminescent mechanism- luminescence due to the energy liberated in a chemical reaction. Adding non-polar solvents (e.g., ethanol and ether) in moderate concentrations or irradiating chlorophyll solutions with ultraviolet (UV) light destroys the luminescence of intact cells, which supports the theory that a chemiluminescent mechanism might be involved in DLE. However, the chemiluminescent mechanism involved in DLE was thought to differ from the mechanism involved in phosphorescence- luminescence that persists after removal of the exciting source due to storage of energy. DLE was believed to involve biochemical reactions that produce excited molecules and to be associated with an enzyme system. Phosphorescence, on the other hand, is purely a physical reemission of trapped light energy (12).

1.16. Factors affecting DLE

All fruits, vegetables, and plant materials undergoing photosynthesis probably produce DLE. Major factors affecting the intensity of emitted light are:

- Wavelength of excitation
- Excitation intensity
- Excitation time and time after illumination
- Dark periods
- Sample thickness and area of excitation
- Temperature and
- Chlorophyll content of plant material

1.17. Quality Evaluation based on DLE

Typically, the process of ripening involves degradation of chlorophyll, which exposes carotenoids, giving characteristic maturity symptom of yellowing (10). Because of the change in pigment composition, fruits exhibit reduced absorbance in the 670 nm region, as they mature. Spectrophotometric methods take advantage of this optical characteristic in indicating maturity and other quality attributes of fruits and vegetables. Because of their simplicity, spectrophotometric methods have been widely used for maturity and quality evaluation of biological materials. However, DLE measurements have the following advantages over spectrophotometric methods:

1. Illumination need not be spectrally well-defined, which permits wider choice and simpler design of the exciting source.
2. Illumination and measurement are not simultaneous, i.e., illumination can be removed in time and position from measurement.
3. Beyond saturation level, additional illumination has little effect on increasing DLE intensity, so small changes in illumination will not interfere with quantitative measurement of DLE.

Measurements of DLE on Bananas indicate that the intensity of emitted light decreased with ripening, and Bananas just beyond optimum ripening had only a trace of DLE (7). The grade of peel colour of Bananas had been related to DLE intensity, which indicates the possibility of using DLE measurements to determine the state of maturity of Bananas. Other quality attributes of Bananas, such as sugar content of the flesh (S, Brix) of Bananas increased from 1 to20% during ripening, the intensity of emitted light (Y volt) decreased almost linearly as

$$Y= -0.773\ S + 3.26$$

Similarly, the relationship between fruit firmness (F, N) and DLE intensity was expressed as:

$$Y= 14F -3.50$$

1.18. Optical Chlorophyll Sensing System

The first visible sign of ripening is a colour change from green to yellow, caused by a reduced chlorophyll content in the peel. There are three major methods for peel colour analysis: visual, chemical, and instrumental (17, 36, 23). Since the change in peel colour from green to yellow during the ripening of Bananas depends mostly on loss of chlorophyll, its content can be used to characterize the change in peel colour. Chemical determination of chlorophyll content involves the extraction and measurement of peel pigment components (8) which is a destructive method and, therefore, undesirable. Recent advances in electro-optical technology make low cost colour measurement possible. An optical chlorophyll sensing system includes a sensing head, an amplifying circuit, and a datalogger. The sensing head was composed of a support structure with two monochromatic illuminating sources and two detectors. Light emitting diodes (LED) and phototransistors were used as illuminating sources and detectors because of their long life, availability and low cost. Flowchart of Banana peel testing program is given in Figure 4 (25).

The output from the prototype sensor was determined by placing a Banana sample and the detector head in a light-tight chamber in order to avoid interference from other light sources. Because of slight colour variations on the fruit surface, the reflectance intensity of each Banana was measured twice on both sides of the main axis near the center region of the Banana and the average was taken.

This optical chlorophyll sensing system rapidly and accurately measures the chlorophyll content of Banana peels. It is also simple to operate, non-destructive, and inexpensive. It is envisioned that this device could be used to monitor Banana peel colour during ripening, replacing the standard manual colour determination. This would eliminate inspection during ripening, thereby improving the quality of ripening and preventing the release of ethylene gas.

1.19. Biochemical Means / Destructive / Non-destructive

A. Water

Water is the most abundant constituent in the pulp and peel of Banana. Water percentage increases in the pulp during ripening due to respiratory breakdown of starch and osmotic movement of water from peel to pulp. Water is also lost from peel externally due to transpiration, until the ripening processes degrade the peel tissue preventing further water loss. In a fully ripe Banana, water occupies about 75% of the pulp mass (33).

B. Carbohydrates

The main change in fruit pulp during ripening is the conversion of starch to sugars, and peel colour is closely correlated with the starch:sugar ratio. Starch declines from about 20-23% at harvest to 1-2% in ripe fruit. Sugars increase in about the same proportions. During the early stage of ripening, the ratio of sugars is about 65:20:15 (sucrose:glucose:fructose), indicating that sucrose appears first, and hexose sugars later. This conversion of starch to sucrose is at maximum 2 days after the ethylene peak. Loss of carbohydrate via respiration is small and total carbohydrate remains fairly constant during ripening. Many enzymes are apperantly involved in the total conversion process, and the initiation of enzymatic starch hydrolysis during the ethylene peak appears to be regulated by fructose 2,6-bisphosphate (33).

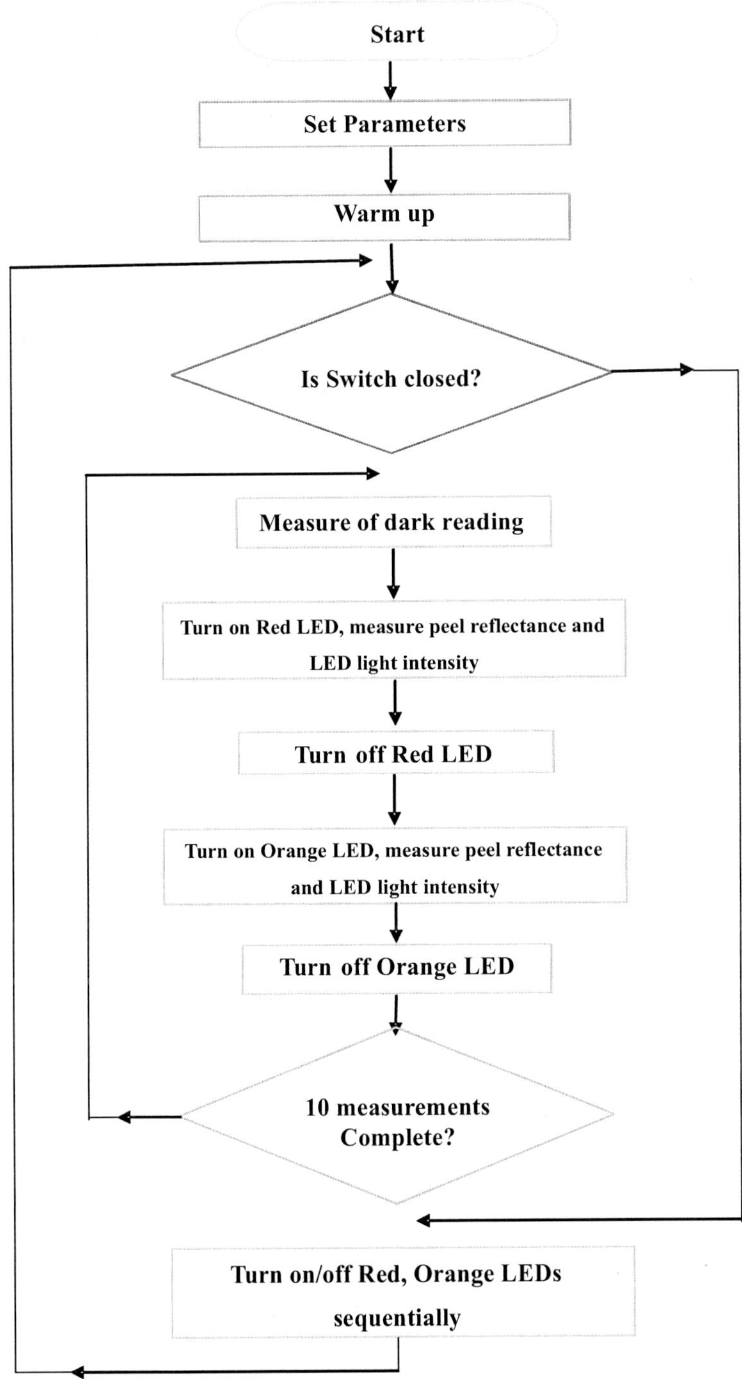

Fig. 4: Flowchart of Banana peel testing program

C. Pigments

During ripening, peel colour changes from dark green to bright yellow and this is due to chlorophyll breakdown which gradually unmasks the carotenoid pigments also present in unripe peel. These processes occurs readily at controlled ripening temperatures of 16-18°C but when ripening occurs naturally at ambient temperatures above 25°C, the degradation of chlorophyll is suppressed while pulp ripens rapidly (33).

D. Aromatic compounds

The characteristic aroma of ripening Banana is due to a combination of many volatile compounds, knowledge of which is still incomplete. These compounds are derived during the ripening period from metabolic pathways involving the formation of esters, acetates, alcohols, and carbonyl compounds. As ripening proceed, aroma emission increases until the peel turns brown after which the aroma subsides. Altogether more than 300 compounds have been observed which could have a role in the aroma component of ripening Banana fruit (33).

E. Phenolics

Phenolics ae responsible for the astringency of Bananas before ripening, and also for certain browning reactions. They are localized mainly in the latex vessels of the pulp and peel. Loss of astringency during ripening is due to polymerization of the phenolics. For example, dopamine represents about 80% of the tannin in the pulp at harvest but it gradually decreases during the ripening process. Browning is linked to enzymatic oxidation of phenolics and polyphenol oxidase is the enzyme. The peel contains double the total polyphenol content of the pulp, at harvest. Excessive browning of phenolics in the latex vessels of the peel, induced by temperature below 13°C, is commercially important (33).

F. Pectins

Ripe pulp contains 0.5-0.7% pectin. As ripening progresses, the water -soluble pectins increase and the insoluble pectins decrease, which is associated with pulp softening. These changes are catalysed by pectin methyl esterase, whose activity remains constant during ripening (33).

G. Lipids

Lipid concentration remains fairly constant during the ripening process. Lipids comprise about 1% of the fresh mass of peel and between 0.2-0.5% of the pulp of bananas. Mostly polyunsaturated fatty acids, particularly linolenic acid constituents are present. Cholesterol content is zero (33).

H. Proteins

Although the protein content of ripe Banana amounts to only 1.1% of its total composition, this comprises some important essential amino acids, such as lysine. In the pulp of Cavendish Bananas, total free amino acids increase from 330-750 mg/100g fresh mass when green, to 2700-3500 mg/100g at over-ripe stage. Histidine, aspargine and glutamine are the predominant amino acids at the fully ripe stage (33).

I. Other constituents

The main organic acids are mallic, citric and oxalic. During ripening, pH decreases and free acidity increases in the pulp (pH 4.0 t fully ripe stage). Bananas contain about 0.8% fibrebin the pulp and somewhat more in the peel. Non-cellulosic insoluble polysaccharides are the most abundant fibre component (33).

2. References

1. Amin, M.N. Hossain, M.M. Rahim, M.A. and Uddin, M.B. (2015). Determination of Optimum Maturity Stage of Banana. Bangladesh J. Agril. Res. 40(2): 189-204.

2. Ara, N. Bashar, M.K. Kabir, W and Kaisar, M.O. (2009). Variety and planting time effect on growth and yield of Banana (Musa sapientum L.). Journal of Agricultural Research, 47(2): 153-163.

3. Artursson, T. Ekloev, T. Lundstroem, I. Martensson, P. Sjoestroem, M and Holmberg, M. (2000). Drift correction for gas sensors using multivariate methods, Journal of Chemometrics. 14: 711-723.

4. Bato, P.M. Nagata, M. Cao, Q.X. Hiyoshi, K and Kitahara, T. (2000). Study on sorting system for strawberry using machine vision (part 2): development of sorting systemwith direction and judgement functions for strawberry (Akihime variety). J Jpn Soc Agrl Mach. 62(2): 101–110.

5. Chadha, K.L. (2001). Hand book of Horticulture. Indian Council of Agricultural Research, New Delhi. 143-153.

6. Chegeh, B.K. Mahinda, M.N. Kinyua, G.K. Kambo, C.M and Kuria, S. (1995). A survey on causes and quantification of post harvest losses in Kenya. Final technical report. KARI, Thika.

7. Chuma, Y. Sein, K. Kawano, S and Nakaji, K. (1977). Delayed light emission as a means of automatic selection of satsuma oranges. Trans Am Soc Agric Eng. 20(5): 996.

8. Clesceri, L.S. Greenberg, A.E and Trussell, R.R. (1989). Chlorophyll Method 10200H. Standard Methods for the Examination of Water and Wastewater, 17th ed. American Public Health Associate, Washington, DC. 31–39.

9. Dull, G.G. (1986). Non-destructive evaluation of quality of stored fruits and vegetables. Food Technol. 40(5): 106-110.

10. Eskin, N.A. Henderson, H.M and Townsend, R.J. (1971). Biochemistry of foods. New york: Academic press. 46.

11. Gowen, S. (1995). Bananas and plantains. Chapman and Hall, London. 404-408.

12. Gunasekaran, S and Panigrahi, S. (2001). Non-destructive Food Evaluation: Techniques to Analyze Properties and Quality. Marcel Dekkar, New York. 99-136.

13. Hewage, K. Wainwright, H and Wijeratnam, R. (1996). Quantitative assessment of chilling injury in Bananas using a calorimeter. Journal of Horticultural Science. 71.

14. Hunter, R.S. (1952). Gloss Evaluation of Materials. ASTM Bulletin, No. 186: 48–55.

15. Loesecke, V.H.W. (1950). Bananas; Chemistry, Physiology, Technology. New York: Interscience.

16. Kajuna, S.T. (1998). Colour changes in Bananas and plantains during storage. Journal of Processing and Preservation. 22: 27-40.

17. Knee, M. (1980). Methods of measuring green colour and chlorophyll content of apple fruit. J. Food. Technol. 15: 493–500.

18. Knowledge Master (2003). University of Hawaii, College of Tropical Agriculture and Human Resources, University of Hawaii-Manoa. Computer Resource Database. http://www.extento.hawaii.edu/kbase/crop/crops/banana.htm.

19. Li, M. Slaughter, D.C. and Thompson, J.F. (1997). Optical chlorophyll sensing system for Banana ripening. Post harvest biology and technology. 12: 273-283.

20. Madrid, M. and Lopez-Lee, F. (1996). Differences in ripening characteristics of controlled atmosphere or air-stored Bananas. International Post harvest Science Conference Post harvest. 96: 464.

21. Marriott, J. and Lancaster, P. (1983). Bananas and plantains. In: Handbook of Tropical Foods, (Eds.) H. T and C. Jr, Marcel Dekker, Inc. New York. 85-143.

22. Marriott, J. and Palmer, J.K. (1980). Bananas—physiology and biochemistry of storage and ripening for optimum quality. Critical Reviews in Food Science & Nutrition. 13: 41-88.

23. Medlicott, A.P. Semple, A.J. Thompson, A.J. Blackbourn, H.R and Thompson, A.K. (1990). Measurement of colour changes in ripening Bananas and mangoes by instrumental, chemical and visual assessments. Trop. Agric. (Trinidad) 69(2): 161–166.

24. Mendoza, F and Aguilera, J.M. (2004). Application of image analysis for classification of ripening bananas. J Food Sci 69:E471–E477

25. Meng Li. David, C. Slaughter and James, F.T. (1997). Optical chlorophyll sensing system for Banana ripening. Post harvest Biology and Technology. 12: 273–283.

26. Mpyshameem. (2010). Banana. Horticulture, www.e-krishi.co.cc/horticulture/Banana, accessed on 03/26/2012.

27. Muchui, M.N. Njoroge, C.K. Kahangi, E.M and Onyango, C.A. (2010). Determination of maturity indices of tissue cultured bananas (Musa spp.)'Williams' & 'Grande Naine'. Acta Hort. 879: 425–430.

28. Narayan, C.K. (2015). Bananas and plantains post harvest management, storage, ripening and processing. New India Publishing Agency, New Delhi. 1-4.

29. Patil, K.B and Rawale, K.P. (2012). Pre and post harvest management of banana for domestic and export market. In: Singh HP (ed) Proceedings of the national conference on adaption to climate change for sustainable production of banana, Jalgaon, India. ICAR, New Delhi. 88–100.

30. Prabha, D.S. and Kumar J.S. (2015). Assessment of banana fruit maturity by image processing technique. J Food Sci Technol. 52(3): 1316–1327.

31. Quevedo, R. Mendoza, F. Aguilera, J.M. Chanona, J and Gutierrez-Lopez, G. (2008). Determination of senescent spotting in Banana (Musa cavendish)using fractal texture fourier image. J Food Eng. 84: 509–515.

32. Ramma, I. Beni Madhu, S.P and Peerthum, P. (1999). Post Harvest Quality Improvement of Banana. Food and Agricultural Research Council, Réduit, Mauritius. 187-194.

33. Riyadi, S. Ashrani, A. Mohd, R. Mustafa, M and Hussain, A. (2007). Shape characteristics analysis for papaya size classification. In: Proceedings of the 5th Student Conference on Research and Development (SCOReD'07), Selangor, Malaysia. IEEE. 371–375.

34. Robinson, J.C and Sauco, V.G. (2010). Bananas and plantains. CAB international. UK. 263-266.

35. Roseleena, J. Nursuriati, J. Ahmed, J and Low, C.Y. (2011). Assessment of palm oil fresh fruit bunches using photogrammetric grading system. Intern Food Res J. 18(3): 999–1005.

36. Strehler, B.L and Arnold, W. (1951). Light production by green plants. *J. Gen Physiol.* 34: 809.

37. Wainwright, H. and Hughes, P.A. (1989). Objective Measurement of Banana Pulp Colour. Int. J. Food Sci. Technol. 24: 553–558.

38. Ward, G. and Nussinovitch, A. (1996). Peel Gloss as a Potential Indicator of Banana Ripeness. Lebensm.-Wiss. u.-Technol. 29: 289–294.

❑❑❑

Chapter 7

Post Harvest Management, Processing and Value Addition of Arid and Semi-Arid Fruits

Yogendra Singh[1], Nirmal Kumar Meena*[1], Kalpana Choudhary[1], Suresh Gurjar[2] and Narednra Verma[1]

[1]Department of Fruit Science, CH&F, Jhalawar, (AU, Kota), Rajasthan-326023, [2]Department of Horticulture, BAU, Birsa, Ranchi, *Email of corresponding author's: nirmalchf@gmail.com

ABSTRACT

India is recognised as a rich heritage of neglected and underutilized fruit species. Many underutilized fruit species are found in desert part of India. These fruit species have great potential to provide nutritional and livelihood security. Among them, Aonla, Annonaceous fruits, Datepalm, Jharberi (*Zizyphus nummularia*), Ker (*Capparis decidua*), Khejri (*Prosopis cineraria*), Lasoda (*Cordia myxa*), etc. are pre-dominant. These fruits provide supplement nutrition in poor diet of rural and tribal peoples. Owing to their nutraceuticals prosperity, these fruits play crucial role in curing of different ailments. Mature fruits of Ker and Pilu are rich source of anti-oxidants, phenolic and carotenoids content. Processing and value addition of these fruits is done by the localized, which helps to earn extra income. Aonla can be processed into preserve, candy, laddu, powder and *Triphala*. Unripe fruits of Ker and Jharberi are used to preparation of pickles which has great demand in market. Khejri pods are consumed fresh as well as in dried form. Fresh pods of Khejri locally known as "*Sangri*" are widely used in the form of chutney and vegetable which fetches higher price in restaurants. Lasoda fruits are rich in gum and this gum is used for coating of fruits and vegetables. Selling of fresh ripe fruits of Lasoda, Jharberi and Ker also added extra income to rural farmers. Due to lacking of technical knowledge, appropriate post harvest management and processing is a major concern. In majority of fruits technical standards are negligible along with important commercial cultivars. So, there is an urgent need for developing post harvest technologies and methodologies, which can reduce the post harvest loss and maximize their processing. In this chapter, post harvest management, processing and value addition of important arid and semi-arid fruit crops are discussed.

1. AONLA (*Emblica officinalis* Gaertn.)

Introduction

Aonla (*Emblica officinalis* Gaertn.) is an important arid fruit crop indigenous to India, belongs to Euphorbiaceae family. It is also known as 'Indian gooseberry', 'Amla', Amalakki and Nelli. It is considered as "Wonder fruit for health" because of its inherent medicinal properties.

Nutritive and medicinal value

The fruits of aonla are highly nutritious, with the rich sources of vitamin C and having many medicinal uses. It has an important therapeutic role from time immemorial and is frequently recommended for its synergistic effects in ayurvedic and unani systems of medicine (Agarwal and Chopra, 2004). It is used to make many ayurvedic tonics, such as Triphala, Chyvanprash, etc. which are found beneficial.

Components	Composition
Moisture	81.8 g
Protein	0.5 g
Fat	0.1 g
Minerals	0.5 g
Fibre	3.4 g
Carbohydrates	13.7 g
Vitamin C	600 mg
Vitamin B	0.03 mg
Calcium	50.0 mg
Phosphorus	20.0 mg
Iron	1.20 mg
Carotene	9 mg
Riboflavin	0.01 mg
Niacin	0.2 mg
Choline	256 mg

Source: Ganachari *et al.*, 2005

Varieties/cultivar

Cultivars like Banarasi, Francis (Hathi Jhool), Chakaiya, Banarasi NA-4 (Kanchan) and NA- 5 (Krishna), NA-6 (Amrit), NA-7 (Neelum), NA-9 and NA-10 are well known for commercial cultivation. Banarasi and Chakaiya cultivars fruits are generally used for making *murabba* (preserve), candy, etc.

Maturity indices and harvesting

The budded plants take less time and produce crop in 4-5 years, while seedlings required a long time 8-10 years to come in the full bearing stage. Fruits are ripen during November-December. It is also vary cultivar to cultivar, such as Banarasi and Chakaiya cultivar mature in the November, whereas Frainsic cultivar fruits are mature in November to first week of December. Fruits should be harvested at proper maturity when the colour of fruits turn dull green or greenish yellow or sometimes brick red colour can also be seen. Change in seed colour from creamy white to brown. For harvesting of aonla growing degree days is also taking in consideration (5000- 5200^{0} days) (Dhillon, 2013). Fruits are harvested individually plucked by climbing on the tree with the help of bamboo ladders. Fruit plucking should be done at morning or evening hours.

Yield

A full mature tree bears 125-200 kg fruit tree^{-1}, however, the fruits yield various variety to variety.

Sorting and grading

For efficient marketing, fruits are generally sorted and graded. Grading can be done on the basis of size and quality. On the basis of size fruits are graded into three grades which are large size fruits (>4 cm in diameter), small size fruits (<4 cm in diameter) and last is defective fruits (Dhillon, 2013).

Post harvest treatments

Aonla fruits can be treated with 1 per cent solution of $CaCl_2$ and 1 per cent solution of $CaNO_3$ for 30 min. each, and with single layer of malted paraffin wax, at 58-60°C to maintain less PLW, higher TSS and ascorbic acid (Pathak *et al.* 2009). Treatment of fruits with 1.5 per cent $CaCl_2$ with storage of fruits in zero energy cool chamber increase shelf-life of fruits (11 days), it also reduces physiological losses in weight (up to 16 per cent) and spoilage losses (upto 16.5 per cent) (Singh *et al.,* 2010).

Packaging

At present proper packaging is inadequate in case of aonla. Generally, fruits are packed in wooden crates with polythene or news-paper liner (to minimize bruising) for distant market. Fruits can also be packed gunny bags and jute bags. The corrugated fiber boxes are better as this provides appropriate ventilation inside the box and can also re-usable. Packaging of fruits, in corrugated fiber board boxes which are lined with newspaper causes minimum spoilage (16.0%) in fruits (Singh *et al.,* 2005).

Storage

Storage conditions should be appropriate to reduce post harvest losses in aonla. Storage and processing methods can curtail the post harvest losses upto 30 per cent (Goyal *et al.* 2008) and increase the shelf-life of fruits which can make the fruit available for longer period. For better and long shelf-life storage, fruits should be stored at ambient temperature and relative humidity.

Processing / Value addition

Fruits of aonla can be processed in many quality products, such as preserve, jam, sauce, pickle, candy, chutney, osmo-air dried aonla slices, pills, shreds and supari due to its nutritious value (Rakesh *et al.,* 2004). To reduce the enzymatic spoilage, blanching with hot water or with potassium meta-bisulphite before drying is preferred, it also improves the colour and texture of the shreds (Prajapati *et al.,* 2009). Low cost preservation technology are applied to prepare products by dehydration, such as dried whole fruit, flakes (Verma and Gupta, 2004), slice (Alam *et al.,* 2010), supari (Damame *et al.,* 2002), shreds (Sagar and Kumar, 2006), and powder (Vijayanand *et al.,* 2007). Aonla is a very important ingredient in the Chyavanprash, and a constituent of Triphala (Boora and Bons, 2015).

Small size aonla fruits, which are not suitable for preparation of preserve, may be utilized for pickle making (Goyal *et al.,* 2008).

Preserve which is also known as murabba, is a very popular product in India. Aonla preserve is prepared from large size (Chakaiya cultivar) mature fruits. These fruits are piercing and impregnated in sugar till the fruits become tender. Aonla preserve should have 65 °B TSS and a minimum 55 per cent of fruit portion. It is effective in reducing cholesterol level, purifying blood also helpful in improving eyesight (Ranote and Singh, 2006).

As like preserve aonla candy is also a popular product of aonla fruits, due to its higher nutritive value and longer storage life. It has 75 °B TSS. The preparation procedure of candy is same as like aonla preserve such, it is also pierced and stepped in sugar solution but drying is additional step in case of candy. A fruit impregnated with sugar, removed from it than drained the sugar solution and dried, is called as candid fruit or fruit candy. It is used to make ayurvedic tonics, such as Triphala and Chyvanprash. Triphala is prepared from mixture of three different types of herbs fruit: Aonla, Bibhitaki or Baheda and Haritaki or Harad. Its powder is also known as 'Triphala Churna'. Chyvanprash is made from aonla fruits, different herbs sugar, honey, etc.

Technological flow-chart for processing of Preserve and Candy

Select mature fruit
↓
Washing
(With clean water)
↓
Blanching
(In 2 per cent solution for 10 min.)
↓
Separation of the fruits
↓
Steeping in 50 per cent syrup solution
(For 24 hrs.)
↓
Steeping in 60 per cent syrup solution
(For 24 hrs.)
↓
Steeping in 70 per cent syrup solution + Cardamom/ginger powder for 3 days
↓ ↓
Draining of excess syrup Preserve
↓
Drying of segments (Cabinet drier) up to 75 per cent TSS
↓
Wrapping with powdered sugar and mixing pieces with fried spices
↓
Packing
↓
Storage

Sources: Nayak *et al.,* 2012

2. Annonacious fruits

Introduction

Annonacious fruits (*Annona spp.*) are belongs to family Annonaceae. Out of the several species of *Annona,* at least five are available in India yielding edible fruits. They are Custard Apple, Cherimoya, Soursop, Ramphal and Atemoya (Vinay *et al.,* 2015). These are thrives well in tropical and warmer subtropical regions of India. Its commercial cultivation is gaining momentum in Gujarat. It is a climacteric fruit.

Botanical or specific, common and vernacular names and their synonyms of the *Annona species* studied.

Common name	Botanical name	Synonyms
Custard apple, Sugar apple, seethapalam	*A. squamosa* L.	*A. asiatica* L., *A. cinerea* Dunal, Guanabamus squamosus Gomez
Bullock's heart, ramphal	*A. reticulata* L.	*A. excelsa* Kunth, *A. laevis* Kunth, *A. longifolia* Moc. and Sesse, *A. riparia* Kunth.
Cherimola, Cherimoya	*A. cherimola* Mill.	*A. tripetala* Aiton, *A. pubescens* Salisb.
Soursop	*A. muricata* L.	Munolla
Wild soursop	*A. senegalensis* Pers.	*A. arenaria* Thonn., *A. chrysophylla* Boj., *A. chrysophylla* var. *porpetac* Bail., *A. senegalensis* var. *porpetac* Bill. Wild

Source: Pinto *et al.* (2005)

Nutritive and medicinal value

Annonacious fruits are highly nutritious and have many medicinal valves. The unripe fruits are rich sources tannin, is dried, pulverized and used against diarrhea. The leaf decoction is used as a vermifuge. Its crushed leaves or a paste of the flesh may be poulticed on boils, abscesses and ulcers. Its bark is very astringent and the decoction is taken as a tonic and also as a remedy for diarrhoea and dysentery. In severe cases, the leaves, bark and green fruits are all boiled together for 5 minutes in a litre of water to make an exceedingly potent decoction. Fragments of the root bark are packed around the gums to relieve toothache. The root decoction is taken as a febrifuge. Relaxing muscles and protecting heart against diseases, helps to grow weight and natural anti-cancerous properties. Lower the risk of arthritis (Morton, 1987)

Nutrient value of 100 g of edible pulp of cherimoya, custard apple, soursop and sugar apple.

Nutrient	Cherimoya	Custard apple	Soursop	Sugar apple
Water	77.3±3	75.8±2.8	81±2.5	72.6±2.4
Protein	1.6±0.6	1.85±0.05	1±0.55	1.68±0.8
Ash	0.7±0.1	0.95±0.15	0.61±0.2	0.7±0.1
Carbohydrate	18.42±4	18.7	17.25±0.1	19.6±1
Fibre	1.64±0.5	2.55±0.35	0.86±0.1	1.4±0.6
Energy	68.6±13.4	75	65±5	96±10
Ca	27.14±5	24	15±7	26.2±6
P	35.2±18	26	28±1	42±14
Fe	0.6±0.2	1.0	0.7-0.1	0.8±0.5
B_1	0.09±0.03	0.07	0.07±0.01	0.1±0.01
B_{12}	0.12±0.2	0.12	0.08±0.035	0.13±0.05
Ascorbic acid (mg)	11.5±5.5	30	19.4±3	37.38±4.62

Source: Pareek *et al.*, 2011

Cultivar/Varieties

Some commercial cultivars of annona species are:

Species	Cultivar
Annona squamosa	Balanagar, British Guinea, Mammoth, Barbados and Red Sitaphal
A. atemoya	Island Gem, Pink's Mammoth
A. cherimola	Golden Russet, Libby, Bronceada, Whaley, Mosman
A. moricata	Blanca, Cerradina, Morada and Lisa

Source : Dhillon, 2013.

Maturity indices and harvesting

Harvesting cannot be done on the basis of flowering, because it can occur during several months in annonaceous fruits. Harvesting of fruits before maturity results low quality fruits, whereas over-ripe fruits are spoiled, so harvesting should be done at proper maturity. The primary maturity index is changes of skin colour from dark green to light green in *A. cherimoya, A. atemoya,* sweetsop, soursop and custard apple (Pinto *et al.,* 2005). On the other hand appearances of cream colour between the segments on the skin (Kader and Arpaia, 2006). Harvesting is done manually by plucking individual fruits.

Post harvest treatments

Some post harvest treatments are given to annona group of fruits to increase shelf-life as well as to reduce flesh browning. Waxing is done in cherimoya fruits to reduce their ethylene production and respiration rate, minimize browning and extend its shelf-life. Different type of coating material can be used, such as caranauba wax, orange oil, sellac gum lac and sodium polyphosphate (Yonemoto *et al.*, 2002). If the fruits of sugar apple are treated with $CaCl_2$ (6%) and stored at 16°C had reduced weight loss due to lower respiration rates, less peroxidase activity, and higher firmness (Lima, 2000).

Packaging

Annona fruits are climacteric and highly perishable in nature, so after picking fruits are kept in boxes or basket with cushioning material to avoid mechanical injury. Fruits of Atemoya can also be packed in individual films to extending their shelf-life (Yamashita *et al.*, 2002). At a temperature of 15°C, atemoya fruit had a shelf-life of 17 days, with wrapping (with PD-955, a co-polymer).

Storage

Annonas are climacteric fruit, generally characterized by high respiration and ethylene production, and are chilling sensitive. Extended fruit storage periods are still not possible, mainly because of the high susceptibility to chilling injury. The fruits of cherimoya can be stored for 21 days, if these are preserved at 11-12°C temperature, whereas fruits of sugar apple can be stored upto 7 days at a temperature of 15-16°C and 85 per cent relative humidity (Dhillon, 2013).

Processing / Value addition

Flow chart of custard apple milk shake

Mature ripe fruit
↓
Washing
↓
Cutting into two halves
↓
Scooping of pulp
↓
Sieving
↓
Removal of seeds
↓
Addition of jaggery into pulp
↓
Addition of milk into pulp (1:1)
↓
Mixing
↓
Addition of vanilla essence
↓
Mixing
↓
Cooling
↓
Ready to serve

Source: Gupta *et al.* (2018)

Other by-products

The leaves have been employed in tanning and they yield a blue or black dye. A fiber produced from the young twigs is superior to the bark fiber from *Annona squamosa*.

3. Date Palm (*Phoenix dactylifera* L.)

Introduction

Date palm (*Phoenix dactylifera* L.) is an important arid and semi-arid region fruit tree. It is commonly known as *Khajoor*. Not only does date palm have an effective role in the ecosystem of the desert, but it has several agricultural and animal husbandry uses. Fresh dates are typically marketed different growth and development stage, such as: *Khalal, Rutab* and *Tamar,* and can be consumed soft, semi-dry or as dry fruits, depending on their water content at harvest when fully-ripe, or in various processed forms (Sawaya, 1986). Date fruits are non-climacteric fruit, with very low metabolic activities.

Nutritional and medicinal value

Date palm fruits are considered highly nutritious and a high-energy food (300 calories/100 g) due to these richness in sugars which providing quick energy intake (Ahmed *et al.,* 1995), minerals (Elleuch *et al.,* 2008), vitamins, phenols, flavonoids, anthocyanins and carotenoids with functional properties (Al-Turki *et al.,* 2010). The soluble fiber helps against constipation (Kulkarni *et al.,* 2008), It also has omega-3, omega-6, and oleic acid, which are heart-healthy.

Parameter g/100g	**Fruit Flesh**
Moisture content	3.50±0.05
Crude protein	17.15±0.15
Crude fiber	0.5
Ash content	1.50±0.17
Carbohydrate	75.85±0.18
Energy kj/100g	337.7±8.82
Vitamin C ìg/100	100.26±1.12
Ca	105.2±3.52
Fe	47.5±1.11

Source: Sadiq *et al.,* 2013.

Varieties

The commercially cultivated varieties of date palm are Halawy, Barhee, Mendjool, Khalas, Khuneizi, Zahidi, Shamran, Sewi, Zaglool, Khadrawy, etc.

Maturity indices and harvesting

Date should be harvest at proper time to reduce the incidence of fruit cracking or splitting, excessive dehydration, insect infestation, and attack by micro-organisms. Date palms are picked several times during the harvest season since all dates do not ripen simultaneously. Dates are harvested at or near maturity. Generally, dates are harvested in July to August at the *Khalal* stage or in September to December at the *Rutab* and *Tamar* stages in northern hemisphere production regions. In the southern hemisphere harvesting takes place in February, March and April. Time of harvest is based on date fruit's appearance and texture, which are related to moisture and sugar content (Yahia, 2004). Fruits harvested, when appearance of a reddish ring at the perianth end, have better storage potential than fruits left on the palm until the ring has faded with more advanced maturity (Rygg, 1975). Harvest is generally done by hand, with access to the crown of the tree being by way of limbing or mechanical lifts.

Yield

Trees of date are start bearing after 5-6 years of planting. Fruits yield is law at initial stage, a full grown tree gives higher yield. Fruit yields 100-200 kg/tree with proper cultural practices (Dhillon, 2013).

Sorting and grading

The fruits consist of defects and abnormalities, such as immature or over-ripe fruits, mechanically damaged fruits during harvesting or on the palm, fruits damaged by birds or insects, and fruits with physiological disorders or diseases. Grading is done manually or mechanically to separate dates based on their size and weight. Grading of fruits according to size is one of the quality criteria for dates. The size of dates depending on cultivar.

Storage

Storage of dates can be done under refrigeration or freezing, mainly to reduce enzyme activity (Biglari, 2009). The optimum storage temperature is 0°C, which allows a storage period of 6–12 months. Storage of dates depends on the variety of dates. Semi-dry dates have a longer storage life than soft dates. Relative humidity of 70–75 per cent in storage rooms will be useful.

Processing / Value addition

Many processed products, including syrup, animal feed, date powder, different types of bread, marmalade, sweet candy, chocolate, date paste, alcohol, and others, can be obtained from date. Other parts of the date palm are also important in the agricultural economy. For example, the stem is used for making boats, covering the roof of rural houses, paper and wood industries, and making fiber. The foliage is used for making handicrafts like fans and straw hats (Jahani, 2002).

Flow chart of drying dates

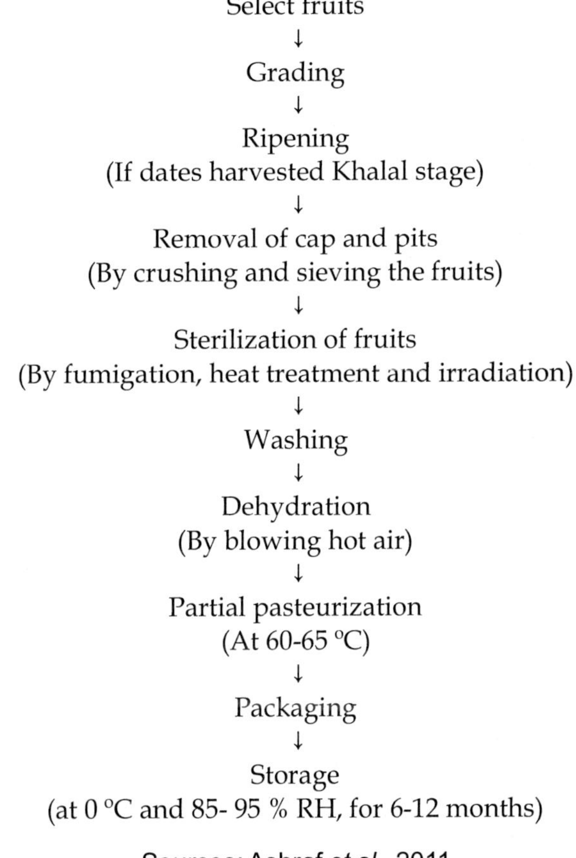

Sources: Ashraf *et al.*, 2011

By-products

The major by-products from date fruit processing plants are low-grade rejected date fruits, date pits, and date presscake (Hui, 2006). Dates with improper appearance that can not be packaged for marketing are mostly utilized in animal feed and as substrate for fermentation products because of their nutritive components, such as sugars, proteins, minerals, and fiber (Shi, 2005).

4. Jhar Ber (*Ziziphus nummularia)*

Introduction

Jhar ber (*Ziziphus nummularia*) is an underutilized, most commonly occurring, a thorny shrub species with a height of 1-2 m and belongs to the buckthorn family Rhamnaceae. Generally, it is found in the west Indian desert where it is locally known by different names, such as Bairi, Badber, Karkanrha, etc. It occupies almost all the habitats including crop lands and grazing lands (Kumar *et al.*, 2005). It is found growing naturally on farmlands and spreads through vegetatively by root suckers and also by seeds.

Nutritive and medicinal value

The fruits of *Jhar ber* are nutritious and a rich source of vitamin C. Generally, fruits are eaten fresh as well as dried. The fruits are contains carbohydrate. Small fruits are used for cold, diarrhoea, indigestion, dysentery and inflammation of gums (Kapoor *et al.*, 2014). The immature fruits of the plant are useful in the management of vomiting, burning sensations and as tonic, while dried fruits are useful as an anticancer, anodyne, refrigerant, sedative, stomachic and also used in the treatment of anaemia, bronchitis, burns, chronic fatigue, diarrhea, hysteria, loss of appetite and pharyngitis (Chopra *et al.*, 1965). The leaves are used to treat scabies and other skin diseases. The fruits are greedily eaten by gerbils and rats and are used as baits for poisoning these rodents.

Varieties/cultivar

This is not grown commercially by the farmers; hence, there is no any standard commercial cultivar available. This is wildly grown plant species, which is predominantly found in barren land.

Maturity indices and harvesting

Fruiting comes in November-December. The fruits are mostly small, round shaped with green fruits in early stage which turn yellow or red at maturity and dark brown after drying. In rural area fruits are harvested manually or by shaking of the plants. For making vegetables and pickles mature green yellowish fruits should be harvested. For fresh consumption ripe fruits should be harvested.

Sorting and grading

The berries are less than one centimetre in diameter, sorted and graded for fetching better price. So, diseased, fungal attacked, damaged, under-sized and over-ripe berries should be sorted out. There is no commercial standard grading method is available, but can be graded on the basis of shape and size for local market and to get better prise.

Post harvest treatments

Post harvest dipping of fruits in cold water reduced respiration, ethylene production and enzymatic activities, whereas hot water treatment hinders the development of pathogens, thereby prolonging the shelf-life and quality of fruits. Hot water dipping of fruits has been reported to reduce water loss and PLW of fruits during storage.

Packaging

Packaging is an essential operation for fruits either for proper storage or for safe transport to markets. The fruits are packed in gunny bags, net bags, cloth packages or wooden or plywood boxes with holes or slits. For local market, cheaper materials can be used, if cushioning and ventilation is provided. Fruits should be packed in non-organic materials to avoid spoilage caused by microbes during storage and transport. For local market sometime bamboo busket can be used.

Storage

The fruits are stored at optimum temperature and relative humidity, to prolong the shelf-life. At higher temperature fruits respire rapidly resulting poor shelf-life, whereas at very low temperature fruits undergone to susceptible to chilling injury. So, it is important to maintain an optimum temperature range for storage of fruits. In normal situation, fruits are stored at room temperature.

Processing / Value addition

The fruits which are small berries ripen during November and December, eaten fresh as well as dried. Fruits used to make a refreshing drink. In India, the berries are collected, when fully ripe, in the beginning of the November months. The dried fruits are grinded and the powder is sieved. The sieved powder consumed after meal as mouth freshner which is very rich in sugar and ascorbic acid (Meghwal and Singh, 2016). In rural areas, fruits are also used to making pickles and chutney.

Other by-products

Leaves of *Jhar ber* called *pala* is very nutritious for animals, used as fodder. It is rich in protein (12.9-16.9 %) and crude fibres are much palatable to animals. For that purpose, the branches along with leaves are collected and left for sun drying. The dried leaves are separated used to feeding of livestocks, while the thorny bushes are also used for fencing purposes. Fodder: The leaves of *Z. nummularia,* average yield of forage was about 1000 kg ha^{-1}. It is also used for fuel and timber wood (Meghwal and Singh, 2016).

5. Kair (*Capparis decidua* Forsk.)

Introduction

Kair (*Capparis decidua* (Forsk., Edgew. Synonym *Capparis aphylla* Roth) is the most under-utilized shrub which belongs to family Capparaceae. It occurs in tropical and warm temperate region (Vyas *et al.,* 2009). *Kair* is a xerophytic plant which made it suitable for cultivation in arid and semi-arid regions of country, especially on wind erosion and sandy wastelands. The immature small fruits are rich sources of phenol content, carbohydrates, protein and mineral elements (Chouhan *et al.,* 1986).

Nutritive and medicinal value

Fruits and flower buds are rich in surface wax, while seeds are rich in oils (Meghwal and Tiwari, 2002). The fruits contain alkaloids, glycosides, terpenoids, sterols, flavanoids, phenols (anti-bacterial) and fatty acids (Sadia Chishty and Monika, 2016). The trees small fruits and its different plant parts are very nutritious and have many medicinal properties, such as pharmacological activities like hypercholesterolemic, anti-inflamatory and analgesic, anti-diabetic, anti-microbial, anti-plaquer, anti-hypertensive, and purgative activities. It is carminative, tonic, emmenagogue, aphrodisiac alexipharmic, and also effective for appetite, rheumatism, lumbago, hiccough, cough and asthma (Satyanarayan *et al.,* 2008). Tender shoots, when chewed, relieves toothache. Root bark is used against intermittent fever.

Composition (per 100 g of Edible Portion)

Unripe fruit of *kair*		Fruit pulp	
Protein	14.88 %	Moisture	56 %
Fat	7.43 %	Crude protein	8.6 %
Ash	5.96 %	True protein	5.02 %
Crude fiber	12.32 %	Total sugar	1.8 %
Carbohydrates	59.41 %	Reducing sugar	1.1 %
Beta-carotene	5.4 mg	Phosphorus	0.057 %
Ascorbic acid	120.7 mg	Potassium	1.026 %
Calcium	90.0 mg	Calcium	0.055 %
Phosphorus	179.0 mg	Magnesium	7.81 mg
Iron	3.5 mg	Ascorbic acid	7.81 mg
Copper	1.1 mg		
Zinc	1.6 mg		
Manganese	1.9 mg		

Source: Chouhan *et al.*, 1986

Maturity indices and harvesting

A full mature shrub comes in bearing after 4-5 years of age. The fruits are harvested in April-May. The fruits should be harvested at green, tender at small pea stage for pickling and vegetable purposes. The stage can be judged by size of the fruits and also by pressing the berries. The fruits should be harvested after 7-10 days of fruit setting with 5-8 mm diameter during March-April to fetch better price in the market (Meghwal, 2002). Harvesting of fruits is done manually by plucking individual fruits or sometimes by beating the shrub with wooden sticks and the fruits are collected from the ground.

Yield

Initially, the yield of immature/unripe green fruits is low, it is about 2-3 kg plant^{-1}, whereas the yield of a full grown plant is upto 10-15 kg fruits plant^{-1} (Meghwal and Tiwari, 2002).

Post harvest treatments

Fruits are sold in the local or distance market after harvesting. Due to their astringency (due to presence of phenols and tannins) fruits are not eaten fresh. Fruits astringent taste can be removed by treating these with 5 per cent salt solution or by dipping for 4-5 days in butter milk in earthen pots. After this treatment, the fruits are used as vegetable or can be dried for further use. Drying can be achieved by keeping fruits in open sunny condition and then preserve these. The dried fruits, with 5-7 per cent moisture, can be preserved for long time in airtight containers (Meghwal and Tiwari, 2002).

Storage

The fruits are stored either in pots or in plastic containers, while processed dried fruits are stored in flexible polybags at ambient conditions of temperature and relative humidity. Dried fruits can be stored in polybags without any deterioration in the quality long time.

Processing / Value addition

The unripe/ripe fruits of *Kair* are generally not eaten fresh due to their acrid taste, but can be converted into a variety of by-products after processing. The pickles are the most commonly and widely utilized post harvest product of *Kair*. The processed fruits can be utilized directly for preparation of pickles or as a vegetable or can be dehydrated for off-season utilization. Because of acrid taste, the fruits are not utilized directly but they are utilized only after processing. The method of processing of *Kair* is given below in the form of flow sheet (Saroj and Awasthi, 2006). *Kair* sangri is one of the most popular dishes incorporating the fruit. Dehydrated *Kair* is one of the important ingredients of '*Panchkuta*', a dehydrated mixture of sangria, *Kair*, *Kachri*, kumat and mango/kamal kakri. The method of processing of *kair* is given below in the form of flow sheet (Saroj and Awasthi, 2006).

Flow chart of *Kair* pickle

Harvesting of tender Fruits
↓
Cleaning and removal of fruit stalk
↓
Store the fruits in earthen pot
↓
Mix curd (250 g) and salt (50 g) kg^{-1} fruit
↓
Add water till the fruits are properly embedded
↓
Keep the pot in sunny place after closing the lid
↓
Drain the water after 4 days
↓
Repeat the process at least 4 times
↓
Taste the fruit, as it gives flat and salty taste and repeat the process once more, if needed
↓
Ready to use

Other by-products

New sprouted shoots are used as fodder for camel and goat during spring summer season (Meghwal and Tiwari, 2002). The plant and its parts are widely used by traditional healers and tribal people for curing variety of ailments. The *kair* branches can be used for fencing, which are not affected by white ants and also as firewood, as it burns quickly. It is very efficacious plant for checking soil erosion by wind and water (Gupta *et al.*, 1989).

6. Khejri (*Prosopis cineraria* L. Druce)

Introduction

Khejri (*Prosopis cineraria* L. Druce) is an evergreen, medium large thorny tree, which is a most important component of arid farming system and plays significant role in the economy of people of Indian Thar desert. It is also called as Shami/*Janti*/kalp vraksha' of Indian desert area. It is very slow growing and has deep root system which makes it able to withstand in the severe drought condition of arid and semi-arid regions. The trees are owing to multiple uses, such as atmospheric nitrogen fixing ability and suitability in agro forestry systems and it has been conserved in arable land since ages. Pods of

Khejri are locally called "sangri" are brown to chocolate colour and having sweet, dry and yellow coloured pulp on ripening. Pods are hang in cluster of upto 12 from the tree, each containing several seeds embedded.

Nutritive and medicinal value

As like, arid foods had are great nutritional values, different parts of *Khejri* plants have long been used by desert people for various purposes. Its nutritious pods both tender and dried are consumed as a vegetable by human beings while, highly nutritious leaves constitute important source of fodder for livestock and also used as traditional medicine. The anti-bacterial activity of the various extracts of the stem bark of *Khejri*. Roots of these plants are benefits for nitrogen fixation (Panwar *et al.*, 2014).

Constituent	Composition
Energy	334.8 Kcal
Protein	23.2 %
Fat	2.0 %
Carbohydrate	56.0 %
Fiber	20 %
Calcium	414.0 $mg100^{1}g$
Phosphorus	400.0 $mg100^{-1}g$
Ferrous	19.0 $mg100^{-1}g$
Vitamin A	0.055 $mg100^{-1}g$
Vitamin B_2 (Riboflavin)	1.2 $mg100^{-1}g$
Vitamin C	523.0

Source : Rathore and Meena, 2004

Varieties/cultivar

Some recently developed cultivar of *Khejri* is Thar Shobha,

Maturity indices and harvesting

Khejri is a slow growing tree, take 7-8 years to come in bearing. Tender pods are harvested for vegetable propose. Its dried pods brown to coffee in colour on ripening and have a sweetish pulp. Harvesting is done by cutting branches or by climbing on the tree, than the individual pods are harvested from the branches.

Packaging

A good packing of fruits has grate importance in reducing spoilage or wastage and it is also increasing shelf-life of fruits. There is not any standardized packaging material for *Khejri* pods packaging, but we can use gunny bags, fibre cartons, baskets and wooden crates, etc.

Storage

The major aim of storing fruits is to increase its shelf-life by giving appropriate environmental conditions (temperature and humidity), by which fruits can be available for long time or in off-season. So, *Khejri* pods should be store at proper temperature and relative humidity.

Processing / Value addition

The ripe dried pods (having 9-14 per cent crude protein and 6-16 per cent sugar (Arya *et al.*, 1991) can be powdered and used in the preparation of bakery items, such as biscuits and cookies.

Other by products

In other by products of *Khejri* tree, its leaves and tender shoots which are used as fodder readily eaten by animals, such as camel, goat, cattle, etc. locally, it is called as *'Loom'*. This is highly nutritious and palatable, when green as well as after its drying. Some products, such as *Khejri* gum and wood are also use as by-product. Its looped branches are used for fencing material due to its thorniness.

7. Lasoda (Cordia myxa L.)

Introduction

Lasoda (*Cordia myxa* L.), locally known by different names, such as Indian cherry/*Gonda/Lasora/lehsua* belongs to family Boraginaceae and is grown all over India except high hills. It has great capacity to tolerate drought and hence, quite widespread in arid and semi-arid regions of north India.

Nutritive and medicinal value

Being a multipurpose plant, it has long been associated with health, nutrition and other diversified uses. Fruits are considered as rich sources of antioxidants *i.e.* carotenoids, ascorbic acid, phenols, *etc.* Besides this fruits are also contains minerals, fiber and vitamins, which provides essential nutrients for the human health (Mala, 2009). Its fruits and other plant parts are used in curing various ailments viz. skin diseases, dropsy, dysentery, dyspepsia, cholera, headache, etc. (Meghwal and Singh, 2016).

Components	Range
Carbohydrate	92 %
Fat	2 %
Energy	394 (Kcal/100 gm)
Protein	2 %
Fibre	2 %
Ca	55 (mg/100g)
P	275 (mg/100g)
Fe	6 (mg/100g)

Source : Tiwari, 2016

Varieties / Cultivar

Some recently developed cultivars are: CAZRI-2021 and CAZARY-2025.

Yield

The farmer got fruit yield from these plants in the second year and in third year, it ranged from 5-15 kg per plant.

Sorting and grading

Considerably variation exists in the quality of the harvested fruits, therefore, grading is required to get suitable prices of the produces. Systematic grading should be done on the basis of shape, size and weight of fruits to get maximum prices in the market.

Post harvest treatments

Packaging

Proper packaging reduces the post harvest losses of any harvested produce, it also make produce convenient to handling and protect it from spoilage during marketing process. The fruits of lasoda are perishable and are normally packed in bamboo baskets for local market and for distance market, it should be packed in cardboard boxes.

Processing / Value addition

The immature fruits of *gonda* are used as vegetable, pickled with raw mango and can be dehydrated for future use. Sometime fruits are dehydrated after blanching for used as vegetable during off-season (Singh, 2001).

Flow chart of lasoda pickle

Wash the lasoda fruits with water

↓

Break the fruit into two pieces and remove the stone

↓

Heat mustard oil in pan, add fenugreek seed, fry till brown and then add lasoda fruit and fry

↓

Add spices like thyme powder, red chilli powder, turmeric powder and salt to the mixture cook for 3-5 minutes

↓

Put the cool mixtures into clay pot for fermentation

↓

Heat mustard oil in pan, cool and put into the clay pot over the mixture and *kair* in sunlight for fermentation

Source: Gupta *et al.*, 2018

Conclusion

The under-utilized fruit crops continue to play an important role in the economy and nutritional security to the poor people throughout the developing World, particularly in the arid and semi-arid conditions. By converting these fruits into various processed and value added products, the loss can be minimized. Their processed products are also gaining popularity not only in the domestic markets but in the abroad also. These products are having high potential of nutrients and more shelf-life, thus can be used during adverse situation. Despite their potential for dietary diversification and the provision of micronutrients, such as vitamins and minerals, they continue to attract little research and development attention. Alongside their commercial potential, many of the underutilized fruits also provide important environmental services, as they are adapted to marginal soil and climatic conditions.

References

1. Agarwal, S. and Chopra, C.S. (2004). Studies on changes in ascorbic acid and total phenols in making aonla products. *Bev. Food World*, 31(5): 32-33.

2. Ahmed, I.A., Ahmed, A.K. and Robinson, R.K. (1995). Chemical composition of date varieties as influenced by the stage of ripening. *Food Chem.*, 54: 305–309.
3. Alam, M.S., Amarjit, S. and Sawhney, B.K. (2010). Response surface optimization of osmotic dehydration process for aonla slices. *J. Food Sci. Technol.*, 47(1): 47–54.
4. Al-Turki, S., Shahba, M. A. and Stushnoff, C. (2010). Diversity of anti-oxidant properties and phenolic content of date palm (*Phoenix dactylifera* L.) fruits as affected by cultivar and location. *J. Food Agric. Env.*, 8: 253–60.
5. Arya, S.S., Toky, O.P., Bisht, R.P. and Tomar, R. (1991). *Prosopis cineraria* - A promising multipurpose tree for arid lands. *Agroforestry Today*, 3(4): 13-14.
6. Ashraf, Zahra and Hamidi-Esfahani, Zohreh (2011). 'Date and Date Processing: A Review', *Food Reviews Int.*, 27(2): 101-133.
7. Biglari, F. (2009). Assessment of Anti-oxidant Potential of Date (*Phoenix dactylifera* L.) Fruits from Iran. School of Industrial Technology. University Sains Malaysia. Penang, Malaysia, 15–105.
8. Boora, R.S. and Bons, H.K. (2015). The minor fruits-health treasure. *Kisan World*, 42(11): 32-33.
9. Chauhan, B.M., Duhan, A. and Bhat, C.M. (1986). Nutritional value of Kair (*Capparis decidua*) fruit. *J. Food Sci. Technol.*, 23(2): 106-108.
10. Chopra, R.N., Nayar, S.L. and Chopra, I.C. (1965). *Glossary of Indian Medicinal Plants.* New Delhi, C SIR.
11. Damame, S.V., Gaikwad, R.S., Patil, S.R. and Masalkar, S.D. (2002). Vitamin C content of various aonla products during storage. *Odisha J. Hort.*, 30: 19–22.
12. Dhillon, W.S. (2013). In: *Fruit production in India*, Narendra Publishing House, Delhi, 11, 17, 27, 223.
13. Elleuch, M., Besbes, S., Roiseux, O., Blecker, C., Deroanne, C., Drira, N. and Attia, H. (2008). Date flesh: Chemical composition and characteristics of the dietary fibre. *Food Chem.*, 111: 676–682.
14. Ganachari, A., Thangavel, K. and Eswaran, S. (2005). Post harvest processing, potential of aonla (*Emblica officinalis*) fruit. *Food Pack.*, 5(5): 34-35.
15. Goyal, R.K., Patil, R.T., Kingsly, A.R.P., Himanshu, W. and Kumar, P. (2008). Status of post harvest technology of aonla in India- A Review. *American J. Food Technol.*, 3(1): 13-23.

16. Gupta, I.C., Harsh, L.N., Shankarnarayanan, K.A. and Sharma, B.M. (1989). Wealth from wastelands. *Indian Farming* 38(11): 18-19.

17 . Gupta, N. Meenakshi Trilokia, Monika Sood, Dogra, J. and Singh, J. (2018). Utilization of Under-utilized Fruits through Value Addition in *Kandi* Areas of Jammu Region. *Int. J. Current Microbiol. Appl. Sci.,,* 7(5): 1965-1977.

18. Hui, Y.H. (2006). *Fruit and Fruit Processing*; Blackwell Publishing: Ames, Iowa, 391–411.

19. Jahani, A. (2002). *A. Date, the life fruit*; Agricultural Sciences' publication: Tehran, Iran, 147.

20. Kader, A.A., and Arpaia, M.L. (2006). Cherimoya: Atemoya and sweetsop. Produce fact sheet. UC Davis, California: Post harvest Technology Research and Information Centre. www.postharvest. ucdavis.edu.

21. Kapoor, B.B.S. and Arora, V. (2014). Ethnomedicinal Plants of Jaisalmer District of Rajasthan used in Herbal and Folk Remedies: *Int. J. Ethnobiol.* Ethnomed, 1: 1.

22. Kulkarni, S.G., Vijayanand, P., Aksha, M., Reena, P. and Ramana, K.V.R. (2008). Effect of dehydration on the quality and storage stability of immature dates (*Pheonix dactylifera*). *Food Sci. Technol.,* 41: 278–283.

23. Kumar, S., Farzana, P. and Narain, P. (2005). Medicinal Plants in the Indian Arid Zone. Technical Bulletin, CAZRI, Jodhpur, 64.

24. Lima, C.L.C. (2000). Uso de ceras e do calico, associados ou nao, na conservacao poscolheita de frutos da pinheira *(Annona squamosa* L.) armazenados ao ambiente ou sob refrigeracao. Tese. (Duotorado em Energia na agricultura) Botucatu: Faculdade de Ciencias Agronomicas, Universidade Estadual Paulista, 119.

25. Mala, R. (2009). Nutrient content of important fruit trees from arid zone of Rajasthan. *J. Horti. Forestry.,* 1(7): 103-108.

26. Meghwal, P.R. and Singh, A. (2016). Underutilized Fruits Research in Arid Regions: A Review. *Ann. Arid Zone,* 56(1&2): 23-36.

27. Meghwal, P.R. and Tiwari, J.C. (2002). *Kair (Capparis decidua* (Forsk.) Edgew.) A multipurpose woody species for arid regions, *Forests Trees Liveli.,* 12: 313-319.

28. Meghwal, P.R. (2002). Judging the culinary maturity of kair fruits (Teent) (*Capparis decidua*). *Current Agric.,* 26(1-2): 75-76.

29. Meghwal, P.R. and Tiwari, J.C. (2002). *Kair* (*Capparis decidua*): A multipurpose woody species for arid region. *Forest, Trees Liveli.*,12(4): 313-319.

30. Morton, I.J. (1987). Soursop. In I.J. Morton (Ed.), Fruits of warm climates. Miami, Florida: Media Incorporated, 69-72.

31. Nayak, P., Tandon, D.K. and Bhatt, D.K. (2012). Study on changes of nutritional and organoleptic quality of flavoured candy prepared from aonla (*Emblica officinalis* G.) during storage. *Int. J. Nutri. Meta.*, 4(7): 100-106.

32. Panwar, D., Pareek, K. and Bharti, C.S. (2014). Unripe Pods of *Prosopis cineraria* used as a vegetable (sangri) in Shekhawati region. *Int. J. Scient. Eng. Res.*, 5(2).

33. Pareek, O.P., Kaushik, R.A. (2011). Post harvest physiology and technology of Annona fruits. *Food Res. Int.*, 44: 1741–1751.

34. Pathak, P.K., Divedi, P. and Kumar, S. (2009). Effect of post harvest treatments on shelf-life of aonla (*Emblica officinalis*) fruits damaged during harvesting. *J. Food Sci. Technol.*, 46(3): 283-285.

35. Pinto, A.C.Q., Cordeiro, M.C.R., de Andrade, S.R.M., Ferreira, F.R., Filgueiras, H.A.C. and Alves, R.E. (2005). Annona species, Southampton: International Centre for Under-utilized Crops, University of Southampton, 263.

36. Prajapati, V.K., Nema, P.K., and Rathore, S.S. (2009). Effect of pretreatment and recipe on quality of solar dried aonla (*Emblica officinalis* Gaertn) shreds. In: Int Conference on Food Security and Environmental sustainability. Indian Institute of Technology, Kharagpur, December 17–19.

37. Rakesh, Arya, S.S. and Moond, S.K. (2004). Processed products of aonla. *Proc. Food Indus.*, 7(4): 20-23.

38. Ranote, P.S. and Singh, S. (2006). Value added aonla products. The Schumacher Centre for Technology and Development Bourton-on-Dunsmore Rugby, Warwickshire, CV23 9QZ United Kingdom. Website: http: //practical action. org/practicalanswers/

39. Rathore, M. and Meena, R.K. (2004). Nutritional Evaluation of Some Famine Foods of Rajasthan Desert, *Indian Forester*, 130(3): 304-311.

40. Rygg, G.L. (1975). Date development, handling and packing in the United States. Agricultural Research Service. US Dept Agric, Washington, DC. Agric Handbook 482. 56p.

41. Sagar, V.R. and Kumar, R. (2006). Preparation and storage study of ready-to-eat dehydrated gooseberry (aonla) shreds. *J. Food Sci. Technol.,* 43(4): 349-352.

42. Saroj, P.L. and Awasthi, O.P. (2006). Advances in Arid Horticulture, Vol. 2 Production Technology of Arid and Semi-arid Fruits, International Book Distributing Company, Lukhnow, 405-433.

43. Satyanarayana, T., Anjana, A. and Mathews, V.P. (2008). Phyto-chemical and pharmacological reviews of some Indian *Capparis* species. *Pharma. Rev. Suppl.* 2: 36-45.

44. Sawaya, W.N. (1986). Date of Saudi Arabia. Riyadh: KSA Safer Press. 75–87.

45. Shi, J., Ho, C. and Shahidi, F. (2005) *Asian functional foods;* CRC Press: New York, 508–516.

46. Singh, V., Singh, H.K. and Chopra, C.S. (2005). Studies on processing of aonla *(Emblica officinalis Gaertn.)* fruits. *Bev. Food World,* 32(12): 50-52.

47. Singh, S., Singh, A.K., Joshi, H.K., Bagle, B.G. and More, T. (2010). Effect of zero energy cool chamber and post-harvest treatments on shelf-life of fruits under semi-arid environment of Western India. Part 2. Indian gooseberry fruits. *J Food Sci. Tech.,* 47(4): 450–453.

48. Singh, I.S. (2001). Minor fruits and their uses. *Ind. J. Horti.,* 58: 178-182.

49. Tewari V.P. (2016). Some Important Fruit Trees and Shrubs of Hot Arid Regions of Rajasthan State in India, Their Uses and Nutritive Values. *J. Plant Chem. Ecophysiol.* 1(1): 1004.

50. Verma, R.C. and Gupta, A. (2004). Effect of pre-treatments on quality of solar-dried amla. *J. Food Eng.,* 65: 397–402.

51. Vijayanand, P., Kulkarnu, S.G., Reena, P., Aksha, M. and Ramana, K.V.R. (2007). effect of processing on gooseberry fruits and quality change and quality changes in dehydrated gooseberry powder during storage. *J. Food Sci. Technol.,* 44: 591-594.

52. Vinay, G.M., Chithiraichelvan, R. and Halesh, G.K. (2015). Nutrients changes during off-season flowering in custard apple (*Annona squamosa* L.) cv. Balanagar, induced by pruning and defoliation. *The Bioscan,* 10(4): 1605-1610.

53. Vyas, G.K., Sharma, R., Kumar, V., Sharma, T.B. and Khandelwal, V. (2009). Diversity analysis of *Capparis decidua* (Forsk.) Edgew using biochemical and molecular parameters. *Gen. Res.Crop Evol.* 56: 905-911.

54. Yahia, E.M. Gross, K., Wang, C.Y. and Saltveit, M. (2004). Date. In: *The Commercial Storage of Fruits, Vegetables and Florist and Nursery Crops*. Agriculture Handbook 66. Beltsville, MD: USDA, 3.

55. Yamashita, F., Miglioranza, L.H.D.S., Miranda, L.D.A., and Souza, C.M.D.A.E. (2002). Effects of packaging and temperature on post harvest of atemoya. *Rev. Brasil. Fruti.*, 24: 658-660.

56. Yonemoto, Y., Higuchi, H., and Kitano, Y. (2002). Effects of storage temperature and wax coating on ethylene production, respiration and shelf-life in cherimoya fruit. *J. Japan. Soci. Hortic. Sci.*, 71: 643-650.

57. Sadia Chishty and Monika (2016). Medicinal and Nutritional Importance of *C. decidua* (Forssk.) Edgew. (Capparaceae): A Review. *Int. J. Sci. Research*, 2319-7064 Vol. 5 Issue 2.

❑❑❑

Chapter 8

Post Harvest Management, Processing and Value Addition of Avocado

Tapas Sarkar, Nilanjana Datta and P.Dutta

Bidhan Chandra Krishi Vishwavidyalaya, Mohanpur, Nadia, W.B., India

ABSTRACT

Avocado (*Persea americana*), a climacteric fruit is used primarily as a vegetable. Avocado is rich in monounsaturated fatty acid (oleic acid), health-promoting phytosterols, and phenolic antioxidants. Avocados are commercially valuable and are cultivated in tropical and Mediterranean climates throughout the world. They have a green-skinned, fleshy body that may be pear-shaped, egg-shaped, or spherical. Commercially, they ripen after harvesting. Avocado trees are partially self-pollinating and are often propagated through grafting to maintain a predictable quality and quantity of the fruit. The word "avocado" comes from the Spanish *aguacate*, which, in turn, comes from the Nahuatl word *ahuacatl*, which goes back to the proto-Aztecan *pawa* which also meant "avocado". Sometimes, the Nahuatl word was used with the meaning "testicle", probably because of the likeness between the fruit and the body part. Avocado is a very versatile fruit. Its creamy flavour goes well with many foods, making it a refreshing and nutritious addition to various recipes. Avocados can also be beneficial when used topically, as they have nourishing and moisturizing properties. Value-added products are defined as a change in the physical state or form of the product (such as milling wheat into flour or making strawberries into jam). The production of a product in a manner that enhances its value, as demonstrated through a business plan (such as organically produced products). Generally large-sized, good quality avocados are sold in retail markets. Avocados are processed as guacamole, frozen slices and sauces, puree, canned, dehydrated, and as avocado oil. Processing of avocado into guacamole or pulp has enabled the exporters to serve far and wide markets.

1. Introduction

The avocado (*Persea americana*) is a tree, long thought to have originated in South Central Mexico, classified as a member of the flowering plant family Lauraceae. The fruit of the plant, also called an avocado (or avocado pear or alligator pear), is botanically a large berry containing a single large seed known as a "pit" or a "stone".

1.1. World Production and Leading Producer Countries

Avocado has a long history of production and consumption, especially in Mexico, and North and Central America. In 2004, the world production of avocado was 3,222,069 metric tonnes valued at $606,608,000 [20]. In 2007, the leading producer was Mexico, followed by Indonesia and United States of America (Table 2) [21].

1.2. Consumption and Market Trends

Mexico is the leading exporter and consumer of avocado in the world.

Table 2: Major world producers of avocado

Rank	Country	Production-2007 (MMT)
1.	Mexico	1.143
2.	Chile	0.250
3.	Indonesia	0.202
4.	United States of America	0.188
5.	Dominican Republic	0.184
6.	Columbia	0.158
7.	Brazil	0.154
8.	Peru	0.122
9.	Spain	0.120
10.	Guatemala	0.114
	World (Total)	3.569

Source: FAO, 2007.

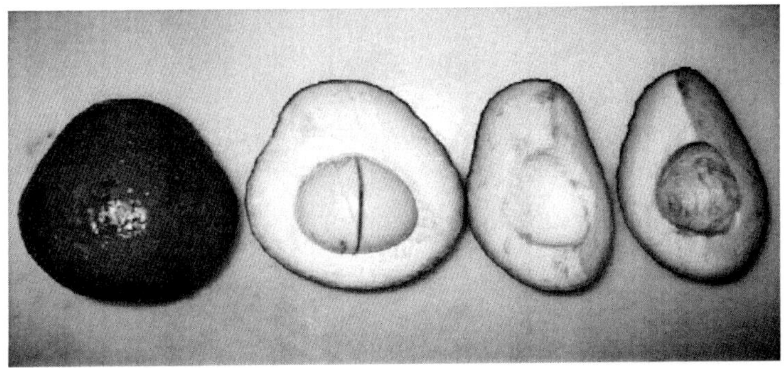

Figure 1. Fully mature, whole and cut avocado fruit

Table 1: Fatty acid profile of avocado oil vis-a-vis olive oil

NZ avocado oil	NZ olive oil	Type of Fatty acid	% Fatty acid
Palmitic, 16:0	12.5–14.0	8.6–12.9	
Palmitoleic, 16:1	4.0–5.0	0.3–0.7	
Stearic, 18:0	0.2–0.4	2.1–2.8	
Oleic, 18:1	70–74"	77.0–82.6	
Linoleic, 18:2	9.0–10.0	4.6–7.5	
á-linolenic, 18:3	0.3–0.6	0.5–0.7	
Arachidic, 20:0	0.1	0.0–0.6	
Gadoleic, 20:1	0.1	0.0–1.4	

Source: Eyres *et al.* (2009).
"Includes 18:1 isomer (5%).

1.3. Postharvest Physiology, Storage, and Shelf-Life

1.3.1. Maturity Indices

Avocado is harvested when fully mature yet green and firm when harvested immature, it will not ripen properly, whereas an over-ripe avocado will decay rapidly after harvest. Typical maturity indices for fruits, such as colour, smell, or texture of the fruit are not reliable indicators for avocados [25]. Both its colour and smell change with ripening after harvest. Although colour change(skin darkening) before harvest is observed in late season avocados, the fruit may still be unripe [13]. Other maturity indices include assessment of oil content, dry matter, or percentage of moisture. Depending on the cultivar, growing conditions, time of harvest, post harvest handling, and method of determination, the oil content of 8–33% and dry matter of 17–21% are considered typical for commercial grade avocado [50 & 41].

In South Africa, percentage of moisture (complementary to dry matter) is used as a maturity index [12]. Blakey *et al.* [8] reported that the percentage of moisture content at harvest correlated well with the post harvest ripening ability of the fruit without shriveling. Firmness correlates well with other parameters of maturity, such as oil content and is a good predictor of expected storage time [57]. As the fruit matures and ripens, the firmness measured by a penetrometer declines accordingly. Use of near infrared spectroscopy (NIRS) method is becoming popular for measuring dry matter and water content of avocados. It enables the producer to rapidly scan and sort not only the harvested fruit into different classes for further handling and processing [12], but also the measurement of dry matter or water content in field before the harvest [8].

1.3.2. Fresh Avocado Packaging and Shelf-Life

Avocados are picked by hand using special clippers and ladders. Once picked, they are pre-cooled overnight to remove field heat in the packing house to ensure the quality. Before packaging, avocados are washed, lightly cleansed, sized, and graded for quality. The highest grade is separated for supply to supermarket for raw consumption; others are used for processing. The packed avocados are maintained at optimum cool temperature to ensure ripeness and quality while awaiting shipment. Once picked, avocado ripens quickly at room temperature because of the production of ethylene gas during storage. All these steps from picking to grading take 5–9 days [23]. Skin of avocado changes colour upon ripening; for example, Hass avocados change colour from green to purple to black. To enhance shelf-life, most producers size avocados by weight or separate by firmness Heterogeneity in firmness can cause variability in ripening and, thus, setting up of the climacteric process. It is important to sort avocados in lots of different ripening stages and employ post harvest measures to extend shelf-life and quality, as ethylene production fromsome avocados in a lot can trigger ripening which once initiated, is hard to stop [49].

1.3.3. Physiological Disorders: Softening, Chilling Injury, Oxidation, etc.

Ripening of climacteric fruit involves a series of metabolic changes initiated by a surge in the production of the plant hormone ethylene after harvesting. High ethylene production is a result of high respiration rate, which is partly responsible for altering the fruit's biochemistry, physiology as well as gene expression [27], thus causing softening of the mesocarp(pulp) as well as colour and flavour change. Softening is caused by several cell wall-degrading enzymes. Early stages of avocado softening is based on the cellulose activity in the cell wall, which disrupts the polygalacturanse in the wall matrix, and releases polygalacturanase for the final fruit softening [71]. It is also associated with embryo growth and ethylene production during cold storage [34]. It is suggested that during chilling stress, some tissues accelerate their ethylene production, resulting in ultra-structural changes in the membrane leading to disrupted ion balance, release of polyphenol oxidase (PPO) enzyme, and leakage of electrolytes.

1.3.4. Post harvest Handling and Storage

As indicated before, ethylene plays an important role in the ripening of avocado and its shelf-life. Storage at low temperature delays ethylene production; however, storage at low temperatures can also cause chilling injury. Low storage temperatures generally range between 5 and 8°C but could be reduced to as low as 2–8°C in order to reach a distant market [24]. To prevent chilling injury in cold storage, ethylene absorption sachets are used to reduce levels of ethylene in the storage atmosphere [1], or the avocados are treated with 1-methylcyclopropene (1-MCP), an inhibitor of ethylene action, before cold storing [36]. [23] measured the potency of 1-MCP in delaying ethylene-induced avocado ripening at various concentrations using cultivars Ettinger, Hass, Reed, and Fuerte. Mature avocados were treated for 24 hours at 22°C with various concentrations of 1-MCP after ventilation and exposing to 300 mL/L ethylene for 24 hours. Subsequently, they were stored in ethylene-free air at 22°C and assessed for ethylene production, fruit firmness colour, cellulose, and galacturonase activity during ripening. It was observed that 1-MCP inhibited ethylene-induced ripening at very low concentration of 30–70 nL/L and delayed ripening upto 13 days. Further, cultivars Fuerte and Hass [6]; [3] were successfully stored for 5–9 weeks without chilling injury at 5–7°C in a controlled atmosphere (CA) storage of 2% oxygen and 10% carbon dioxide. [47] reported that a combination of 3% oxygen and 8% carbon dioxide extended shelf-life of Hass avocado at 5°C upto 9 weeks. This reduced ratio of oxygen in storage atmosphere resembles anaerobic respiration of fruit tissues producing acetaldehyde (AA) and ethanol. Accumulation of these endogenous respiration metabolites gives aroma to the fruit and inhibits ethylene production in plants. Application of exogenous AA vapour to avocados before storage also inhibited fruit ripening by reducing ethylene production and activities of cell wall-depolymerizing enzymes and chilling injury symptoms at low temperature of 2°C [58]. Waxing is commonly used to prevent post harvest water loss, [36] investigated a combined effect of ethylene inhibitor (1-MCP) treatment and waxing on avocado's shelf-life. The heat shock treatment of avocados not only prevented unfavourable effects of chilling injury but also protected against Mediterranean fruit flies eggs [35]. Pre-treatments of moderate heat (i.e., 38^{0}C) also induce tolerance to much higher temperatures required for insect killing. This induced tolerance of fruits is suggested to be due to the induction of heat shock proteins (HSPS) that are stimulated in fruit by post harvest heat treatments of avocado [74]; [73]. Similar induction of HSPS gene expression was seen in avocados that were picked from trees facing the sun, which, therefore, exhibited reduced external chilling injury in cold storage, as compared to those picked from trees facing shade [72].

1.3.5. Avocado Enzymes: Polyphenol Oxidase, Lipooxygenase

Peroxidase or oxidase are present in most of avocado varieties but are absent in the green fruit. Fuerte has a negligible amount of oxidase and it does not turn black readily when cut surfaces are exposed to oxygen[37].

The activity of lipoxygenase enzyme is associated with fungal rotting, if the fruit is infected. In a mature fruit, flavonoid epicatechin, a natural anti-oxidant, decreases while the fruit is ripening. Epicatechin is an inhibitor of lipoxygenase enzyme, thus its activity increases when the level of its inhibitor declines. These metabolic changes in the fruitalso reduce the level of an antifungal compound, 1-acetoxy-2-hydroxy-4 oxo-heneicos12, 15 (antifungal diene, AFD) resulting in activation of quiescent *Colletotrichumgloeosporioides. Colletotri chumgloeosporioides* is a phytopathogenic fungus that attacks a wide variety of tropical and subtropical fruits including avocado. Unripe avocado isresistant to this fungus because of its high content of the epicatechin and lower concentration of lipoxygenase enzyme [32].

PPO is a copper-containing enzyme which catalyzes two different reactions in the presence of oxygen, such as hydroxylation of monophenols to o-diphenols and subsequently o-diphenols to o-quinones. Oquinones then polymerize non-enzymatically with other o-quinones, phenolic compounds or amino acids and form red to brown to black pigments [19]. Inhibition ofbrown coloration can be achieved by inactivation of these enzymes by high temperature, pressure, or both or other treatments, such as removal of oxygen, etc.

1.3.6. Bioactive Compounds

Together with fiber (5.2 g/100 g), high â-sitosterol (a phytosterol) [16]; [53]; [59] and fatty acid, oleic acid, avocado is suggested to improve lipid profile in healthy and mildly hyper cholesterolemic subjects by reducing total cholesterol, low-density lipoprotein (LDL) cholesterol, and triglycerides while increasing high-density lipoprotein (HDL) cholesterol[39]. Research studies on Mexican populations show that people consuming avocado have decreased serum total cholesterol, LDL cholesterol, and triglycerides while increased HDL cholesterol compared to control diet [2]; [39].

Adding avocado to meals increases the feeling of satiety resulting in reduced food intake and, thus, better body weight management. Body weight correlates positively with reduced risk of coronary heart diseases or stroke and diabetes [45]. Partial replacement of carbohydrate foods withavocado extract in the diet of type-2 diabetic subjects showed a remarkable improvement in the lipid profile and plasma glucose concentration[26].

Avocado is rich in antioxidants, such as tocopherol and carotenoids including zeaxanthin, cryptoxanthin, α-carotene, and β-carotene) and lutein; lutein is 70% of the measured carotenoids, which is highest in avocado among fruits. The acetone extract of avocado containing tocopherol and carotenoids was capable of inhibiting growth of both androgen-dependent (LNCap) and androgen independent (PC-3) human prostate cancer cell lines in vitro [45]. Lutein has been shown as an anticarcenogenic carotenoid; however, acetone extract of avocado had a greater inhibitory effect of the above-mentioned prostate cancer cell lines compared to purified lutein, which suggests the involvement of a combination of factors rather than lutein alone [45]. Other studies showed that chloroform extract of avocado selectively induced apoptosis in human oral cancer cells but not in normal oral epithelial cell lines [14]; [15] or human breast cancer cell line [61]. Therapeutic value of avocado is realized in Mexico folk medicine. Its fruit, oil, and leaves are used to treat a wide variety of diseases. For example, the oil from its seed is considered a useful treatment for rashes, scars, or dysentery. Avocado leaves are not only used by Mexicans but also by people in Brazil, Panama, Jamaica, Nigeria, and Indonesia to treat hypertension.

1.3.7. Nutritional Quality

Avocado contributes minerals including magnesium, potassium, iron, copper, and phosphorus, and vitamins, such as vitamin C, thiamin, riboflavin, niacin, folic acid, vitamin E, and β-carotene. The nutritional composition of avocado is given in Table 3.

Using avocado as a spread on sandwiches replacing butter, cream cheese, and fatty mayonnaises can help reduce intake of calories, saturated fat, sodium, and cholesterol. Phytonutrients are better absorbed in the presence of fat, and avocado is abundant in both fat and phytonutrients. Research study in humans showed that avocado added to salsa enhanced lycopene and β-carotene absorption by 4.4 and 2.6 times compared to control salsa without avocado. Addition of either 150 g avocado or 24 g avocado oil to salad enhanced α-carotene, β carotene, and lutein absorption by 7.2, 15.3, and 5.1 times, respectively compared to avocado-free salad [67].

Table 3: Avocado raw edible portion nutrient composition and percent daily values for vitamins and minerals

Nutrients	Per 100 g edible portion
Energy	160 Kcal (670 KJ)
Carbohydrates (g)	8.53
Sugars (g)	0.66
Dietary fiber (g)	6.7
Fat (g)	14.66
Saturated (g)	2.13
Monounsaturated (g)	9.80
Polyunsaturated (g)	1.82
Protein (g)	2.0
Vitamins:	
Thiamin (mg)	0.067
Ribofl vin (mg)	0.130
Niacin (mg)	1.738
Pantothenic Acid (mg)	1.389
Vitamin B6 (mg)	0.257
Folate (µg)	81
Vitamin C (mg)	10
Minerals:	
Calcium (mg)	12
Iron (mg)	0.55
Magnessium (mg)	29
Potassium (mg)	485
Phosphorus (mg)	52
Zinc (mg)	0.64

Source: USDA nutrient database.

1.3.8. Value Added Products

I. Fresh-Cut Avocado

Enzymatic browning of sliced avocados is a serious problem in the preparation of fresh-cut products. Using Hunter colour values as indicators of browning, [29] investigated various browning inhibitors. Ascorbic acid and sodium sulphite were effective browning inhibitors when combined with citric acid. Other reports [65] indicated retention of texture (firmness when avocado slices were dipped in citric acid and ascorbic acid solutions, and stored under modified atmosphere. [46] reported that pectin-based coating extended shelf-life of avocado for more than 1 month at 10°C.

II. Frozen Avocado

[38] Reported that commercial processing of avocados started in the United States in 1964 at the Frigid Foods, Inc., Escondido, California, by Calavo Growers. The company developed frozen avocado halves and slices by immersing in liquid nitrogen (-196°C). Later on, frozen avocado sauce (guacamole dip) was also developed [68]. The processing potential of avocado cultivars at different stages of fruit maturity was investigated by [56]. The composition of edible part of avocado fruit (mesocarp) varied depending upon the cultivar and harvest season. They reported various constituents, such as dry matter (19.18–30.29%), fat (8.3–16.75%), protein (2.1–2.3%), carbohydrates (6.8–8.1%), and mineral ash (0.7–1.2%). Fruits harvested near to the end of harvesting season gave the highest values for these constituents. [54] have also reported the effect of cultivar and maturity on the quality of frozen (–40°C) avocado pulp stored at "18°C. Of the five cultivars tested, Edranolgave is the best quality frozen product. A minimum of 15% oil content was required in these cultivars at their maturity level to obtain an acceptable quality frozen avocado pulp product. [66] have prepared an acceptable frozen guacamole base using 100 parts of avocado flesh with 5 parts of lemon juice, 4 parts of chopped onions, and 1 part of salt. This product when packed with minimum headspace in glass jars or tin cans had a shelf-life of 7 months at -18°C (0°F). [9] developed a process to produce both fresh-cut and frozen avocado halves, slices, and chunks. The major problems related to product browning, as well as taste and texture of the fruit, were addressed. They suggested the use of anti-oxidants (ascorbic acid) to prevent browning during defrosting process. Calcium ions play an important role in the browning of avocado fruit. To prevent enzymatic browning during frozen storage and thawing. The correct use of anti-oxidants, modified atmosphere packaging (MAP), and frozen storage conditions will decrease the potential for microbial growth and will enhance the marketability of fruit products in a wider area.

In case of fresh avocado fruit, off-taste develops due to tissue collapse. Even the frozen product on defrosting develops offalvow. [10] investigated the use of MAP with decreased oxygen and increased carbon dioxide levels. In another study, [11] developed a process to prevent browning of frozen avocado halves and chunks. They ripened the avocado fruit to correct softness, cleaned it externally with a disinfectant, peeled, and sliced. The slices were dipped in boiling water for sufficient time to inactivate the PPO, as well as to leach phenolics from the cut surface of the slices. The slices were frozen in liquid nitrogen. These steps enhanced the flavour of the product and eliminated bitterness associated with the storage of frozen avocado products. [60] reported that ascorbic acid-treated frozen avocado chunks did not darken, when kept at -26°C. The darkening of colour of frozen chunks correlated well with the disappearance of most phenolic compounds, except the simple cinnamic acid derivatives.

[69] investigated the use of 60Co gamma rays (doses 0.5–2.5 kGy) on the quality of frozen avocado pulp. The *Listeria monocytogenes* was reduced by 1–4 log cycles without adversely affecting the hydrolytic or oxidative rancidity at any of the doses for 120 weeks of frozen storage. Even the highest radiation doses did not produce any off-flavour or affect the green colour (chlorophyll) of the product.

III. Dehydrated Avocado

During dehydration, many quality characteristics, such as colour, texture, flavour, porosity, and the rehydration ratio, are affected. Drying of avocado has a few special problems. Avocado is dried either by the spray drying or by the drum drying technique. These drying techniques produce a product that has only pale green colour and has a chalky off-flavour, which does not have a big market. By adding a few spices, the off-flavour can be masked and a ready mix of guacamole is produced.The dehydrated avocado may be in use in pet food manufacture as an ingredient rich in oil. A number of patents have been issued for dehydrated avocado pieces, powdered compositions. avocado concentrate, and osmotic drying. [30] Spray dried avocado paste and studied the effect of synthetic and natural anti-oxidants on the oxidative stability and palatability of stored avocado powders at temperatures ranging from 6 to 40°C. During storage, peroxide value was measured and the development of rancidity was determined by sensory evaluation. Tertiary butylated hydroxyquinone (0.05%) + citric acid (0.1%) mixture coupled with nitrogen packaging were found to be the most effective in extending the shelf-life of spray dried avocado powder during ambient storage. [40] developed a mathematical model based on one-dimensional, steady state to suitably describe the rehydration behaviour. The effect of variety, storage time, temperature, and atmosphere on the chemical and organoleptic quality of freeze-dried avocado puree and guacamole has been studied by [28]. They found nouse of peroxide value in predicting product acceptability, as deterioration in sensory quality was observed without significant increase in peroxide value. [70] reported manufacture of high quality freeze-dried guacamole and [52] investigated the hygroscopic properties of freeze-dried guacamole during storage.

IV. Guacamole

The value-added product, guacamole, is prepared from a fully ripe fruits by blending the pulp with herbs, spices, lemon juice, and salt for flavouring. The process outline for preparing guacamole from ripened avocado fruit is described in Figure 2. After the ripening is completed, the fruit is chilled to 5°C. The fruit is dipped in 200 ppm hypochlorite solution to lower the microbial load on the surface. Avocado Ripening of mature fruit 85–90% RH, 20–22°C Washing and dipping in 200 ppm hypochlorite solution Stem removal, cutting into halves, seed removal, Separation of pulp from peel, with machine or Manually Formulation (spices, etc.) of guacamole Blending of ingredients Guacamole Packaging, sealing with vacuum or nitrogen flushing Frozen storage at -22°C.

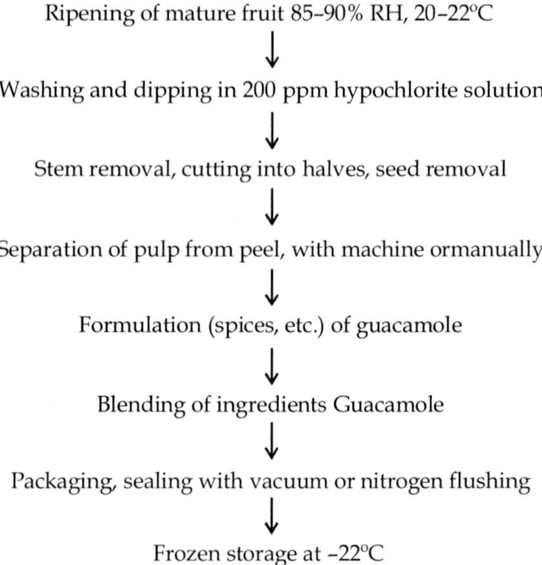

Figure 2: Typical process for the production of avocado guacamole

At first stems are removed, fruit is cut into halves, and the seed is removed. The pulp is removed from peel either by ricing machine or manually. All the remaining spices and other ingredients (as per formulation) are added to the pulp and blended to obtain a uniform-flavoured guacamole. The product is filled into containers, vacuum-sealed (or nitrogen-flushed) and stored frozen (–22°C). Low initial counts of micro-organisms of the ripened fruits are very important for the production of guacamole, as any heat given to the pulp causes colour and flavour deterioration. [4] collected samples of guacamole from a number of restaurants in the Queretoro city, Mexico and tested for the presence of pathogenic organisms. They also inoculated the guacamole pulp with *Salmonella* spp., *Staphylococcusaureus*, and *Escherichia coli* O150:H7 to study the behaviour of these pathogens. A wide variation observed for the pathogen counts in guacamole samples correlated well with the sanitary conditions observed in catering establishments, with higher numbers from street vendors. No growth of pathogens was observed in guacamole stored at a temperature of 4–7°C. A few patents have been issued for the preparation of guacamole [31], use of hurdle technology for shelf stable guacamole, and avocado fruit flesh processing.

V. Avocado Puree (Sauce)

The process for avocado puree or sauce is similar to guacamole up to pulp extraction step. The next step consists of high shear blending of pulp with water, gums, thickeners, and spices. A thick sauce should have an apparent viscosity of 8,000–12,000 centipoises as measured by a Brookfield viscometer using spindle number 4. The sauce is then filled into polyethylene jars, and frozen stored at –18°C. The use of traditional heating of avocado puree produces off-flavours and, therefore, use of an alternative process of high temperature short time (HTST) heating using microwave energy has been investigated by Guzman *et al.* (2002). In another study, use of microwave heating of avocado puree at high energy level for 180 seconds brought only 60% reduction in PPO activity [63] have investigated the kinetics of PPO activity and browning of avocado puree preserved by anti-browning agents (ascorbic acid, or EDTA) and anti-microbial agents (sorbic acid). The avocado puree, when added to the sausage formulation, had no adverse effect on the acceptability of this meat product. [18] investigated effect of α-tocopherol and sorbic acid on development of oxidative rancidity in avocado puree. As expected, α-tocopherol reduced the rancidity of lipid fraction in avocado puree stored under different MAP conditions. Sorbic acid used as an anti-microbial agent in puree had a moderate pro-oxidant effect and enhanced the oxidative rancidity of lipid fraction of puree. Lipid oxidation is one of the major phenomena that limits the shelf-life of avocado products. Use of 100 ppm of β- tocopherol and 200 ppm of ascorbic acid has been reported to have beneficiary effect on the stability of lipid fraction of avocado puree [17]. The use of microwave energy in the preparation of avocado puree also has a tremendous potential to retain better colour and flavour in the finished product. The effects of microwave heating of avocado puree on the volatile components as influenced by the heating times, pH values, and addition of avocado leaves have been studied by [43]. The addition of avocado leaves had a positive effect on the flavour of puree. According to their results, the optimum values for the processing of avocado puree were microwaving for 30 seconds at pH 5.5 and addition of 1% avocado leaves [33] have also studied the effect of microwave heating time, pH, and addition of avocado leaves on the volatile profile of puree and obtained similar results [64] studied role of storage temperature, pH, water activity (AW), MAP, and addition of chemical preservatives to extend the shelf-life of avocado puree. According to their findings, 300 ppm of sorbicacid was sufficient to inhibit yeast and fungi responsible for the spoilage of avocado puree.

VI. Avocado Oil

Avocado oil is known to be rich in many bioactive substances that can prevent and control hyperlipidemia. Because of its unique nutritional quality, avocado oil has been named as the new wonder oil [48] developed a rapid method for the sequential extraction and subsequent estimation of fatty acids and sugars from avocado mesocarp tissues. The fatty acid composition and phytosterols present in avocado oil are similar to olive oil [62]. Depending upon the season and location of orchard, the oil content in Hass variety of avocado can vary from 16 to 30%. Not only the late season fruits have more oil, but also it is easier to extract oil from the idioblast cells that contain fat. A typical composition of the avocado fruit is: (1) flesh 65%; (2) seed, 20%; and (3) peel, 15%. However, several avocado varieties from Venezuela have been reported to have low (6.73–8.07%) to very low (3.0–6.70%) oil content [7] developed avocado oil extraction technology suitable for the developing countries. Highest recoveries (70–80%) were obtained at water to avocado ratio of 5:1, pH 5.5, and centrifugal force of 12,300×g. Use of 5%calcium carbonate or calcium sulphate eliminated the requirement of organic solvents during extraction process. Processes for extra virgin oil from avocado pulp and for the manufacture of refine avocado oil rich in triglycerides have been granted US patents. [51] studied the effect of fruit ripeness and method of fruit drying on the extract ability of avocado oil with hexane and super critical carbon dioxide. Oil extraction yield was higher from ripe than unripe fruits. Oil extraction from freeze-dried avocados was higher than oven-dries samples. Hexane proved to be better solvent for oil extraction than super critical carbon dioxide. Besides triglycerides, avocado oil also contains a small but significant amount of un-saponifiable components. [22] have identified nine un-saponifiable components through chromatographic fractionations and spectroscopic analysis. These products have a homogeneous structure, having a furylnucleus with position 2 substituted by an aliphatic, mono-, or polyunsaturated chain of 13–17 carbon atoms (always odd carbon numbers). The presence, identification and characterization of pigments, glycolipids and phospholipids, and furan lipid-rich un-saponifiable materials in avocado oil have been reported.

A. High-Pressure Processing of Avocados

As described elsewhere in this book, in this method, the food is subjected to a very high pressure in the range of 200–900 MPa with or without heating. The objective is to inhibit spoilage micro-organisms and pathogens, inactivate enzymes, and retain sensory and nutritional qualities. The ultra-high pressure (UHP) food processing technology by Avomax (TX, USA) has been applied for the production of avocado pulp and guacamole. [55] studied the application of HPP with pressures of 689 MPa for the processing of guacamole by measuring its effect on microbiological quality, sensory properties, colour, catechol oxidase, and lipoxygenase activities. Although the HPP-treated avocado guacamole had fresher taste than the frozen; the HPP-treated product was inferior in appearance and colour, as the enzyme PPO was still active in the product. The HPP treatment is not capable of eliminating the PPO activity completely, which causes the avocado products to turn brown, once the vacuum packed product is opened.

B. Standards for Processed Avocados

The Codex (Stan 197-1995) gives definition and classification of avocados, size codes, and requirements for quality, contaminants, hygiene, packaging, marketing, and labelling. For more information, the readers are referred to the list of Codex Standards on the Codex Alimentarius(http://www.codexalimentarius.net).

C. Avocado By-products

The avocado lipids contain small amounts of un-saponifiable components, which have been found useful in the treatment of osteoarthritis [42] investigated the ant oxidative activity (*in vitro*) of polyphenols extracted from the sarcocarp, seed, and peel of avocados. The methanol extracts from the peel of avocado showed higher free radical scavenging activities than the sarcocarp and seed and increased apoptosis in MDA-MB-231 cells. A patent [5] reported the production of fats rich in ω-3 unsaturated fatty acids by growing microalgae, *Thraustochytrium*, *Schizochytrium*, or their mixture, on flax seed, rapeseed, soybeans, and avocado meal. Avocado honeys produced in Spain showed high pH, ash, and total mineral content, potassium being the most abundant (73% of the total minerals). Perseitol, a sugar, was suggested as an indicator of the degree of purity of avocado honey.

2. Conclusions

Avocados are commercially valuable and are cultivated in tropical and Mediterranean climates through out the world. They have a green-skinned, fleshy body that may be pear-shaped, egg-shaped, or spherical. Commercially, they ripen after harvesting. Avocado trees are partially self-pollinating and are often propagated through grafting to maintain a predictable quality and quantity of the fruit.Avocado has healthy lipid composition. In addition, it is rich in minerals, vitamin E, and health promoting phytochemicals. As indicated in this review many processed forms of avocado can extend use of this important produce and improve the nutritional quality of our diet.

3. References

1. Adkins M.F., Hofmanb P.J., Stubbings B.A., Macnish A.J. (2005). Manipulating avocado fruit ripening with 1-methyl cyclopropene. *Postharvest Biol Technol* 35: 33–42.
2. Alvizouri-Munoz M., Carranza-Madrigal J., Herrera- Abaraca J.E., Chavez-Carbajal F., Amezcua-Gastelum J.L. (1992). Effects of avocado as a source of monounsaturated fatty acids on plasma lipid levels. *Arch Med Res* 23: 163–167.

3. Arpaia M.L., Faubian D., Mitchell F.G., Mayer G. (1990). The use of controlled atmosphere for long-term storage of "Hass" avocados. In: Shepherd JS (editor), *Yearbook.*
4. Arvizu-Medrano S.M., Iturriaga M.H., Escartin E.F. (2002). Indicator and pathogenic bacteria in guacamole and their behaviour in avocado pulp. *J Food Saf* 21(4): 233–244.
5. Barclay W.R. (2008). Process of the heterotrophic production of microbial products with high concentrations of -3 highly unsaturated fatty acids. Patent No. US 2008/0166780 A1.
6. Barmore C.R., Rouse A.H. (1976). Pectinmethylesterase activity in controlled atmosphere stored avocado. *J Am Soc Hort Sci* 100: 294–296.
7. Bizimana V., Breene W.M., Csallany A.S. (1993). Avocado oil extraction with appropriate technology for developing countries. *J Am Oil Chem Soc* 70(8): 821–822.
8. Blakey R.J., Bower J.P., Bertling I. (2009). Influenc of water and ABA supply on the ripening pattern of avocado (*Persea americana* Mill.) fruit and the prediction of water content using Near Infrared Spectroscopy. *Post harvest Biol Technol* 53: 72–76.
9. Bower J.P., Dennison M.T. (2003). Progress in the development of avocado products. *South African Growers" Assoc Yearbook* 26: 35–39.
10. Bower J.P., Dennison M.T. (2004). Alternative avocado products. *South African Growers Assoc Yearbook* 27: 32–34.
11. Bower J.P., Dennison M.T. (2005). A process to prevent browning of frozen avocado halves and chunks. *South African Growers Assoc Yearbook* 28: 40–41.
12. Clark C.J., McGlone V.A., Requejo C., White A., Woolf A.B. (2003). Dry matter determination in "Hass" avocado by NIR spectroscopy. *Post harvest Biol Technol* 29: 300–307.
13. Cox K.A., McGhie T.K., Anne White A., Woolf A.B. (2004). Skin colour and pigment changes during ripening of "Hass" avocado fruit. *Postharvest BiolTechnol* 31: 287–294.
14. Ding H., Chin Y.W., Kinghorn A.D., D'Ambrosio S.M. (2007). Chemo preventive characteristics of avocado fruit. *Semin Cancer Biol* 17: 386–394.
15. Ding H., Han C., Guo D., Chin Y.W., Ding Y., Kinghorn A.D., D'Ambrosio S.M. (2009). Selective induction of apoptosis of human oral cancer cell lines by avocado extracts via a ROS-mediated mechanism. *Nutr Cancer* 61(3): 348–356.

16. Duester K.C. (2001). Avocado fruit is a rich source of betasitosterol. *J Am Diet Assoc* 101: 404–405.

17. Elez-Martinez P., Soliva-Fortuny R.C., Gorinstein S., Martin-Belloso O. (2005). Natural antioxidants preserve the lipid oxidative stability of minimally processed avocado puree. *J Food Sci* 70(5): 5325–5329.

18. Elez-Martinez P., Soliva-Fortuny R.C., Martin-Belloso O. (2007). Oxidative rancidity in avocado puree as affected by-tocopherol, sorbic acid and storage conditions. *Eur Food Res Technol A* 226(1–2): 295–300.

19. Espi'n J.C., Trujno M.F., Tudela J., Garci'a-Ca'novas F. (1997). Monophenolase activity of polyphenol oxidase from Hass avocado. *J Agric Food Chem* 45(4): 1091–1096.

20. FAO, (2000/09). Avocado production in Asia and Pacific Food and Agriculture Organization of the United Nations Regional Office for the Asia and the Pacific Bangkok and Thailand. Rap Publication 2000/09.

21. FAO, (2007). *FAO Stat Book*, web site: http://faostat.fao.org.

22. Farines M., Soulier J., Rancurel A., Montaudoin M.G., Leborgne L. (1995). Influence of avocado oil processing on the nature of some unsaponifiabl constituents.*J Am Oil Chem Soc* 72(4): 473–476.

23. Feng X., Apelbaum A., Sisler A.C., Goren R. (2000). Control of ethylene responses in avocado fruit with 1-methylcyclopropene. *Post harvest Biol Technol* 20: 143–150.

24. Flitsanov U., Mizrach A., Liberzon A., Akerman M., Zauberman G. (2000). Measurement of avocado softening at various temperatures using ultrasound. *Post harvest Biol Technol* 20: 279–286.

25. Gaete-Garreton L., Vargas-Hernandez Y., Leon-Vidal C., Pettorino-Besnier A. (2005). A novel non-invasive ultrasonic method to assess avocado ripening. *J Food Sci* 70(3): E187–E191.

26. Gallagher A.M., Flatt P.R., Duffy G., Abdel-Wahab Y.H.A. (2003). The effects of traditionl anti-diabetic plants on *invitro* glucose diffusion. *Nutr Res* 23: 413–424.

27. Giovannoni J. (2001). Molecular biology of fruit maturation and ripening. *Ann Rev Plant Physiol Plant Mol Biol* 52: 725–749.

28. Gomez R.F. and Bates R.P. (1970). Storage deterioration of freeze-dried avocado puree and guacamole. *J Food Sci* 35(4): 472–475.

29. Gomez-Lopez V.M. (2003). Inhibition of surface browning, cut avocado. *J Food Qual* 25: 369–379.

30. Grajales-Lagunes A., Garcia-Galindo H.S., Angulo-Gerrero O., Monroy-Rivera J.A. (1999). Stability and sensory quality of spray dried avocado paste. *Drying Technol* 17(1/2): 317–326.

31. Griebel J.M., Kargel B.C. (2004). Method of treating avocados and method of preparing guacamole therefrom. Patent No. WO 2004/ 100670 A1. (USA).

32. Guetsky R., Kobiler X., Wang N., Perlman N., Gollop G., Avila-Quezada I., Hadar P.D. (1995). Metabolism of the flavonoid epicatechin by laccase of Colletotrichumgloeosporioides and its effect on the pathogenecity on avocado fruits. *Phytopathology* 11:1341-1348.

33. Guzman-Geronimo R.I., Lopez M.G., Dorantes-Alvarez L. (2008). Microwave processing of avocado: volatile flavor profilin and olfactometry. *Innovative Food Sci Emerg Technol* 9(4): 501 506.

34. Hershkovitz V., Friedman H., Eliezer E. Goldschmidt and Edna Pesis, (2009). The role of the embryo and ethylene in avocado fruit mesocarp discoloration. *J Exp Bot Vol* 60(3): 791–799.

35. Jang E.B., Chan H.T. Jr, Nishijima K.A., Nagata J.T., McKenney M.P., Carvalho L.A., Schneider E.L. (2001). Effect of heat shock and quarantine cold treatment with a warm temperature spike on survival of Mediterranean fruit fly eggs and fruit quality in Hawaii-grown "Sharwil" avocado. *Post harvest Biol Technol* 21: 311–320.

36. Jeong J., Huber D.J., Sargent S.A. (2003). Delay of avocado (*Persea americana*) fruit ripening by 1-methylcyclopropene and wax treatments. *Post harvest Biol Technol* 28: 247–257.

37. Kahn V. (1976). Polyphenol oxidase isoenzymes in avocado. *Phytochemistry* 15: 267–272.

38. Kurlaender A. (2004). Avocados. In: Barrett D.M., Somogyi L., Ramaswamy H. (editors), *Processing Fruits-Scienceand Technology*. New York: CRC Press, Taylor & Francis Group, pp. 739–750.

39. Ledesma R.L., Munari A.C.F., Domi'nguez B.C.H., Montalvo S.C., Luna M.H.H., Juarez C., Lira S.M. (1996). Monounsaturated fatty acid (avocado) rich diet for mild hypercholesterolemia. *Arch Med Res* 27: 519–523.

40. Lee K.T., Farid M., Nguang S.K. (2006). The mathematical modelling of the rehydration characteristics of fruits. *J. Food Eng* 72(1): 16–23.

41. Lee S.K., Young R.E., Schiffman P.M. and Coggins C.W. Jr. (1983). Maturity studies of avocado fruit based on picking dates and drying weight. *J Am Soc Hort Sci* 108(3): 390–394.

42. Lee S.G., Yu M.H., Lee S.P., Lee I.S. (2008). Anti-oxidant activities and induction o apoptosis by methanol extracts from avocado. *J Kor Soc Food Sci Nutr* 37(3): 269–275.

43. Lopez M.G., Guzman G.R., Dorantes A.L. (2004). Solid phase microextraction and gas chromatography-mass spectrometry of volatile compounds from avocado puree after microwave processing. *J Chromatogra* 1036(1): 87–90.

44. Lopez-Malo A., Palou E., Barbosa-Canovas G.V., Welti-Chanes J., Swanson B.G. (1998). Polyphenol oxidase activity and colour changes during storage of high hydrostatic pressure treated avocado puree. *Food Res Int.* 31(8): 549–556.

45. Lu Q.Y., Arteaga J.R., Zhang Q., Huert S., Go V.L.W., Heber D. (2005). Inhibition of prostate cncer cell growth by an avocado extract: role of lipid-soluble bioactive substances. *J Nutr Biochem* 16: 23–30.

46. Maftoonazad N., Ramaswamy H.S. (2009). Effect of pectin based coating on the kinetics of quality changes associated with stored avocados. *J Food Process Preserv* 32(4): 621–643.

47. Meir S., Akerman M., Fuchs Y., Zauberman G. (1995). Further studies on the controlled atmosphere storage of avocados. *Post harvest Biol Technol* 5: 323–330.

48. Meyer M.D., Terry L.A. (2008). Development of a rapid method for the sequential extraction and subsequent quantificatio of fatty acids and sugars from avocado mesocarp tissue. *J Agric Food Chem* 56(16): 7439–7445.

49. Mizrach A., Galili N., Rosenhouse G. (1991). Determination of fruit and vegetable properties by ultrasonic excitation. *Trans ASAE* 34: 2135–2138.

50. Morris R., O'Brian K. (1980). Testing avocados for maturity. *AgrGaz New South Wales*, pp. 42–44.

51. Mostert M.E., Botha B.M., du Plessis L.M., Duodu K.G. (2007). Effect of fruit ripeness and method of fruit drying on the extractability of avocado oil with hexane and super critical carbon dioxide. *J Sci Food Agric* 87(15): 2880–2885.

52. Mujica-Paz H., Valdez-Fragoso A., Vergara-Balderas F., Rangel-Marron M., Welti-Chanes J. (2005). Hygroscopic properties of freeze-dried guacamole. *IFTAnnual Meeting, July 15–20, New Orleans, Louisiana.* Abstract No. 99C-14.

53. Naveh E., Warman M.J., Sabo E., Neeman I. (2002). Defatted avocado pulp reduces body weight and total hepatic fat but increases plasma cholesterol in male rats fed diets with cholesterol. *J Nutr* 132(7): 215–218.

54. Olaeta J.A., Rojas M. (1987). Effect of cultivar on the qualityof frozen avocado pulp. *South Afric Avocado Growers Assoc Yearbook* 10: 163-164.

55. Palou E., Hernandez-Salgado C., Lopez-Malo A., Barbosa-Canovas G.V., Swanson B.G., Welti-Chanes J. (2000). High pressure-processed guacamole. *Innovative Food Sci Emerg Technol* 1(1): 69–75.

56. Pauker R., Bernstein S., Popelf G., Rosnthalf I. (1992). An assessment of processing potential of avocado fruit. *Calif Avocado Soc Yearbook* 76: 137–144.

57. Peleg K., Ben-Hanan U., Hinga S. (1990). Classification of avocado by firmnes and maturity. *J Text Stud* 21: 123–139.

58. Pesis E., Ackerman M., Ben-Arie R., Feygenberg O., Feng X., Apelbaum A., Goren R., Prusky D. (2002). Ethylene involvement in chilling injury symptoms of avocado during cold storage. *Post harvest Biol Technol* 24: 171–181.

59. Plaza L., Nchez-Moreno C.N.S., De Pascual-Teresa S., De Ancos B.A., Cano P.M. (2009). Fatty acids, sterols and anti-oxidant activity in minimally processed avocado during refrigerated storage. *J Agric Food Chem* 57: 3204–3209.

60. Ramirez-Martinez J.R., Luh B.S. (2006). Phenolic compounds in frozen avocados. *J Sci Food Agric.* 24(2): 219–225.

61. Roberts C.G., Gurisik E., Biden T.J., Sutherland R.L., Butt A.J. (2007). Synergistic cytotoxicity between tamoxifen and the plant toxin persin in human breast cancer cells is dependent on Bim expression and mediated by modulation of ceramide metabolism. *Mol Cancer Therp* 10: 2777–2785.

62. Salgado J.M., Daniell F., Regitano-D'Arce M.A.B., Frias A., Niero Mansi D. (2008). The avocado oil (*Perseaamericana*Mill) as a raw material for the food industry. *Clencia e Tecnologia de Alimentos* 28(Suppl.): 20–26.

63. Soliva-Fortuny R.C., Elez-Martinez P., Sebastian-Caldero M., Martin-Belloso O. (2002). Kinetic of polyphenol oxidase activity inhibition and browning of avocado puree preserved by combined methods. *J Food Eng*. 55(2): 131–137.

64. Soliva-Fortuny R.C., Elez-Martinez P., Sebastian-Caldero M., Martin-Belloso O. (2003). Effect of combined methods of preservation on

the naturally occurring microfloro avocado puree. *Food Control* 15(1): 11–17.

65. Soliva-Fortuny R.C., Martin-Belloso O. (2003). New advances in extending the shelf life of fresh-cut fruits: a review. *Trends Food Sci* 14: 341–353.

66. Stephens T.S., Lime B.J., Griffith F.P. (1957). Preparation of a frozen avocado mixture for guacamole. *Proc Rio Grande Valley Hort Soc* 11: 82–89.

67. Unlu N.Z., Bohn T., Clinton S.K., Schwartz S.J. (2005). Carotenoid absorption from salad and salsa by humans is enhanced by the addition of avocado or avocado oil. *J Nutr* 135(3): 431–436.

68. Urbanek J. (1966). Delicate avocado yields to liquid nitrogen freezing process. *Canner/Packer* 135(5): 31–33.

69. Valdivia M.A., Bustos M.E., Ruiz J., Ruiz L.F. (2002). The effect of irradiation on the quality of the avocado frozen pulp. *Radiat Phys Chem* 63(3–6): 379–382.

70. Vergara-Balderas F., Rangel-Marron M., Mujica-Paz H.,Valdez-Fragoso A., Welti-Chanes J. (2005). Freeze drying of guacamole. IFT Annual Meeting, July 15–20, New Orleans, Louisiana. Abstract No. 99C-13.

71. Weemaes C., Ludikhuyze L., Van Den Broeck I., Hendrickx M. (1999). Kinetic study of anti-browning agents and pressure inactivation of avocado polyphenol oxidase. *J Food Sci* 64(5): 823–827.

72. Woolf A.B., Cox K.A., White A., Ferguson I.B. (2003). Low temperature conditioning treatments reduce external chilling injury of "Hass" avocados. *Post harvest Biol Technol* 28(1): 113–122.

73. Woolf A.B., Bowmen J.H., Ferguson I.B. (1999). Pre-harvest exposure to sun influence post harvest responses of "Hass" avocado fruit. *Post harvest Biol Technol* 15: 143–153.

74. Woolf A.B., Lay-Yee M. (1997). Pre-treatments at 38°C of "Hass" avocado confer thermotolerance to 50æ%C hot water treatments. *HortScience* 32: 705–708.

□□□

Chapter 9

Post Harvest Diseases of Tomato and its Management

Shalini Yerukala, Sumalatha Nallabeema and Suresh Veera

Department of Plant Pathology, College of Agriculture, Rajendranagar, Professor Jayashankar, Telangana State Agricultural University, Hyderabad-500030, Telangana, India

ABSTRACT

Tomato fruit has wide array of nutrients and often challenged with diverse plant diseases at pre- and post-harvest stages of crop growth. General plant diseases that attack and infect the tomato at post-harvest stage include fungal, such as black mould, anthracnose, b*otrytis gray mold, early blight, late blight, r*hizopus rot, phoma rot, southern blight, fusarium rot and phomopsis fruit rot, etc. Pseudofungi, such as l*ate blightand buckeye fruit rot*. Bacterial diseases include bacterial speck, *bacterial spot, b*acterial soft rot and hollow stem etc. To reduce and minimize the post harvest loses caused by different phytopathogens. Identification of causal agent responsible for the plant diseaseis very important. Hence, the knowledge on post harvest diseases are very essential, to reduce and in prevention of post harvest loses caused by plant pathogen infection. Further efficient post harvest management practices must be followed for sustainable tomato production and to reduce potential health risk from infected fruits, for healthy living, and to increase commodity value in market, and to reduce fruit wastage.

1. Tomato Post Harvest Diseases

Tomato (*Lycopersicon esculentum* Mill.) known as "golden apple" (Chohan *et al.*, 2017) belongs to the family Solanaceae (nightshade) is one of the major horticultural crop with an estimated global production of 177 million tonnes (F.A.O. 2016) with world area harvested about 4 million hectares (F.A.O. 2016). Further, tomato world production share by region-wise include Asia occupying highest 52.9%, followed by Americans (18.5%), Europe (16.4%), Africa (11.8%) and least by Oceania (0.4%) (F.A.O. 2016). However, botanically tomatois a berry fruit but considered as a vegetable (Passam *et al.*, 2007).

Tomatoes have high nutritional value because of major source of anti-oxidant include lycopene, **beta-carotene, chlorogenic acid** and vitamins, such as C, K, B-6, B-12, A, E, D, folate, niacin, pyridoxine, thiamin, riboflavin.Minerals include Ca (Calcium), Fe (Iron), Mg (Magnesium), P (Phosphorus), K (Potassium), Na (Sodium), Zn (Zinc), etc. Tomatoes are also rich in carbohydrates and fiber. Other phytonutrients include carotene-β, carotene-α and lutein-zeaxanthin, etc. (USDA National Nutrient database, 2018; Bjarnadottir, 2015). During pre and post harvest phase, tomato fruits are subjected to several plant diseases. Some of the well known post harvest diseases are detailed in the chapter (Table 1).

1.1. Fungal Disease

A. **Alternaria Canker:** *Alternaria alternate* f.sp.*lycopersici*

Symptoms: On green fruits, the disease develop as sunken tiny spots surrounded by yellow halo known as freckle. Later, the spots enlarge and become darker with concentric rings. When infection occur on ripe fruit, black large sunken lesions develop and generally known as black mould. Usually the fruit lesions are not clear at harvest but may appear clearer after the fruit harvest (Fig. 1).

Dispersion: Plant pathogen over winters on decaying plant debris and seed borne. Spread by wind or rain water.

Management: Disease resistant cultivars, discard old plant debris, good drainage and irrigation, especially by drip irrigation to avoid plant surface wetting, fungicides such as application of azoxystrobin (Amistar), copper hydroxide (Kocide) and Mancozeb (protective application).

B. Anthracnose:*Colletotrichumgloeosporoides, C.coccoides*and *C. dematium*

Symptoms:Disease lesions develop on fruit as circular water soaked and slightly sunken regions with back spots at center (Fig. 2).

Dispersion: Over winter on infected plant debris, soil and seed borne.

Management: Discard the infected diseased tomatoes and old plant debris, use clean healthy seeds, crop rotation, proper plant spacing, proper irrigation practices, plant staking, soil mulch to prevent fruit touch on soil, appropriate harvest time, weeding. Chemicals include protective and curative fungicide mancozeb + carbendazim (Dithane M-45 + Bavistin, pyraclostrobin (cabrio), protective-chlorothalonil (kavach), azoxystrobin (amistar), curative-trif loxystrobin + tebuconazole (native).

C. Bacterial speck: P*seudomonas syringae*pv. *tomato*

Symptoms: Very small raised back specks or lesions often surrounded by dark green ring (Fig. 3).

Dispersion: Over winter on infected debris or infected seed or cool temperature or rainfall.

Management: Disease free seed, clean and healthy seed, crop rotation, sanitation, chemicals include streptomycin sulphate+ tetracycline hydrochloride (Agrimycin), soil drench at root zone with copper oxychloride (Blitox), copper hydroxide (kocide), Cuprous oxide (Nordox), Copper sulphate (cuproxat).

D. Bacterial spot: X*anthomonascampestris*pv. *vesicatoria*

Symptoms: Tiny, raised, scabby lesions develop on fruits (Fig. 4).

Dispersion: Warm and wet weather favourable for the disease, plant pathogen over winters on infected plant debris or seed.

Management: Use clean and healthy seed, crop rotation, discard infected and old plant debris.

Chemicals include streptomycin sulphate + tetracycline hydrochloride (Agrimycin), soil drench of copper oxychloride (Blitox) at the plant root zone, copper hydroxide (kocide), cuprous oxide (Nordox), copper sulphate (cuproxat).

E. Botrytis grey mould: Botrytis cinerea

Symptoms: White circular ring spots develop on green fruits known as ghost spots. Initial lesions develop as water-soaked spots with light brown center later become soft and water mass (Fig. 5).

Dispersion: Cool or wet weather favourable for the disease, plant pathogen over winters in soil or on perennial hosts.

Management: Proper plant spacing, plant staking, discard infected and old plant debris.

F. Buckeye fruit rot: Phytophthora nicotianae var. parasitica, P. capsici, and P. drechsleri

Symptoms: Initially, symptoms develop as water-soaked lesions on fruit. Later develop greyish brown discolouration on fruit skin and fruit pulp and becomes brown, watery, soft and decays. Greyish-greenish or brownish, water-soaked spots develop on fruits. Later lesions enlarge and develop as concentric rings or target spot (Fig. 6).

Dispersion: Warm and wet weather favours the disease. Phytopathogen is soil borne, disease develop when fruits touch soil, by rain splash, plant pathogen over winters on infected plant debris.

Management: Plastic mulch or straw mulch helpful in prevention of fruit contact to ground soil, plant staking, crop rotation,discard infected and old plant debris, appropriate fruit harvest time. Chemicals include protective fungicide application mancozeb(Dithane M-45), chlorothalonil (kavach), Fosetyl-Al (Aliette), Dimethomorph (Acrobat), Metalaxyl +mancozeb, drenching with Azoxystrobin (Amistar)–curative. Seed treatment with agrosan or ceresin @2gm/ kg of seed.

G. Early blight of tomato: Alter*naria solani*

Symptoms: Fruit lesions appear as black or brown, leathery, frim and concentric ringscoveredwith black spores (Fig. 7).

Dispersion: Cool weather favours the disease; plant pathogen over winters in soil and on infected plant debris.

Management: Clean and disease-free seed usage, crop rotation, discard infected and old plant debris, minimize plant injury, avoid dense plant populations grown in field for proper air ventilation, proper plant spacing, use drip irrigation to reduce plant surface wetness.

Chemicals include Azoxystrobin(Amistar), protective - Chlorothalonil (kavach) and Mancozeb (Dithane M-45), Copper hydroxide (Kocide), Propineb (Antracol), curative fungicide Difenoconozole (Score), Pyraclostrobin (Cabrio).

H. Late blight of tomato: Phytophthora infestans

Symptoms: Greenish brown lesions on fruits (Fig. 8).

Dispersion: Cool and wet weather favours the disease development, phytopathogen spread by wind or over winter on infected plant debris.

Management: Discard and destroy infected and old plant debris, crop rotation, proper sanitation measures, plant stalking, minimize leaf wetness by drip irrigation usage.

Chemicals include protective application of Mancozeb (Dithane M-45) and Chlorothalonil (kavach), Propineb (Antracol), Dimethomorph (Acrobat), Metalaxyl +mancozeb (Ridomil MZ gold)-protective and curative application, Azoxystrobin (Amistar) as curative application.

I. Rhizopus rot: *Rhizopus stolanifer*

Symptoms: Water-soaked lesions appear on fruit, and lesion surface covered with thin cotton-like fungal mycelium, later stage the white tuft mycelium will be covered with dark fungus sporulation's (Fig. 9).

Dispersion: Infected fruit residues.

Management: Proper fruit washing after field harvest, early and quick detection of disease symptoms to reduce loses, handle carefully to avoid wounds and injury to the fruits, use clean storage containers to reduce the pathogen load and heat treatment such by hot air.

J. Bacterial Soft Rot and Hollow Stem: *Erwiniacarotovorapv.carotovora*

Symptoms: On fruit soft watery tiny spots develop and later lesions enlarge and fill whole fruit and turn the fruit into soft waterywith liquefying fruit tissues (Fig. 10).

Dispersion: Wet and high humid weather favours the disease.

Management: Handle carefully to avoid wounds and injury to the fruits. Disinfects tools and equipment's used, proper pruning tomato plants, crop rotation, well drainage of soil during planting, use chlorinated water to reduce the risk of infection during washing, dry weather suitable for harvesting.

K. Phomarot: *Phoma destructive*

Symptoms: Black lesions develop on fruit and later stage lesion cover entirely either by fungus pycnidia or fruiting bodies (Fig. 11).

Dispersion: Moderate temperature and high humidity are favourable for the disease development. Seed borne disease

Management: Seed treatments are essential in controlling seed-borne diseases.

L. Southern blight: *Sclerotium rolfsii*

Symptoms: When fruit comes in contact with soil,develops slightly sunken, yellow lesions develop, later the lesions collapse and cover by a white fungal mass with sclerotia (Fig. 12).

Dispersion: Soil borne, disease infested soil, weeds and crop hosts.

Management: Use clean equipment's in field, discard diseased and infected plants and plant debris,crop rotation, soil solarization effective in reduce of viable sclerotia.

M. Fusarium rot: *Fusarium oxysporum*f.sp.*lycopersici races 1-3*

Symptoms: Disease characterized by fruit vascular tissue discoloration (Fig. 13).

Dispersion: Soil borne pathogen, spread by infected transplants and chlamydospores in soil particles on contaminated shoes or plant stakes or farm machinery or farm equipment's.

Management: Selection of disease resistant cultivars, clean equipment'susage, crop rotation, soil solarization.

N. Fruit rot: *Phomopsis vexans*

Symptoms: Infected fruits become soft, watery later turn black and mummified. The lesion center with grey and black pycnidia.

Dispersion: Spores spread by rain splash or insects or contaminated farm equipment, high moisture on plant surface. The pathogen survives on crop debris, seeds, and soil.

Management: Selection of disease free quality seeds, disease resistant varieties based on available, crop rotation, destroy crop residues, proper timing of irrigations and proper plant spacing.

Table 1: List of post-harvest diseases of tomato

Tomato post harvest diseases	
Fungal Diseases	**Plant Pathogen**
Freckle or Black mold.	*Alternaria alternate* f. sp.*lycopersici*
Tomato anthracnose	*Colletotrichumcoccoides,* *C. gloeosporoides, C. dematium*
Botrytis gray mold	*Botrytis cinerea*
Early blight of tomato	*Alternaria solani*
Rhizopus rot	*Rhizopus stolanifer*
Phoma rot	*Phoma destructive*
Southern blight	*Sclerotium rolfsii*
Fusarium rot	*Fusarium oxysporum* f. sp. *lycopersici races 1-3*
Phomopsis Fruit rot	*Phomopsis vexans*
Pseudo fungi (EPPO-European and Mediterranean Plant Protection Organization, 2018)	
Late blight of tomato	*Phytophthora infestans*
Buckeye fruit rot	*Phytophthoraparasitica*and *P. capsici*
Buckeye rot	*Phytophthora nicotianae* var. *parasitica, P. capsici,* and *P. drechsleri.*
Bacterial Diseases	
Bacterial speck	*Pseudomonas syringae pv. tomato*
Bacterial spot	*Xanthomonascampestris* pv. *vesicatoria*
Bacterial soft rot and Hollow stem	*Erwiniacarotovora* pv.*carotovora*

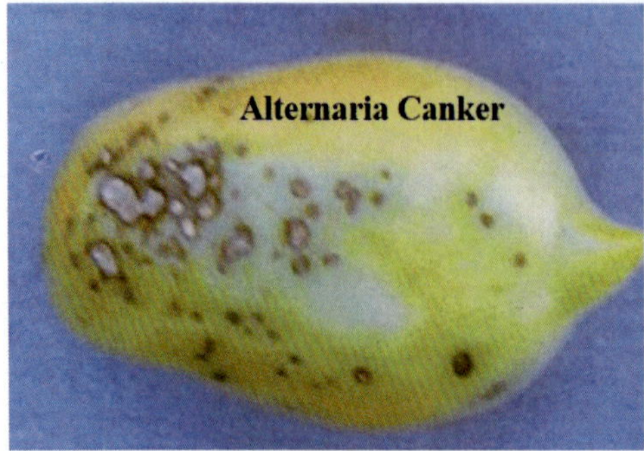

Photo Credit: Texas A&M Agri Life Extension, USA

Fig. 1: Tomato Alternaria canker

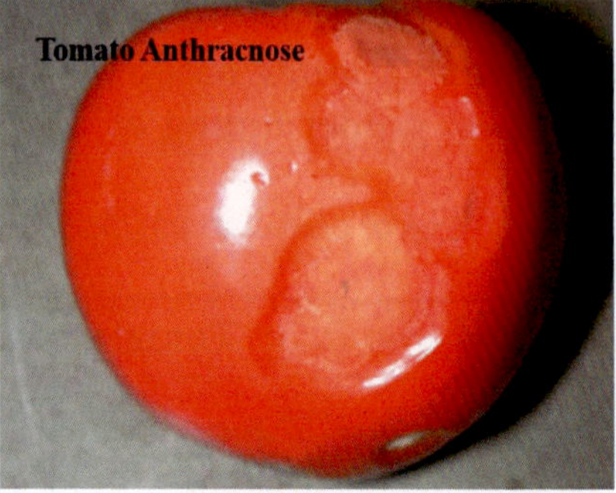

Photo Credit: University of Minnesota Extension, USA

Fig. 2: Tomato anthracnose

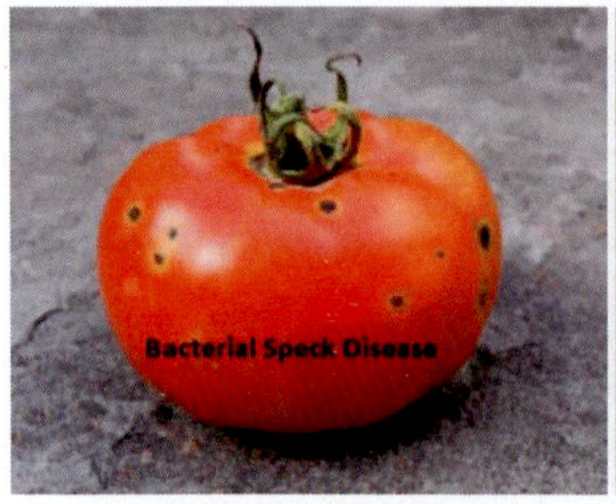

Photo Credit: Gregory Martin, Boyce Thompson Institute, USA

Fig. 3: Tomato bacterial speck

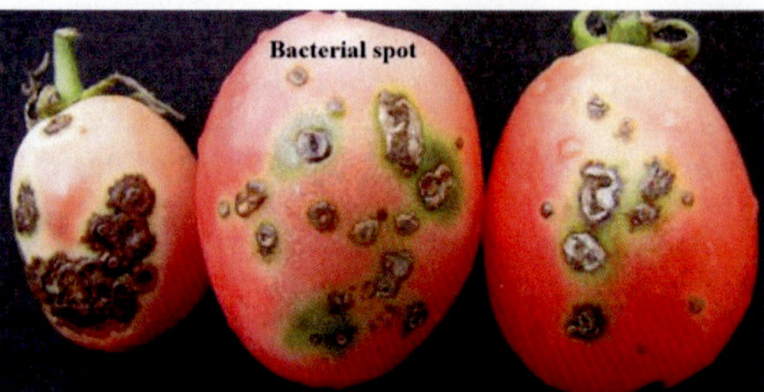

Photo Credit: Greenlife Crop Protection Africa

Fig. 4: Tomato bacterial spot

Photo Credit: Shawn Butler, NCSU PDIC, USA

Fig. 5: Tomato grey mold

Photo Credit: Cornell University, USA

Photo Credit: D. Ferrin, LSU AgCenter, Bugwood.org, USA

Fig. 6: Tomato Buckeye fruit rot

Photo Credit: Dr. J. Raja, Annamalai University, India

Fig.7: Tomato early blight

Photo Credit: Claudia Nischwitz, Utah State University, USA

Fig. 8: Tomato late blight

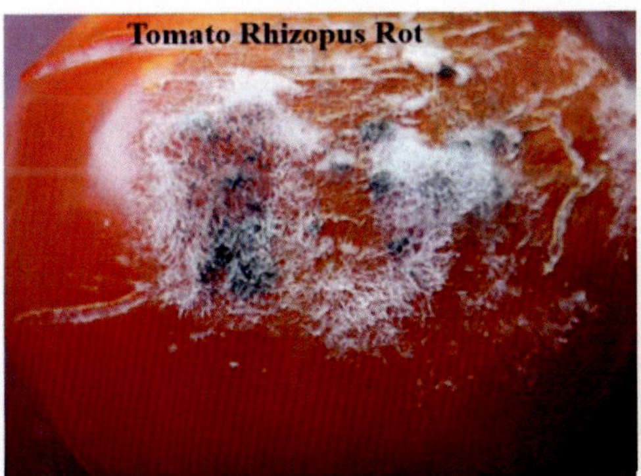

Photo Credit: Oscar Mario Castro Solano, Costa Rica, APS

Fig. 9: Tomato rhizopus rot

Photo Credit: Dennis Van Dyk @ONvegetables

Fig. 10: Tomato bacterial soft rot

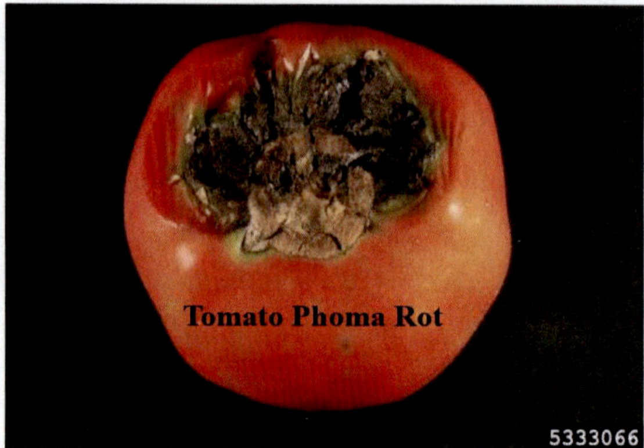

Photo Credit: OllieMartin@WikiGardener

Fig. 11: Tomato phoma rot

Photo Credit: Clemson University - USDA Cooperative Extension, Bugwood.org

Fig. 12: Tomato southern blight

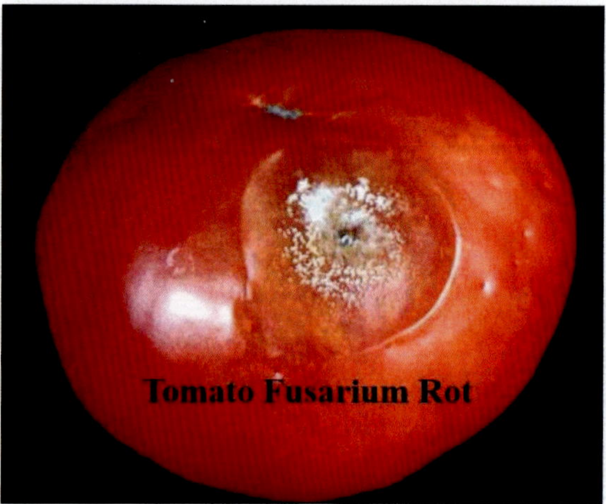

Photo Credit: Edwards, Don, University of California, USA

Fig. 13: Tomato fusarium rot

2. References

1. Bautista-Baños, S., Velaquez-Del Valle, M.G., Hernandez-Lauzardoa, A.N. and Barka, E.A. (2008). The Rhizopus stolonifer-Tomato interaction. Research Signpost 37/661 (2),Fort P.O., Trivandrum-695 023,Kerala, India. https://goo.gl/YDRXye
2. Bjarnadottir, A. (2015). Healthline Media. Health line newsletter. Retried on May 15, 2018. https://goo.gl/4TLi3d
3. Chauhan, S., Parveen, R., Abid, M., Naqvi, A.H. and Naz, S. (2017). Management of Seed Borne Fungal Diseases of Tomato: A Review. *Pakistan Journal of Phytopathology*, 29(1): 193-200. https://goo.gl/gDPhmD
4. Coates, Lindy and Greg Johnson. (1997). "Post harvest diseases of fruit and vegetables." *Plant pathogens and plant diseases*: 533-548. https://goo.gl/Mr1RM6
5. EPPO, (2018). European and Mediterranean Plant Protection Organization (EPPO). Global database. https://gd.eppo.int/taxon/PHYTIN
6. FAO (Food and Agricultural Organization), (2016). FAOSTAT, core production, 2016. Retried on May 15, 2018. Available online: http://www.fao.org/faostat/en/#data/QC/visualize
7. Karen, D. and Walt, S. (2004). Tomato disorder: Post harvest fruit diseases. University of Wisconsin Extension Publication, Division of Co-operative Extension of the University of Wisconsin-Extension, 432, N. Lake St. Rm. 103, Madison, WI 53706. I-05-04. A3799. https://learningstore.uwex.edu/Assets/pdfs/A3798.pdf
8. Magray, M.M., Hussain, K., Aijaz, M., Singh, P.K. and Nab, A. (2017). Post harvest management of vegetable crops: Introduction. Journal of Pharmacognosy and Phytochemistry; 6(6): 182-185. https://goo.gl/NnsLmd
9. Mullen, J. (2001). Southern blight, Southern stem blight, White mould. *The Plant Health Instructor*. DOI: 10.1094/PHI-I-2001-0104-01. Updated 2006. *APS Americana Phytopathological Society*.https: //goo.gl/tqNzma
10. Ozbay, N. and Newman, S.E. (2004). Fusarium crown and root rot of tomato and control methods. *Plant Pathol Journal*, 3: 9-18. https://scialert.net/fulltextmobile/?doi=ppj.2004.9.18

11. Passam, H.C., Karapanos, I.C., Bebeli, P.J. and Savvas, D. (2007). A review of recent research on tomato nutrition, breeding and post-harvest technology with reference to fruit quality. *The European Journal of Plant Science and Biotechnology*, 1(1): 1-21. https://goo.gl/N722t2

12. Peter, N. and Andy, W. (2005). Diagnosing and Controlling Fungal Diseases of Tomato in the Home Garden. Extension Specialist in Vegetable Pathology, Rutgers University. New Jersey Agricultural Experiment Station Rutgers, The State University of New Jersey, 88 Lipman Drive, New Brunswick, NJ 08901-8525.Co-operative Extension Fact Sheet FS547. https://njaes.rutgers.edu/fs547/

13. Ray Cerkauskas, (2005). Cornell University. Published by AVRDC-2005. The World Vegetable Centre Fact Sheet, Tomato Diseases, Soft Rot, AVRDC Publication 05-632, 2005.www. avrdc.org. The World Vegetable Center; P.O. Box 42, Shanhua; Taiwan 741; ROC. http://203.64.245.61/web_crops/tomato/soft_rot.pdf

14. Schwartz, H.F. and Gent, D.H. (2007). Eggplant, Pepper, and Tomato. *Phomopsis Fruit Rot (Phomopsis Blight). High Plains IPM Guide, a cooperative effort of the University of Wyoming, University of Nebraska, Colorado State University and Montana State University*, 157-168. https://goo.gl/RVtXpf

15. Shankar, R., Harsha, S. and Bhandary, R. (2014). A practical guide to identification and control of tomato diseases. Tropica Seeds Pvt. Ltd No. 54, South End Road, 1st Floor, Nama Aurore Building, Basavangudi, Bangalore-560004. INDIA. PP 1-49. https://goo.gl/PPif9T

16. Srivastava, M.P. and Tandon, R.N. (1966). Post harvest diseases of tomato in India. Mycopathologia et MycologiaApplicata (1966) 29: 254. https://doi.org/10.1007/BF02128452.

17. USDA National Nutrient database (2018). United States Department of Agriculture Agricultural Research Service. Retried on May 15, 2018. https://goo.gl/5RgBmw

❑❑❑

Chapter 10

Post Harvest Management, Processing and Value Addition of Betelvine (*Piper betle* l.)

Tanuja Priya, B.[1], Syed Sadarunissa[2], Rama Devi, P.[3] Rama Krishna[4] and Sireesha, K.[5]

[1,5]Assistant Professor, [2,3]Associate Professor, [4]Associate Dean, Dr. Y.S.R. Horticultural University, V.R. Gudem, West Godavari, Andhra Pradesh-534101, *corresponding author's email: tpriyahort@gmail.com

ABSTRACT

Betel vine is a perennial climber harnessing leaves rich in anti-oxidants and many medicinal properties. It has several beneficial effects, which increased its demand and market price in the social community. However, huge loss is witnessed due to wastage of leaves during storage, transport and perishability ranging from 35-70%. The betel leaves presumed to store for 4 to 12 days in summer and 12 to 21 days in winter by adopting various methods. Further, the value added products of betel leaves developed, elevates the national and international economy as the major growers primarily derive their livelihood through this crop. Thus, this chapter throws a light on the post harvest management, processing and preparation of value added products.

Keywords: Betel leaf, storage, grading, packing, value added products

1. Introduction

Betel vine (*Piper betle* L.), the dioecious shade loving perennial climber that belongs to the family Piperaceae is cultivated across the world with about hundred known varieties [9]. All the varieties are broadly classified under three major groups *viz.*, Kapoori, Bangla and Meetha. The Kapoori leaves are ovate in shape, light green in colour, smooth in texture and less pungent. The pungency in Kapoori type is attributed to presence of terpenyl acetate content. Bangla leaves are cordate in shape, green to dark green in colour, thick with broad lamina, coarse to touch and pungent due to high eugenol content. While, meetha leaves are dark green, cordate to broadly ovate and possess characteristic fennel like aroma and sweet palatable taste due to the presence of an ethole content (Plate 1). It is proclaimed that the most probable place of origin for betel vine is Malaysia and known to thrive best in tropical forest ecosystem with a rainfall ranging 2250-4750 mm, relative humidity 40-80% and temperature 15-40°C [5]. A well-drained fertile sandy loam soil with pH range of 5.6-8.2 is considered suitable for its cultivation [14]. It is propagated vegetatively through cuttings with 3 nodes buried in the soil and three above the soil under the shades of *Sesbania* crop in open conditions (Plate 2) or under partial shade in a small hut like structure called *Boroj*, which is constructed with the locally available materials like bamboo stems, jute sticks, paddy straw, petioles and leaves of banana, *etc.* (Plate 3).

It is presumed that the betel vine, commonly known as paan is considered as most neglected crop in the world, but designated as green gold in India since most of the famers derive their livelihood directly or indirectly from production to marketing [11]. Further, huge capital and labour intensive characteristics of this crop paved to offer perennial employment and income to small and marginal farming community. In fact, right from the dawn of civilization, this edible leaf has achieved an esteemed position in the human society particularly in the countries like Bangladesh, Burma, China, India, Indonesia, Malaysia, Nepal, Pakistan, Philippines, South Africa, Sri Lanka, Thailand, *etc.* where leaves are traditionally used for chewing in their natural raw condition along with many other ingredients. Every sphere of human life including social, cultural and religious gatherings also bestows betel leaves as a significance of benevolence. Besides, betel leaves are known to harbour moisture, protein, fats, minerals, vitamins, phytochemicals and anti-oxidants that aid in curing of various diseases. Therefore, owing to its importance, the area and production under betel vine in recent years has accelerated remarkably, however, decline in market price is manifested due to huge post harvest losses. Severe losses occur because of poor transportation facilities, poor management and improper market facilities, careless handling of the produce by farmers, market intermediaries and consumers and lack of know-how in processing method and value added products. Pursuance of the fastest and most practical route to minimize the post harvest losses of betel leaves is the need of hour for sustainability. Hence, the chapter on post harvest management, processing and value addition of betel vine is focused and presented.

Plate 1: Various groups of betel vine

Plate 2: Betel vine garden under open condition

Plate 3: Betel vine garden under closed condition

1.1. Uses

Betel leaves are chewed as raw along with areca nut, slaked lime, cardamom and clove in many Asian countries. Various properties *viz.*, anti-oxidant, anti-fungal, anti-platelet, anti-diabetic, anti-amoebic, anti-inflammatory, anti-filarial and anti-microbial [15], anti-fertility [23], anti-hyperglycemic [22], anti-dermatophytic [26], anti-nociception [1], anti-cancerous [18] and radio-protective properties [3]in betel leaves was reported. The leaves are also considered to contain vitamins and significant amount of all essential amino acid except lysine, histidine and arginine which occurs in traces. In addition to the medicinal properties, the betel leaves also known to have a good nutritional value. The essential oil of the betel leaves reported to have high marketability, as it is used in the production of perfume, medicine, talc, beverages, food additives and mouthwash [9]. The leaf found to produces aromatic volatile oil that contains phenol called chavicol, which has powerful antiseptic properties. Pharmacological effects of betel chewing included abundant flow of saliva, temporary dulled of taste perception, stimulation of muscular and mental efficiency [7].

1.2. Composition of Betel Leaf

The phytochemical analysis of *Piper betel* leaf revealed that it possesses starch, sugars, tannins, carbohydrate, amino acids, diastases, steroidal components, alkaloids, piperol-A,piperol-B, methyl-piperbetol, and an essential oil composing of terpinen-4-ol, safrole, allylpyrocatecholmono acetate, eugenol, eugenyl acetate and hydroxyl chavicol. It is also reported that the betel oil contains cadinene-

carvacrol, allyl catechol, chavibetol, chavicol, p-cymene, caryophyllene, chavibetol, cineole, estragol, *etc.* as the key components [25]. The nutritional value of raw betel leaf (100g) is presented below [8].

Nutrition in 100 g (raw)	Energy: 44 kcal
Water	85 g
Proteins	3g
Fat	1g
Minerals	2 g
Dietary fibre	2g
Carbohydrates	6 g
Calcium	230 mg
Phosphorus	40 mg
Iron	11 mg
Nicotini acid	0.8 g
Vitamin A	2 mg
Vit B1 (Thiamin)	0.05 mg
Vit B2 (Riboflavin)	0.02 mg
Tannin	1 g
Nitrogen	5 g
Vitamin C	1 mg
Magnesium	23 mg
Potassium	3 g

1.3. Harvesting

Plucking of betel leaves requires an expert hand, with the help of an artificial nail made of iron. The matured leaves are cut along with the petiole about 1cm in length (Plate 4 & 5). Harvesting of betel leaves is done throughout the year, three to four months after plantations and one to one half months after lowering. During rainy season, leaves are plucked continuously at demands.

Betel leaves being highly perishable in nature causes heavy loss to farmers, as they have to sell the products at a throw away price. Bangla and meetha type of betel leaves could be maintained on the vine without harvesting for more than three months to overcome the glut situation in market, while Kapoori types should be harvested immediately on maturity. Hence, extension of shelf-life to maintain the freshness of the green leaves principally Kapoori type is inevitable for the benefit of growers.

Plate 4: Harvesting of betel leaves

Plate 5: Iron nails used for harvesting

1.4. Grading

It is stated that the betel leaves are graded based on the harvest; the leaves harvested from laterals called *Kalli*, leaves harvested from lower part of the vine called *Hakkal* and leaves harvested from main stem called *Fapada* are marketed separately. The Fapada leaves fetches high market price for its bigger size and shape compared to *Kalli* land *Hakkal* eaves. It is also presented that the grading in some parts of Deccan region is done according to the plucking of leaf *viz*., the new betel leaves were called as *Navati*, the second plucking of leaves as *Parati* and third plucking as *Terti*. Furthermore, grading based on age of vine is also done, where leaves from first year garden fetch more prices, compared to matured gardens [16].

1.5. Post Harvest Storage and Packaging

Consumption of betel leaves is mainly in the fresh form and thus, it is essential that they are stored in a system that could maintain their freshness until consumption without any noticeable physiological disorder. Packaging of betel leaves are done for year-round availability, maintenance of taste, marketing, distribution and long shelf-life. Diverse array of options for packing and package materials exists and it varies according to region and consumer preferences.Refrigeration and cold storage systems often used for fresh produce storage are not always suitable due to their high cost and energy requirement. It is also reported that betel leaves are subjected to chilling injury, when stored at refrigeration temperature. Some of the methods for minimizing post harvest losses of betel leaves are given below:

a) **Dehydration**: Reducing the amount of moisture content in the leaves through solar, shade, mechanical and microwave drying to extend the shelf of betel leaves [2].

b) **De-petiolation and de-midribbing**: De-petiolation is removal of the petioles from the leaves to delay senescence [10], while de-midribbing is removal of midrib batta that could also result in delay of senescence. De-petiolation in combination with partial or total removal of midrib, delayed senescence to a greater extent than with de-petiolation alone [12].

c) **Heat treatment**: The leaves subjected to heat treatment at 45°C for one hour in traditional packaging also showed better performance as compared to other treatment [19].

d) **Bleaching and curing**: Green leaves are treated with smoke, high temperature and pressure in a closed room with little or no ventilation to improve organoleptic quality by changing the colour from green to yellow/white [6].

e) **Chemical treatments:** The storability of chemically treated betel leaves were on par with heat treated and the colour of the chemically treated leaves were relatively higher than heat treated leaves [24]. Storage of the betel leaves could be extended on packaging with a mixture of sodium bicarbonate and tartaric acid or benzyl adenine or kinetin [8], while it was reported that the leaves treated with ascorbic acid 100 ppm extended shelf-life up to 11 days [17] (Plate 6).

f) **Modified atmospheric storage**: Post harvest dipping of betel leaves for 6 h in solutions of 25 ppm benzyladenine (BA) and 50 ppm kinetin and packed in vented polythene bags prolonged the shelf-life of the leaves under refrigerated conditions [24]. It was also reported that betel leaves with the relative humidity inside the polythene bags ranged from 92-96% increased the shelf-life of betel leaves upto 21 days under temperature ranging from 10 to 15^{0}C [9].Betel leaves packed in 300 guage polythene bag extended shelf-life for twelve days irrespective of ventilation [17 & 21] (Plate 7). The shelf-life of the betel leaves was found higher (4 days) in thermocol box and bamboo basket at ambient condition with great freshness, no yellowing and higher retention of anti-oxidants [4].

g) **Combined treatments:** Dipping of leaves in 25 ppm of BA for 6 h in combination with de-petiolation and de-midribbing resulted in the increase of shelf-life by about 10 days with negligible spoilage [24].Extended storage was demonstrated, when the leaves were subjected to chemical treatment (5 ppm benzyladenine for 8 h), followed by heat treatment (45°C for 1 h) [19].De-petiolation, de-midribbing and chemicals treatments on combination observed to be best suited to prolong the shelf-life of betel vine leaves.

Plate 6: Chemically treated betel leaves on 11th day of storage (T_1:Ascorbic acid 100ppm, T_2: CaCl2 0.5%, T_3:Citric acid 100ppm, T_4:Potassium meta bisulphate (KMS) 50ppm, T_5:sucrose 2%, T_6:salicylic acid 100ppm, T_7:vitamin K (Maladian) 100ppm, T_8:Sucrose 2% + Salicylic acid 100ppm + KMS 50ppm and T_9: Control)

Plate 7: Betel leaves in various guages of polythene bags on 12th day of storage

Majorly, betel vine growers use bamboo baskets padded with either paper shavings, fresh plantain leaves, moist dry plantain leaves, fresh neem leaves or moist paddy straw for easy transportation of leaves. Leaves packed in bamboo baskets using the hollow method (space left in the middle of the basket) lined with fresh plantain leaves and stored at room temperature extended its shelf-life and 50% rotting reported after 14 days of storage [24].It is further reported that three thousand betel leaves are packed in bamboo baskets called *Karandi* in Maharashtra using Banana fiber and leaves as a packing material (Plate 8), while six and twelve thousand leaves are assembled in packages called *Dappa* and *Dag*, respectively. Here, local mulberry branches are used for making rings and placed on either sides of the packing. Sugarcane or Banana leaves as a padding material and Banana fiber, plastic fiber and strings are used to fasten the package for safe transportation (Plate 9). In *Karandi* and *Dappa* package, mostly *Kalli* leaves are packed, while *Fapada* leaves are packed in *Dag* type of packaging system.

Plate 8: Betel leaves packed in *Karandi*

Plate 9: Betel leaves in *Dappa* or *Dag* package

1.6. Processing and Value Added Products

The various drying methods *viz.*, shade drying, solar drying and mechanical drying of betel leaves are followed. Maximum nutrient could be preserved in betel leaves under shade drying and no significant loss of quality was observed at 40^0C and could be recommended as a controlled method of drying, compared to other methods for further use [19].Betel leaves have several benefits but still a huge amount of wastage of leaves takes place every year, which is ranging from 35-70% due to storage, transport and high perishable nature. Hence, development of value added products to minimize the losses of betel leaf is inevitable to fetch high superior prices in market. Some of the products developed are Gulpaan, Paanlata, Paanchur and Paan squash, Paan shot, betel leaf oil and betel quid [13,20 & 27].

1. **Gulpaan:** Gulpaan is a tasty and nutritious preparation of fresh betel leaf. The betel was washed and pat dried using kitchen towel and finely chopped the betel leaves. Rolled the leaf from one edge to form a proper cone to keep all the ingredients, mixed the chopped betel leaf, gulkand, mukhawas, coconut flakes, cherry and pinch of cardamom powder, placed in the betel leaf, closed, kept the glazed cherry on top of gulpan and then insert the toothpick [20].

2. **Paanlata:** Paanlata is a sweet of fresh betel leaf and mawa. Washed the betel leaves with the help of lime water for about three days, chopped the dry fruits and mixed with the mava and gulkand.One betel leaf was taken, filled the mava mixture, folded the four sides of leaf and inserted the clove. Sugar syrup was prepared and dipped the covered betel [20].

3. **Paanchur:** Paanchur is like a mouth freshner and it also works as a substitute of tobacco. Betel leaves were clean dried in dehydrator and crushed, the fennel seeds combined with betel leaf powder, gulkand, crushed bale seed, mishri, pipermint and cardamom powder was mixed well, stored in air-tight container and used, when required [20].
4. **Paan squash:** Paan squash is a drink of boiled fresh betel leaf. Sugar syrup was prepared by dissolving the sugar in water in a large saucepan or pot. The fresh betel leaf mash was boiled in the sugar syrup. Potassium metabisulphite was mixed in paan squash and stored in a bottle [20].
5. **Paan shot:** All the ingredients *viz.*, four calcutta betel leaves (paan) torn into small pieces, four tablespoons gulkand, two and half table spoons betel leaves (paan) masala, one table spoons fennel seeds (saunf) and two and half cups vanilla ice-cream and ten ice cubes were taken in a mixer, blended till smooth and refrigerated for at least 1 hour. Later, equal quantities into 12 shot glasses were poured and served chilled [13].
6. **Betel leaf oil:** The common method of extraction of the oil is reported as steam distillation. Fresh betel leaves, which were previously dried for two days were steam distilled to get the oil. The aroma of the betel leaf oil is found sharp, bright brown to yellow in colour and tastes bitter [13].
7. **Betel quid:** The term "betel quid" refers to a combination of ingredients that most typically includes areca nut, piper betel leaf (a common vine), slaked lime (calcium hydroxide), and tobacco. The ingredients of betel quid vary considerably by region, country, ethnicity, and personal preference [27].

Gulpaan

Paanlata

Paanchur Paan squash

Paan shot Betel leaf oil

2. Conclusion

Betel vine, being a cash crop is cultivated by large number of small and marginal farmers across the country. However the highly perishable nature of crop, forced the growers to shift the cultivation from this crop due to heavy post harvest losses. Hence, it is further envisaged that adopting various techniques minimizes the losses during storage, processing, transportation and marketing and further assists to generate the revenue adequately through this leafy crop cultivation. Therefore, a well-coordinated effort by the farmers, traders, scientists, technologists, extension workers, physicians, administrators and policy-makers is required to be initiated to boost up the national economy as well as the national employment generation through proper exploitation of this green gold.

3. References

1. Arambewala, L.S.R., Arawwawala, L.D.A.M., Ratansooriya, W.D. (2005). Antinocepative activities of aqueous and ethanol extract of Piper betle leaves in rats. Journal of ethanopharmacology; 102(2): 239-245.

2. Balasubramanian, S., Sharma, R., Gupta, R.K. (2010). Validation of drying models and rehydration characteristics of betel (*Piper betel* L.) leaves. *J. Food Sci. Technol.* 48(6): 685-691

3. Bhattacharya, S., Subramanium, M., Roychaudhary, S., Bauri, A.K., Chattopadhyay, S., Bandyopadhyay, S.K. *et al.* (2005). Radio protective property of ethanol extract of Piper betle leaf. *J. Radiat. Res.* 46(2): 165-171.

4. Bhuvaneshwari, S., Hima Bindu, K. and Kusuma, D.K. (2017). Effect of storage temperature, packaging on the shelf-life extension of betel leaves. International Journal of Research in Applied, Natural and Social Sciences; 5(1): 2347-4580.

5. Chattopadhyay, S.B. and Maity, S. (1967). Diseases of Betel vine and Spices. ICAR, New Delhi.

6. Dastane, N.G., Patil, R.G. and Chaugule, B.A. (1958), This business of bleaching betel leaves, Indian Farming, 7(10): 10-12.

7. Fazal,F., Mane, P.P., Rai, M.P., Tilakchand, K.R., Bhat, H.P., Kamble, P.S, Palatty, P.L., Baliga, M.S. (2014). The phytochemistry, traditional uses and pharmacology of piper betel. linn (betel leaf): a pan-asiatic medicinal plant. Chinese Journal of Integrative Medicine. 1–11

8. Gopalan, C., Rama Shastri, B.V. and Balasubramanian, S.C. (2004). Nutritive Value of Indian Foods, National Institute of Nutrition, ICMR, Hyderabad.

9. Guha, P. (2006). Betel Leaf: The Neglected Green Gold of India. *J Hum Ecol.* 19: 87-93.

10. Imam, M.N. and Pariari, A. (2012). Keeping quality of betel leaves (Piper betle L.) as influenced by different methods and seasons. Journal of Spices and Aromatic Crops; 21(2): 106–112.

11. Jane, N.S., Deshmukh, A.P., Joshi, M.S. (2014). Review of study of different diseases on betel vine plant and control measure. Int. *J. Appl. Innov. Eng. Manag.* 3(3): 560-563.

12. Madan, A., Balan, N., and Barma, R.D. (2014). Reducing post harvest losses of betel (Piper betel L.) leaves by various preservation techniques. Journal of Agri. Search, 1(4): 251-256.

13. Mazumdar, S., Roychaudhary, A., Banerjee., S. (2016). An overview of betel leaf (Piper betle L.): a review. Annals Food Science and Technology, 367-376.

14. Mishra, D., Mishra, S.K., Acharya, A. and Das. J.N. (1997). Annual Report of All India Coordinated Research Project on Betel vine. Department of Plant Pathology, College of Agriculture, Orissa University of Agriculture and Technology, Bhubaneshwar, India.

15. Nagori, K., Singh, M.K, Alexander, A., Kumar, T., Dewangan, D., Badwaik, H. *et al.* (2014). Piper betle L.: A review on its ethnobotany, phytochemistry, pharmacological profile and profiling by new hyphenated technique DART-MS (Direct Analysis in Real Time Mass Spectrometry). Journal of Pharmacy Research; 4(9): 2991-2997.

16. Patil, B.V. (2016). A critical study of irrigation practices for betel vine in the selected area, 4(29): 4419-4425.

17. Priya, B.T., Vani, S., Sireesha, K. and Rama Devi, P. (2017). Enhancement of shelf-life in betel leaves. Trends in bio-sciences; 10(21): 4103-4105.

18. Rai, M.P., Tilakchand, K.R., Palatty, P.L., Rao, P., Rao, S., Bhat, H.P., *et al.* (2011). Piper Betel Linn (Betel vine), the maligned South-east Asian medicinal plant possesses cancer preventive effects: Time to reconsider the wronged opinion. Asian Pac. J. Cancer Prev. 12: 2149-56.

19. Rayaguru, K., Md. Khan, K., Mohanty, S.N., Sahoo, G.R. (2007). Effect of Drying and Storage on Quality of Betel Leaves, Journal of Agricultural Engineering, 44(1): 64-68.

20. Richa and Neetu Singh (2017). Betel leaf product preparation and its nutritional analysis. Asian Journal of Science and Technology; 08(11): 487-6489.

21. Sadarunnisa, S., Madhumathi, C. and Hribabu, K. (2006). Post harvest behaviour of betel leaves as influenced by polythene package, 34(4): 105-107

22. Shantakumari, P., Prakasam, A., Pugalendi, K.V. (2006). Anti-hyperglycemic activity of P. betle on streptozotoan induces diabetic rats. Journal of Medicinal Food; 9(1): 108-112.

23. Sharma, J.D, Sharma, L., Yadav, P. (2014). Anti-fertility efficacy of Piper betle Linn (petiole) on male albino rats. Asian J. Exp. Sci.; 21(1): 145-150.

24. Singh, K.K., Balasubrahmaniam, V.R. and Kochhar, V.K. (1990). Effect of different packing methods, temperature conditions, treatment with chemicals on the senescence and storage behaviour of betel (Piper betle L.) leaves. *J. Plant. Crops,* 18(1): 23-28.

25. Sugumaran, M., Poornima, M., Venkatraman, S., Lakshmi, M., Srinivasan sethuvani, (2011). Chemical composition and anti-microbial activity of sirugamani variety of *Piper betle* Linn Leaf oil. Journal of Pharmacy Research; 4(10): 3424-3426.

26. Trakranrungsie, N., Chatachawanchonteera, A., Khunkitti, W. (2008). Ethanoveterinary study for anti-dermatophytic activity of P. betel, Alpiniagalanga and Allium ascalonicum extract *in vitro*. Research in veterinary sciences; 84(1): 80-84.

27. Zain, R.B., Ikeda, N., Gupta, P.C. *et al.* (1999). Oral mucosal lesions associated with betel quid, areca nut and tobacco chewing habits: consensus. *J. Oral Pathol Med*, 28: 1–4.

❑❑❑

Chapter 11

Nutritional Status of Chekurmanis (*Sauropus androgynous*)

Subhrajyoti Chatterjee[1], Debmala Mukherjee[1], Arghya Mani[2] and Partha Choudhuri[1]

[1]Department of Vegetable Science, Bidhan Chandra Krishi Vishwavidyalaya, Mohanpur, Nadia- 741252, West Bengal, [2]Department of Post Harvest Technology of Horticultural Crops, Bidhan Chandra Krishi Vishwavidyalaya, Mohanpur, Nadia- 741252, West Bengal

ABSTRACT

Promotion of perennial crops is a good alternative for food security as well as nutritional security since they offer more consistent and year round production without any break. Chekurmanis or sweet leaf, [*Sauropusandrogynus* (L.) Merrill] belonging to family Euphorbiaceae, is such a perennial green, which is considered as power house of multivitamins and proteins. Leaves are rich in proteins (6-8%), high compared to other leafy vegetables like methi and palak; Vitamin A (more than mangoes and papayas), and Vitamin B, C, Nicotinic acid, mineral matters, phosphorus and iron. It is the reason for which the plant is nicknamed as 'multivitamin green' and 'multimineral packed leafy vegetable'. The crop is extensively grown in hot and humid climatic situation. The leaves and succulent young tips of the plant are normally eaten like tropical asparagus, raw in the form of salads or steamed or alternatively added to stirfry, rice and egg dishes, soups or casseroles and sometimes, blanched to serve with chilli paste. The flowers and the small purplish fruits of the plant are also reported to be eaten.

Keywords: Chekurmanis, leafy vegetable, multi vitamin green, food security, nutritional security.

1. Introduction

Green leafy vegetables occupy an important place among the food crops, as these provide adequate amount of many vitamins and minerals for humans. They are rich source of carotene, ascorbic acid, riboflavin, folic acid and minerals like calcium, iron and phosphorus. In nature, there are many under-utilized greens of promising nutritive value, which can nourish the ever increasing human population. Many of them are resilient, adaptive and tolerant to adverse climatic conditions. Although, they can be raised comparatively at lower management costs even on poor marginal lands, they have remained under-utilized due to lack of awareness and popularization of technologies for utilization. Now-a-days, under-utilized foods are gaining importance as a means to increase the per capita availability of foods[1].Chekurmanis is one such most important, highly nutritious, shrubby, perennial leafy vegetable crop belonging to family Euphorbiaceae, growing widely in Indonesia, Malaysia and Singapore i.e. South and South-east Asia but remain under-utilized in India except some parts of Southern India.It is known as 'Tropical asparagus' and can be eaten as either raw or cooked like spinach. Besides vegetable, it is also grown as a hedge or fence around vegetable plots or kitchen garden [2].

1.1. Botany

It is a slow growing glabrous shrub; attaining a height of 2-3.5 m. Main branches are terete and flaccid, while lateral branches are thin. It bears dark green oval leaves which are 5-6 cm long. The plant bears small reddish monoecious flowers. Flowers are found in the month of April-July. Fruits are sessile, white or pinkish, 0.2 cm in diameter with a fleshy epicarp. Fruit setting occurs in July-December. It is usually maintained as a perennial plant at a height of 1-1.5 m by frequent harvesting of leaves [2].

Fig. 1: Chekkurmanis plant at IIVR, Varanasi

Fig. 2: Reddish monoecious flowers of Chekkurmanis

Fig. 3: Oval leaves and white-pinkish fruits

1.2. Distribution and Propagation

Chekurmanis grows rapidly in hot humid conditions but becomes relatively dormant in cooler environments. It appears to have originated from the hot humid lowland rainforest of Borneo and its present distribution spans from India and South China, through Indo-China and Malaysia to the Philippines, New Guinea and Solomon islands. It is more vigorous at lower altitudes but also grow up to an altitude of 1300 m in Malaysia and Indonesia. In India, the shrub is reported to be found in Sikkim Himalayas, Khasi, Abor and Aka hills and in the Western Ghats of Kerala at altitudes of 600-1200 m. It is said to have got introduced into Kerala from Malaysia in 1953. Reports of its occurrence in the peninsular India from the evergreen forests, clearings, scrub vegetation, roadsides and on waste grounds in Coimbatore, Madurai, Nilgiri, Tiruchirappalli, Tirunelveli and several other parts are available in literature. Vernacular names of this crop in different parts of India are provided in Table 1.

Table 1: Vernacular names of chekurmanis in different parts of India

Indian States	Local name (Ethnic group/ Location)
Andman & Nicobar Islands	Chakramani
Assamese (Assam)	Bari sundari
Bengali (W. Bengal)	Chekurmani
Khasi (Meghalaya)	Diengsoh pit
Kannada (Karnataka)	Chakraniberu, Chinese soppu
Lepcha (Sikkim)	Sengtungrung
Malayalam (Kerala)	Malay cheera, Elacheera (Muthuvan tribes of Idukki district)
Tamil (Tamil Nadu)	Thavasaimurungai (thavasimurungai)

The plant is usually propagated vegetatively, since the seeds have low germination power, but grows readily from cuttings (specially from semi-hard wood cuttings) [3].

1.3. Nutritive value

It has been reported that an analysis of the proximate composition of *S. androgynus* leaf has shown that its nutritive value is superior to other commonly consumed leafy-vegetables in India [4]. Proximate composition of Chekurmanis leaves is reported in Table 2.

Table 2: Nutritional values of chekurmanis leaves

Constituent	Composition								
	(per cent)			(g/100 g EP)			(g/100 g FL)		
Moisture	85.4	88.32	79.4	-	79.8	89.9	88.46	78.19	88.0
Carbohydrate	-	5.36	6.9	11.6	6.9	54.5	3.86	-	-
Protein	5.25	4.84	7.6	6.8	7.6	15.8	4.19	8.31	3.4
Lipids	0.58	0.19	1.8	-	1.8	4.0	0.19	-	-
Fibre	1.75	1.11	1.9	-	1.90	30.6	1.20	-	-
Ash	5.25	0.17	2.0	-	2.0	12.9	1.38	-	-
Energy (kJ)	-	-	100.5	-	310.0	1327.0	169.1	-	117.0
Reference	5	6	7	8	9, 10	11	12	13, 14	1,15
EP = edible portion; FL = fresh leaf material; – = not recorded									

- Leaves of chekurmanis are rich in minerals. Minerals are considered essential for physical and mental development, immune system and various other metabolic processes.

Table 3: Mineral content of chekurmanis leaves.

Minerals	Contents						
	(ppm DL*)	(mg/100 g DL*)	(mg/100 g FL)	(mg/100 g EP)			
Magnesium	1587.0 ± 2.01	664.9 ± 38.1	-	0.55	-	-	-
Potassium	269.1 ± 0.28	45.7 ±1.5	-	2.77	2610.0	-	-
Sodium	-	306.3 ± 52.8	-	-	-	-	386.0
Calcium	206.82 ± 4.79	84.4 ± 8.1	313.0	2.77	234.0	118.8	70.5
Phosphorus	81.42 ± 2.71	61.2 ± 3.5	-	0.61	64.0	-	128.2
Iron	20.49 ± 0.89	212.5 ± 20.8	10.1	0.01	3.1	13.5	0.90
Zinc	11.02 ± 0.15	15.9 ± 6.6	-	-	-	-	0.90
Vanadium	7.41 ± 0.05	-	-	-	-	-	-
Manganese	4.76 ± 0.82	25.6 ± 5.8	-	-	-	-	-
Copper	1.54 ± 0.06	768.7 ± 11.4	-	-	-	-	0.29
Selenium	1.03 ± 0.02	-	-	-	-	-	-
Chromium	0.52 ± 0.08	-	-	-	-	-	-
Cobalt	0.06 ± 0.04	1.6 ± 00.1	-	-	-		-
References	16	5	1, 13	9	9,7, 10	11	12
*Mean ± standard deviation of three determinations; DL = dried leaf material; EP = edible portion; FL=fresh leaf material; – = not recorded							

- Chekurmanis leaves are rich in vitamins. It is very rich in Vitamin A (9510 IU) and Vitamin C (247 mg/ 100 g fresh leaves) [17]. Evidences have suggested that the plasma levels of Vitamin C in large sections of the population are sub-optimal for the health protective effects. As it cannot be synthesized by human system, our body entirely depend upon dietary sources to meet the needs of this vitamin. Apart from the well-known roles in vision, Vitamin A is also important for several other physiological processes including foetal development, cell growth and regulation of immune system [3]. So, chekurmanis can be a potent source of these two vitamins for mankind. Vitamin content of chekurmanis leaves as reported by different authors are given here under in Table 4.

Table 4: Vitamin content of chekurmanis leaves

Constituent	Content						
	(100 g EP)			(100 g FL)			
Vit A[a]	10000.00	9510.00	-	-	-	-	-
Vit B_1[b] (Thiamine)	0.23	0.48	-	-	-	-	-
Vit B_2[b](Riboflavin)	0.15	0.32	-	-	-	-	-
Vit B_3[b](Nicotinic acid)	-	2.60	-	-	-	-	-
Vit C[b] (Ascorbic acid)	136.00	247.00	56.10	190.83	-	22.00	314.30
Vit E[b](a-tocopherol)	0.43	-	-	-	42.68	-	-
Provitamin A[a] (Retinol equivalent)	-	1889.00	-	-	-	-	-
References	9, 10	8,18	11	13, 14	12, 19	1	5
[a]IU; [b]mg; – = not recorded.							

A number of studies have conclusively established the rich polyphenolic content of the edible leaf of chekurmanis (Table 5). Accumulating biochemical, clinical and epidemiological evidences support the chemoprotective potentials of these phenolic anti-oxidants against oxidative stress-mediated disorders. About 80.5-95.7% of the total leaf-phenolics have been determined to be flavonoids. Anthocyanins and tannins are the other classes of plant phenolics reported.

Table 5: Polyphenolic contents of chekurmanis leaves

Tannins		Total Phenolics		Flavonoids	Anthocyanins	Reference
(mg/100 g)	(mg GAE/100 g)	(mg CE/100 g)	(mg CA/100 g)	(mg/100 g)	(mg/100 g)	
88.68	1150.95[a]	-	-	-	82.94[§]	5
-	1150.00[b]	-	-	1040.00[¶]	-	11
-	149.00[a]	-	-	143.00	-	13
-	138.01[a]	-	-	142.64	1.53[a]	14
-	-	-		576.00	-	20
-	1900.00[a]	-	-	1530.00	-	21
-	2300.00[b]	-	-	-	-	22
-	-	2930.00[c]	-	-	-	23
-	-	2840.00[d]	-	-	-	23
-	-	-	-	785.00	-	24

GAE = Gallic acid equivalent; CA = chlorogenic acid equivalent; CE = catechin equivalent; [§]cynidine-3glucoside equivalent; [a]fresh leaf material;[b]dried leaf material; [¶]Rutin equivalent; [c]aqueous extract;[d]MeOH extract; – = not recorded.

- Chekurmanis leaves are rich in carotenoids. As carotenoids are also pigments that cannot be synthesized by humans, they must be essentially derived through the diet and green leafy-vegetables appear to be the best dietary sources. The total carotenoids content of the edible leaves, recorded in literature, are summarized in Table 6.

Table 6: Carotenoid contents of chekurmanis leaves

Carotenoids	Content					
	(µg/g DL)	(mg/100 g FL)	(mg/100 g DL)	(mg/100 g FL)	(µg/100 mL FL)	(µg/100 g EP)
Total carotenoids	358.9 ± 21.4	5.1 ± 0.07	-	-	19.4 ± 0.5	-
Beta-Carotene	96.40± 7.7	1.6 ± 0.02				
All-*trans*-alpha-carotene	-	-	-	-	-	1335 ± 878
All-*trans*-beta-carotene	-	-	-	-	-	10010 ± 219
Cis-beta-carotene	-	-	-	-	-	1312 ± 349
Lutein	51.59 ± 01.11	-	110.6 ± 4.1	19.5 ± 0.8	-	-
Neoxanthin	93.27 ± 4.54	-	-	-	-	-
Violaxanthin	117.63 ± 15.10	-	-	-	-	-
Zeaxanthin	-	-	48.1 ± 3.7	8.5 ± 0.8	-	-
Retinol equivalents	-	-	-	-	-	1889 ± 466[a]
References	25	14	26	26	5	18

*Mean ± standard deviation; DL = dried leaf material; EP = edible portion; FL = fresh leaf material; – = not recorded; [a]fresh edible portion; RE calculated as (1/6 all-trans-beta-carotene) + (1/12 all-trans-alpha-carotene) + (1/12 cis-beta-carotene).

Health benefits (Medicinal properties)

- A decoction of its roots is often recommended in case of fever in rural areas [17].
- Pounded roots and leaves are reported to be used as a poultice for ulcers in the nose [17].
- In some areas of South India, the juice of the leaves pounded with the roots of pomegranate and the leaves of jasmine is used in eye troubles [17].
- The dark green leaves provide a rich source of chlorophyll which is avaluable blood building element, cell rejuvenator, beneficial to blood circulation and for regular bowel elimination [27].
- As discussed earlier, they are rich in various micronutrients and phytochemicals having anti-oxidant properties which offer protection against heart disease and certain types of cancer [28]. For example, Beta-carotene prevents lung and skin cancer, niacin prevents recurrent heart attacks, ascorbic acid helps in improving immune system and riboflavin for lesion treatment, etc.
- High blood pressure is lowered by eating raw leaves [27].
- Leaves aregiven as vegetable to nursing mothers to stimulate breast milk production. In lactating sheep also, it induced milk production [29].
- It is also beneficial to cure anaemia [27].

- It is suitable to prevent tiredness, to promote absorption from alimentary tract and to prevent chronic cardiovascular diseases [27].
- Leaves are reported to have anti-bacterial activity against *Klebsiella pneumoniae* and *Staphylococcus aureus* [27].
- The leaves are also found to contain the amino acids like lysine, methionine, tryptophan, phenylalanine, threonine, valine, leucine and isoleucine [3].
- Polyphenols present in the leaves have protective effects on human carcinogenesis, cardiovascular and renal disorders, memory and cognitive function, age-related neurological dysfunctions, such as Alzheimer'sdisease, ulcers and several other human ailments [3].
- Substantial amount of flavonoids present in the leaves are known to act at different development stages of malignant tumours by protecting DNA against oxidative damage, inactivating carcinogens, inhibiting the expression of the mutagenic genes and enzymes, which are responsible for activating procarcinogenic substances and activating the systems responsible for xenobiotic de-toxification [3].
- *Sauropusandrogynus* leaf is reported to contain considerable amounts of the alkaloid papavarine (580 mg per 100 gm fresh leaf). Papaverine is an alkaloid anti-spasmodic drug, used primarily in the treatment of visceral spasm and vasospasm (especially those involving the intestines, heart, or brain) and occasionally in the treatment of erectile dysfunction. It is used in the treatment of acute mesenteric ischemia [4].

Uses and foods prepared from chekurmanis

- The leaves, tender shoots and fruits are used either as raw or as cooked vegetable. Cooked leaves are acidic and are used in soup preparation in Java. A sweet meat is also prepared from the fruits there [17].
- In Indonesia, the leaves are reported to be used for giving light green colour in pastry and fermented rice [17].
- Leaves are used in sandwiches, salads, curries, meat, rice & curry dishes, scrambled eggs, omelettes, pickles, casseroles, stir fries and as a garnish [1].

- Vietnamese are said to cook it with crab meat, minced pork or dried shrimp to make soup [3].
- In the Malaysian multi-racial cultures, any vegetable usually eaten raw in the form of salads is called an 'ulam' and chekurmanis is a notable ulam there [3].
- The inhabitants of north-eastern regions of Thailand are reported to eat the fresh tender or soft boiled leaves with sauce of chilli [3].
- Chekurmanis leaves are used for preparing cutlet in southern part of India [30].
- The leaves and succulent young tips of the plant reportedly possess a pleasant taste, similar to fresh garden peas and slightly nutty and these are normally eaten like tropical asparagus, raw in the form of either salads or steamed in Kerala and Karnataka [3]. In Kanyakumari district of Tamil Nadu, leaves are generally used as cattle or poultry feed [17].

Anti-nutritional factors

As the leaves contain significant amount of 'pipevarine', sometimes excessive consumption of raw leaves may cause dizziness, drowsiness, constipation and respiratory disorders [4].

Conclusion

Climate change and population growth impede progress toward achieving food and nutritional security. Production of an expanded range of food crops can reduce risk and enhance food security. Diversified diets, based on a range of crop species, are essential for nutritional security. Leafy and under-exploited vegetables like ckekurmanis in particular have appreciable levels of 'macro', 'micro' and 'phyto' nutrients which can help in alleviating malnutrition and significantly contribute to nutritional security. So, the consumption of this green leafy vegetable should be promoted, especially among the vulnerable sections of the population because of its cheap price. This could be a highly feasible strategy to combat problems of malnutrition in our country in future.

References

1. Sheela K., Nath K.G., Vijayalakshmi D., Yankanchi G.M. and Patil R.B. (2004). Proximate composition of under-utilized green leafy vegetables in southern Karnataka. *J. Hum. Ecol.* 15: 227-229.
2. Verma A. and Sood S. (2018). Chekurmanis is a mineral rich leafy green. *Indian Hort.* 63: 22-23.
3. Petrus A.J.A. (2013). *Sauropus androgynous* (L.) Merrill- a potentially nutritive functional leafy vegetable. *Asian J. Chem.* 25: 9425-9433.
4. Padmavathi P. and Rao M.P. (1990). Nutritive value of *Sauropus androgynous*leaves. *Plant Foods Hum. Nutr.* 40: 107-113.
5. Singh S., Singh D.R., Salim K.M., Srivastava A., Singh L.B. and Srivastava R.C. (2011). Estimate of proximate composition micronutrients and phytochemical compounds in traditional vegetables from Andman and Nicobar Islands. *Int. J. Food Sci. Nutr.* 62: 765-773.
6. Benjapak N., Swatsitang P. and Tanpanich S. (2008). Determination of anti-oxidant capacity and nutritive values of Pak-Wanban (*Sauropusandrogynus* L. Merr.). *K.K.U. Sci. J.* 36: 279-289.
7. Hoe V.B. and Siong K.H. (1999). The nutritional value of indigenous fruits and vegetables in Sarawak. *Asia Pac. J. Clin. Nutr.* 8: 24-31.
8. Khare C.P. (2007). Indian Medicinal Plants-An Illustrated Dictionary. Springer-Verlag, Berlin/Heidelberg, Germany, p. 407.
9. Medicinal plants of Sarawak, SC-Nature Biotech SdnBhd (2011). URL:http://sarawakmedplants.com.my/
10. AVRDC, Know your indigenous vegetables, Leaflet, Sauropus, AVRDC-The World Vegetable Center, Shanhua, Taiwan No. 5.
11. Maisuthisakul P., Pasuk S. and Ritthiruangdej P. (2008). Relationship between anti-oxidant properties and chemical composition of some Thai plants. *J. Food Comp. Anal.* 21: 229-240.
12. Tangkanakul P., Trakoontivakorn G., Auttaviboonkul P., Niyomvit B. and Wongkrajang K., Kasetsart J. (2006). Anti-oxidant Activity of Northern and North-eastern Thai Foods Containing Indigenous Vegetables.*Kasetsart J. Nat. Sci.* 40: 47-58.
13. Andarwulan N., Batari R., Sandrasari D.A., Bolling B. and Wijaya H. (2010). Flavonoid content and antioxidant activity of vegetables from Indonesia.*Food Chem.* 121: 1231-1235.

14. Andarwulan N., Kurniasih D., Apriady R.A., Rahmat H., Roto A.V. and Bolling B.W. (2012). Polyphenols, carotenoids, and Ascorbic acid in under-utilized vegetables.*J. Functional Foods*. 4: 339-347.

15. Yarger L. (2007). Lagos Spinach. ECHO Technical Note. pp 1-8.

16. Suguna G. (2005). Further investigations on the polyphenolic constituents of*Cleomegynandra* and *Sauropusandrogynus*. Dissertation, Pondicherry University, Puducherry. pp. 49-61.

17. Ramachandran C., Peter K.V. and Gopalkrishnan P.K. (1980). Chekurmanis a multi-vit. leafy vegetable. Indian Hort. 25: 17-18.

18. Hulshof P.J.M., Xu C., Bovenkamp P.V.D., Muhilal B. and West C.E. (1997). Application of a validated method for the determination of pro-vitamin A carotenoids in Indonesian foods of different maturity and origin. *J. Agric. Food Chem*. 45: 1174-1179.

19. Ching L.S. and Mohamed, S. (2001). Alpha-tocopherol content in 62 edible tropical plants. *J. Agric. Food Chem*. 49: 3101-3105.

20. AVRDC Report, (2008). AVRDC Publication Number 08-702. AVRDC The World Vegetable Center, Shanhua, Taiwan, p. 75.

21. Subhasree B., Bhaskar R., Laxmi Keerthana R., Lijina Susan R. and Rajashekharan P. (2009). Evaluation of anti-oxidant potential in selected green leafy vegetables. *Food Chem*. 115: 1213-1220.

22. Wong S.P., Leong L.P. and Koh J.H.W. (2006). Anti-oxidant activities of aqueous extracts of selected plants. *Food Chem*. 99: 775-783.

23. Mai T.T., Thu N.N., Tien P.G. and Chuyen N.V. (2007). Alpha-glucosidase inhibitory and anti-oxidant activities of Vietnamese edible plants and their relationships with polyphenol contents. *J. Nutr. Sci. Vitaminol*. 53: 267-276.

24. Miean K.H. and Mohamed S. (2001). Flavonoid (myricetin, quercetin, kaempferol, luteolin, and apigenin) content of edible tropical plants.*J. Agric. Food Chem*. 49: 3106-3112.

25. Fatimah A.M.Z., Norazian M.H. and Rashidi O. (2012). Identification of carotenoid composition in selected 'ulam' or traditional vegetables in Malaysia. *Int. Food Res. J*. 19: 527-530.

26. Liu Y., Perera C.O. and Suresh V. (2007). Comparison of three chosen vegetables with others from South-East Asia for their lutein and zeaxanthin content. *Food Chem*. 101: 1533-1539.

27. Paul M. and Beena Anto K. (2011). Antibacterial activity of *Sauropusandrogynus* (L.) Merr. *Internat. J. Plant Sci*. 6: 189-192.

28. Saxena R. (1999). How green is your diet? *Nutrition,* 33: 9.

29. Suprayogi A. (2000). Studies on the biological effects of *Sauropu-sandrogynus* (L.) Merr: Effects on milk production and the possibilities of induced pulmonary disorder in lactating sheep. 2000. ISBN:3 89712-941-8, Göttingen (DE): Cuvillier Verlag.

30. Krishnendu J.R. (2006). Promoting consumption of green leafy vegetables among rural women through participatory approach. M.Sc. thesis, Department of Home Science, Kerala Agriculture University, Thrissur, p. 31.

❑❑❑

Chapter 12

Bioprospecting Wild Plant Resources to Ensure Food and Nutritional Security

Gopal Shukla, Vineeta, Nazir A. Pala, Abha Manohar K., Anju Puri[*] and Sumit Chakravorty

Department of Forestry, Uttar Banga Krishi Vishwavidyalaya, Pundibari, Cooch Behar, West Bengal, India, [*]Baring Union Christian College, Batala Punjab, c_drsumit@yahoo.com; gopalshukla12@gmail.com

ABSTRACT

Wild edible plant (WEP) resources not only provide vital food supplies, employment, health care and raw materials for billions in the tropical world but also valuable raw materials, cash crops and timber for the developed world as well. Most of these are cheap and readily available with vibrant taste appeal. Their potential in nutritional, medicinal, therapeutic and industrial values is now well-recognized. WEPs are used or available for income smoothing that determines vulnerability to poverty. Gathering wild plants for food is a major livelihood strategy of indigenous and rural poors in tropical countries. The WEPs rescue these people by providing life-sustaining food during times of food crisis or famine which ensures food and nutritional security. Many value added products can also be obtained from WEP species. However, many of these plant species are now threatened and in the verge of extinction due to over extraction, deforestation, and pollution. This warrants sustainable conservation, documentation of indigenous knowledge base and its subsequent sustainable utilization. Today, almost all modern human food is based on a limited number of crops and since food and phyto-resources are shrinking globally, it is imperative to find new alternatives. Although these species continue to be maintained by cultural preferences and traditional practices but they still remain inadequately characterized and neglected by research and conservation. Lack of attention indicates that their potential value is under-estimated and under-exploited. Research on the utilization aspect will help to identify new uses, improve production and also promote welfare of the local community. Unfortunately, little research has been done on identification and utilization. It also places them in danger of continued genetic erosion and disappearance. This would further restrict development options for poor. Therefore, exploration and listing of these plants with their ethno-biological values are important for knowing and evaluating their economic potential. Thus, bio-prospecting the WEP resources can be a potential way of ensuring food and nutritional security in near future.

Keywords: Wild edible plant (WEP), Indigenous community, Bio-prospecting, Food and nutritional security, Conservation, Utilization, Value addition

1. Introduction

Since time immemorial human society is totally depending on the biodiversity. Wild biodiversity supplements the needs of residing community even though their primary reliance is on agriculture [1-2]. Wild plants are gathered and consumed by indigenous people all around the world [3-10]. Each major geographical and agro-ecological region is bestowed with a wide range of wild edible plant (WEP) or indigenous food plants species, some of which though quite important locally, are seldom known outside that region [11-13]. Most of the wild edibles are collected from sacred groves and other forest areas [14-16]. WEPs are the species those are neither cultivated nor domesticated, however, are available in their wild natural habitat and used as sources of traditional food for children and for the poor in rural areas [17-18].

Different WEPs have been significant roles in all geographical regions of world throughout human history maintaining socio-economic development and livelihood security for many people in developing countries providing income and closing food gaps during periods of drought and scarcity [19-22]. In South, South-East and East Asian region WEPs significantly contributes to people's livelihoods supporting household income, food, employment, traditional medicine, timber, livestock fodder and also plays a role in the stability of ecosystems [23]. They form the major supplier of subsidiary nutrition to the indigenous and rural communities, as the common cultivated plants are less familiar and inaccessible to them [24]. These wild edibles are also sometimes nutritionally superior to some cultivated species [25].

Traditional foods are those foods originating locally in an area with respect to the country, region, district or even smaller zone. The definition includes all indigenous food plants found in an area with regard to food products, vegetables, fruit trees and yield products, carbohydrates, vitamins and minerals [26]. As WEPs have high nutritional and neutraceutical properties, they are now globally significant for their potential to biodiversity conservation, food security, nutrition, dietary and culinary diversification, health and income generation but unfortunately their use is still limited relative to their economic potential as their value still remain unknown [27-40]. The information on wild edible plants is scattered in botanical monographs, informal notes and tribal oral tradition. The useful properties of non-domesticated crops known in local communities, require proper study and documentation in order to validate and spread this useful knowledge. Such information can then be utilized to address food insecurity and formulate strategies for pro-poor low cost development program [35].

1.1. Exploration and Status of Wild Edibles

WEPs have been on focus in recent decades globally and also have interest in documenting ethnobotanical information on these neglected WEP resources [41]. Of 2, 50,000-3, 00,000 higher plant species known to man, only a few hundred species have been fully domesticated and many thousands are still gathered from the wild. Only 103 species of food plants contribute 90% of national per capita supplies while 20-30 of these species are regarded as the staples [42]. Wild edibles are consumed mainly by indigenous and rural poors and therefore, also known as "poor people's food". The wild plants consumed by the indigenous and rural communities are not familiar to the urban communities [43, 44]. Approximately, 75, 000 species of plants world-wide are believed to be edible [45]. Ethnobotanical surveys of wild plants reveals more than 7000 species were used as human diet [46, 47]. In India, about 600 plant species are known to be WEPs [48] which were documented from different parts of India [49-55] along with 373 wild relatives of crop species [56].

The Himalayan region is rich in wild growing plants, those are used for food and other subsistence needs by the local communities [57]. Over 675 wild edible species belonging to 384 genera and 149 families were reported from Indian Himalaya [58, 59]. The Sikkim Himalayas was reported to have 190 food plants growing in wild habitats [19]. There are 3000 tropical fruits worldwide with an untapped potential [60]. South-east Asia was listed with 795 species of wild edible fruits and nuts [61]. Many publications have emphasized on the diversity, value, conservation and propagation of wild edible plants since 19th century [62-91]. Table 1 enlists some common WEPs.

Table 1: Some common wild edible plants

S. No	SN/F/Fo	Pu	P	R
1	*Adenanthera pavonina* L. Mimosaceae, T	Sd	Tannin, protein, alkaloid, glycoside, lignin, cellulose	56
2	*Abrus precatorius* L. Fabaceae, S	L	Alkaloid, glycoside, flavonoid, phenol	56
3	*Abutilon indicum* L. Malvaceae, S	L	Steroid, sapogenin, carbohydrate, flavonoid	56
4	*Aegle marmelos* L Rutaceae, T	Ft	Alkaloid, phenylpropanoid, terpenoid	35, 56, 92-95, 96, 97
5	*Aerva lanata* L. Amaranthaceae, H	L	Alkaloids, flavonoid, methyl grevillate, lupeol, lupeol acetate benzoic acid, β-sitosteryl acetate, tannic acid.	56
6	*Agave americana* L. Agavaceae, S		Carbohydrate, reducing sugar, alkaloid, volatile oil, tannin, steroid, glycoside, flavonoid	40
7	*Angelica glauca* Edgew. Apiaceae, H Endangered*	L	Coumarin, saponin, alkaloid, tannin, carbohydrate, protein, fat, steroid glycoside, flavonoid	40
8	*Artocarpus heterophyllus* Lam. Moraceae, T	Ft	Flavonoid, sterol, prenylflavone	56, 94, 95
9	*Asparagus filicinus* Buch. -Ham ex Roxb. Liliaceae, H		Essential oil, asparagine, arginine, tyrosine, flavonoid, resin, tannin	40
10	*Averrohoea carambola* L.Oxalidaceae, T	Ft	Saponin, alkaloid, flavonoid, tannin	92, 95-97
11	*Baccaurea sapida* Muell Arg. Euphorbiaceae, T		Phenol, flavonoid	
12	*Bambusa bambos* L. Poaceae, B	Cu	Tannins, steroids, phenols, glycosides, flavanoids, carbohydrates and proteins annins, steroids, phenols, glycosides, flavanoids, carbohydrates and proteins Tannin, steroid, phenol, glycoside, flavonoid, carbohydrate, protein	56
13	*Bauhiniapurpurea* L. Caes alpiniaceae, T Least concern*	Fl	Carbohydrate, protein, alkaloid, saponin, phenol, flavonoid, fat, anthocyanin, terpenoid, steroid	40, 56,

14	*BauhiniaVariegata* L. Caesalpiniaceae, T Least Concern*	Fl	Steroid, glucoside, tannin, carbohydrate, amide, reducing sugar, vitamin C, crude protein, fiber, calcium, phosphorus, quercetin, rutin, quercitrin	40, 56,
15	*Bauhinia retusa Roxb* Caesalpiniaceae, T	Fl	Steroid, glucoside, tannin, carbohydrate, amide, reducing sugar, vitamin C, crude protein, fiber, calcium, phosphorus, quercetin, rutin, quercitrin	56
16	*Boerhavia diffusa* L. Nyctaginaceae, H	L	Flavonoid, glycoside, alkaloid, rotenoid, steroid, triterpenoid, lipid, lignin, carbohydrate, protein, glycoprotein	56
17	*Capparis zeylanica* L. Capparaceae, C	Ft	Fatty acid, flavonoid, tannin, E-octadec-7-en-5-ynoic acid, alkaloid,saponin glycoside, terpenoid, p-hydroxybenzoic,saponin, syringic, vanillic, ferulic, p-coumanic acid	35, 56
18	*Carissa sp.* Apocynaceae, S		Tannin,2-phenly ethanol, linalool, β-caryophylline, isoamyl alcohol, benzyl acetate	95
19	*Cassia tora* L. Caesalpiniaceae, S	L	Saponin, glycoside, protein, tannin, carbohydrate	56
20	*Centella asiatica* Apiaceae, H Least concern*	L	Carbohydrate, tannin, steroid, terpenoid, alkaloid, flavonoid, cardiac glycoside, saponin	56
21	*Chaerophyllum villosum* Wall exDC Apiaceae, H		Essential oil	40
22	*Chenopodium album* L. Chenopodiaceae, H	L	Flavonoid as phenolic amide, saponin, cinnamic acid amide, alkaloid chinoalbicin, apocortinoid, xyloside, phenol, lignin	56
23	*Colocasia esculanta* L.Schott. Araceae, H	L, T	Alkaloid, glycoside, flavonoid, terpenoid, saponin, phenol	40, 56,
24	*Cordia myxa* L. Boraginaceae, S		Carbohydrate, tannin, protein, glycoside, alkaloid, flavonoid, phenolic compound, saponin,	92
25	*Dillenia indica* L Dilleniaceae, T	Ft	Alkaloid, glycoside, steroid, flavonoid, tannin, saponin, phenol	96-98
26	*Dioscorea bulbifera* L. Dioscoreaceae, H	T	Flavonoid, terpenoid, saponin, steroid, cardiac glycoside, carbohydrate	40
27	*Diospyros melanoxylon* Roxb. Ebenaceae, T	Ft, Sd	Alkaloid, flavonoid, tannin, terpenoid, essential oil	56, 92

28	*Diplazium esculentum* Retz. Dryopteridaceae, Fr Least concern*	L	Alkaloid, reducing sugar, anthraquinones, anthranol glycoside, cyanidin, phenol, saponin, protein	40
29	*Elaeocarpus floribundus* Bl. Elaeocarpaceae, T	Sd	Saponin, tannin, cardiac glycoside, flavonoid, steroid,	96, 97
30	*Emblica Officinalis* Gaertn. Euphorbiaceae, T Vulnerable*	Ft	Fixed oils phosphatide, essential oil, tannin, mineral, vitamin, amino acid, fatty acid, glycoside	56, 94
31	*Eriobotrya japonica* Lindl. Rosaceae, T **(Reference)**	Ft	Essential oil, triterpene, flavonoid	
32	*Fagopyrum cymosum*Trev. **(Family and form)**	Sd	Flavonoid, steroid, organic acid	40
33	*Ficus hispida* L. Moraceae, T	Ft	Alkaloid, carbohydrate, protein, amino acid, sterol, phenol, flavonoid, gum, mucilage, glycoside, saponin, terpene	35, 56, 94
34	*Ficus racemosa* L. Moraceae, T	Ft	Flavonoid, triterpenoid alkaloid, glycoside, sterol, tannin	35, 94, 97
35	*Flacourtia indica* Merr Flacourtiaceae, S	Ft	Phenol, tannin, carbohydrate, saponin, alkaloid, protein, flavonoid, phytosterol, furanoid, coumarin, quinine	35,94, 96, 97
36	*Garciniacowa* Clusiaceae	Ft	Flavonoid, phenol	56,94-96
37	*Gmelina arborea* Roxb.Verbenaceae, T	Ft	Steroid, triterpenoid, saponin, protein, phenol, flavonoid, carbohydrate	55
38	*Grewia asiatica* L. Tiliaceae, T	Ft	Mucilage, flavonoid, alkaloid	35, 97
39	*Grewia subaequalis* Baill. Tiliaceae, T		Polyphenol, anthocyanin, flavonoid, tannin	95
40	*Indigoferapulchella* Roxb. Fabaceae, S	Fl	Terpenoid, flavonoid, steroid	40
41	*Lantana camara* L. Verbenaceae, H	Ft	Alkaloid, phenol, flavonoid, tannin, saponin, terpenoide, phlobetanin, coumarin	56
42	*Lepidium sativum* L. Cruciferae, H	L	Carbohydrate, phenol, flavonoid, alkaloid, protein, saponin, lipid	40
43	*Madhuca indica*J.F.Geml.Sapotaceae, T	Fl, Sd	Alkaloid, tannin, protein, carbohydrate	35, 95
44	*Malpighia punicifolia* L. Malphigiaceae, S	Ft	Phenol, saponin	

45	*Melia azedarach* L. Meliaceae, T	L, F	Alkaloid, tannin, saponin, phenol, glycoside, steroid, terpenoid, flavonoid	56
46	*Mesuaferrea* L. Cluslaceae	Ft	Alkaloid, phenol	56
47	*Mimus opselengi* L. Sapotaceae	Ft		56, 92, 97
48	*Musa balbisiana* Musaceae, H	Ft, Fl, L, Ps	Saponin, glycoside, fatty acid	95
49	*Nasturtium officinale* R. Br. Cruciferae	L	Phenol, flavonoid	40
50	*Nipa fruticans* (Thumb.) Wur mb Arecaceae	L	Phenol, flavonoid	
51	*Oxalis Corniculata* L. Oxalidaceae, H	L	Flavonoid, tannin, phytosterol, phenol, glycoside, fatty acid, galacto-glycerolipid, volatile oil	56
52	*Paspalum scrobiculatum* L. Poaceae, H Least concern*	Sd	Tannin, phenol, saponin, protein, carbohydrate	40
53	*Phytolac caacinosa* Roxb. Phyt olaccaceae, H	R	Saponin	40
54	*Passiflora foetida* L. Plassifloraceae, C	Ft	Carbohydrate, glycoside, flavonoid, resin, balsam, alkaloid, phenol	56
55	*Rhododendron arboreum* Smith. Ericaceae, T	Fl	Polyphenol, quinone, sterol, triterpene, fatty acid	40
56	*Schleichera Oleosa* Lour. Sapindaceae, T	Ft, Sd	Phenol, flavonoid, tannin, fatty acid, sterol	56
57	*Shorearobusta* Gaertn. Dipter ocarpaceae	Ft, Sd	Phenol, tannin, saponin, glycoside, alkaloid	55
58	*Sida cordifolia* Burm. f Malvaceae, H	L	Alkaloid, flavonoid, tannin	56
59	*Simarouba glauca* D.C. Simaroubaceae, T	S	Carbohydrate, protein, alkaloid, phenol, flavonoid, saponin, tannin, phytosterol, glycoside	95
60	*Spondias cytherea* Sonn. Anacardiaceae, T	Ft	Phenol, sterol, triterpene, saponin, essential oil, amino acid, polysaccharide	92, 97
61	*Streblusasper* Lour Moraceae, T	Fl, Ft	Alkaloid, glycoside, sterol	56, 94
62	*Terminalia Bellerica* Gaert. Combretaceae, T	Ft	Flavonoid, alkaloid, terpenoid, tannin	56, 94, 96
63	*Urtica parviflora* Roxb. Urticaceae, H	L	Flavonoid, tannin, scopoletin, sterol, fatty acid, polysaccharide, isolectin, sterol	
64	*Ziziphus mauritiana* Rhamnaceae, T	Ft	Glycoside, saponin, phenol, lignin, tannin	95, 96
65	Mushrooms of Ascomycetes and Basidiomycetes, Fn	Fb	Phenol, polypeptide, terpene, steroid	95

SN/F/Fo: Scientific name/Family/Form; T: tree; S: shrub; H: herb; B- bamboo; C- climber; Fr- fern; Fn- fungus; Sd- seed; L- leaf; Ft- fruit; T- tuber; Cu- culm; Ps- pseudostem; R- root; Fl- flower; Fb- fruiting body; *Conservation status according to IUCN Red Data Book; Pu- part used; P- phytochemical present; R- reference

The unrecognizable nutritional, therapeutic and other values (Fig. 1&2) needs to be explored, as it can an alternative source of food security and can preserve the cultural diversity of a region as well [92, 98].

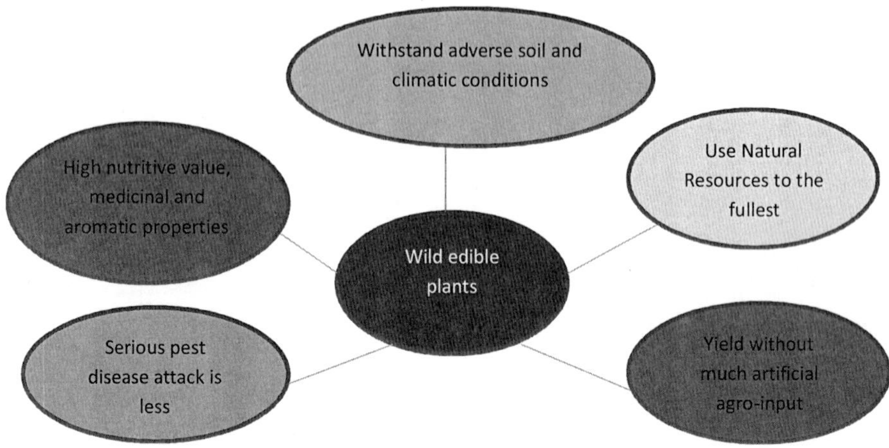

Fig. 1: Potential values of wild edible plants

It can be helpful to overcome the constraints of wider production to ensure dignified living of poor people combating hunger and malnutrition. Now there has been an increasing concern to diversify agriculture by exploring the possibilities of wild plant resources and promote their utilization. Many international research organizations like International Centre for Under-utilized Crops (ICUC), Global Facilitation Unit (GFU), Convention on Biological Diversity (CBD) and Crops for the Future (CFF) now have initiated research and gather information on these wild edibles[95]. The strategy is now to utilize the lands which are unsuitable for growing high demanding crops to domesticate and popularize these plants to bring under cultivation in order to diversify the present day agriculture.

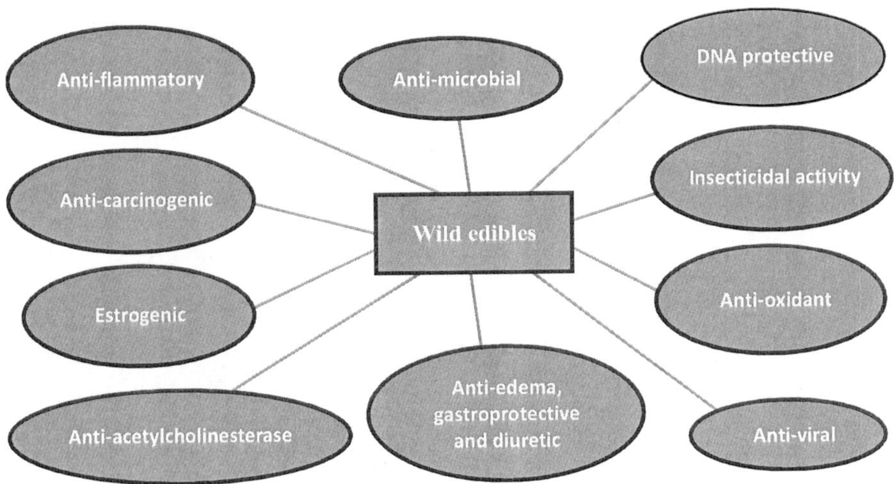

Fig. 2: Bioactivities of wild edibles

It can be helpful to overcome the constraints of wider production to ensure dignified living of poor people combating hunger and malnutrition. Now there has been an increasing concern to diversify agriculture by exploring the possibilities of wild plant resources and promote their utilization. Many international research organizations like International Centre for Under-utilized Crops (ICUC), Global Facilitation Unit (GFU), Convention on Biological Diversity (CBD) and Crops for the Future (CFF) now have initiated research and gather information on these wild edibles [95]. The strategy is now to utilize the lands which are unsuitable for growing high demanding crops to domesticate and popularize these plants to bring under cultivation in order to diversify the present day agriculture.

1.2. Folk Knowledge and Awareness of Wild Edibles

Folk or indigenous knowledge on utilization of wild edible plants as a food source is an integral part of the culture of indigenous and rural people [46, 99-101], who gather and consume wild edible plants as snacks and at times of food scarcity [5, 9, 102-105]. They are knowledge able of traditional plant uses, one of which is the use of wild edible plants as a food source [9, 106-108]. Folk knowledge of plants and their environment have a great value in identifying species for domestication, utilization and commercialization [109, 110]. It is important for people to know the prevailing traditional food in their areas and how they can be improved for better sustainable food security/nutrition [26]. Awareness and knowledge of traditional foods is also important for sustaining the development and utilization of WEP [111, 112]. For instance, knowledge of traditional food plant is important to promote the cereal staples in a more sustainable way. This is because,such know-how will enable for better food security/ nutrition at household level [26]. These staples should be mixed with other food sources to have a balanced diet [113].

1.3. Importance, Utilization and Value Addition of Wild Edibles

Sustainable utilization of WEPs depends on existing folk knowledge and the economic pursuit of the people [47, 114-115]. It is also important for people to know the prevailing traditional food plants in their areas and how they can be improved for sustainable food security/nutrition [111,116, 117]. About 3000 plants species are used as food but only about 150 species have been cultivated and less than 10 plant species are meeting over 90% of the world food demand [118-120]. Many such food resources and valuable plants are still to be explored [121]. The rich diversity of wild plant species in India and elsewhere offers opportunity to indigenous population a variety in family diet and contributes to household food security [120, 122-124].

FAO estimated around one billion people use wild foods in their diet [125]. The leaves, fruits, roots & other parts wild plant species are gathered mostly for home consumption by forest dwellers, indigenous people and marginalized rural communities for food, medicinal and other use like house construction and fuel wood [126, 127]. These collections contribute a major proportion of indigenous dishes globally [58, 128-131]. The collection of WEPs is location specific depending on its availability at different regions but is not nutrition specific, collected only to satisfy hunger [24, 128, 132, 133]. The contributions of wild species to household food supply vary according to region and groups [111, 134-139]. The benefits accrued from the collection of the WEPs mainly depend on the quantity and extent of its consumption [140-147]. It was also reported that among other factors, the degree to which WEPs are used or available for income smoothing determines vulnerability to poverty [148]. Most of the forest dwellers in Bihar, Odisha, Madhya Pradesh, West Bengal and Himachal Pradesh depend on forest for half of their annual food requirements [149, 150].

Collecting wild plants for food is prime livelihood strategy of many rural poors, forest frontier cultivators, continuous croppers or hunter gatherers, forest dwellers and indigenous communities in tropical world [38]. These people meet their subsistence needs from their forest collection besides growing a few crops [151]. Even Mediterranean Europe collecting WEPs is a common practice [152]. These wild plants are especially important for these poor people to sustain their food requirement during food crisis or famine ensuring their food security [2, 38, 132, 153-168]. In drier areas of the world, WEPs are critically important, as they are source of life-sustaining food especially during famines caused by extended drought periods [169-172]. In India, it is reported that WEPs are significant to ensure food and nutrient security of indigenous people and rural poors [128].

More than a billion people are under-nourished globally, of which mostly are rural residents chronically malnourished and threatened with food insecurity [164, 173-177]. WEPs served as "buffer food" as they had rescued people during floods, drought and major famines in Asia, Africa and Latin America [34, 38, 42, 156, 160, 163, 165, 178-186]. These reports are not a historical anecdote or aberration but a present day reality for people across the globe as was indicated in studies from Bosnia-Herzegovina, Congo, Burma, North Korea, Colombia and many other countries [166, 187]. In exceptional cases, WEPs were the only source of food available [42].

WEP species provide nuts, fruits, spices, leafy vegetables, edible oils and beverages which are richly fortified with minerals (iron, zinc), vitamins (A, C, E, folic acid), anti-oxidants, phenols, flavonoid, anthocyanin, β-carotene, proteins and carbohydrate and sometimes even reported with higher nutritional value than several known common vegetables and fruits [34, 69, 93, 94, 133, 188-192]. Many studies conducted on WEPs all around the world concluded critical importance of biodiversity to provide low-cost nutritional supplement to ensure food and nutritional security globally [7, 33, 34, 47, 48, 59, 67, 103, 111, 129, 133, 181, 193-237].

WEPs are also nutritionally important for the collectors but more importantly can be potential contributor to their cash economy [136, 137, 238-245]. In today's climate changed world, the nutritional importance of WEPs is duly recognized because of frequent agricultural crop failures [142], they are pesticides free [246] and available year round [136, 137, 147. 162]. WEPs especially the wild relative of the agricultural crop can be beneficially used in crop improvement program needed to adapt climate change conditions [81]. These plants are also used for other than food purposes like medicines, cosmetics, soap, essential chemicals, fuel wood, building materials, poles, fodder, live fencing, bee forage and soil fertility improvement [163]. Many WEP species especially the wild edible fruit trees are contributing in climate regulation by affecting rainfall cycle and carbon sequestration. Moreover, they are easier to transport than timber or wood products because of higher value-to weight ratio [247].

Traditional ways of processing of WEPs exist across the globe [248] as a way of maximizing its utility [249]. Processing and value addition can also ensure employment and generate income [250-254]. WEPs can be used to make by-products, as they are nutritionally rich which can feed food, beverage and pharmaceutical industries [19, 255]. Their value addition can fetch premium price in the Indian and overseas aiding in income generation [103, 151, 256-258]. The term 'bioprospecting' is commonly used to assess the economic potential of different plant species and their value-addition [259]. Value addition at local level in potential wild edibles has begun to attract attention as being one of the income generating components of the non-farm part of the rural economy. These wild plant resources can be alternative source of economy to many small scale and large scale based industries. Many value added products can be prepared from these wild edibles as they rich in neutraceutical. The main processed products can be dried products, chips, jam, jelly, leather, fruit drinks, chutneys, candies, pickles, squashes, nectar, concentrate, fermented or non-fermented beverages, candy, flour, medicines and nutrient supplements [251, 260-263].

1.4. Threats to Wild Edibles

WEP resources and their indigenous use are in danger of being lost in areas, where environmental and cultural transformations have led to changes in feeding practices. Traditional plant knowledge has vanished over time as many indigenous communities have abandoned or changed their traditional customs [264-266]. It is being lost at an accelerated pace with changes in lifestyle, decreased subsistence use of wild species, changing occupational patterns of household members and disappearance of village elders [267]. Changes in land-use due to urbanization and habitat destruction, as well as the slash and burn system of traditional system with its associated shifting cultivation have been causing forest destruction and degradation [268]. Increased destruction of natural habitats will lead to disappearance of wild plants which will result in the loss of indigenous knowledge with their use

[132]. Modern market economy also contribute to the loss of folk knowledge of plant uses as markets allow access to substitutes for products made from plants [269].

The WEP species are now facing threats due to high population growth which is exerting high pressure on the natural resources [270, 271]. Humans have been using wild plants without managing or propagating them. This, coupled with lack of *ex situ* conservation strategies has led to loss of WEPs in it's the natural habitat [270]. It is no longer possible to leave it to nature to provide the genetic resources. Present utilization strategies of WEPs are not sustainable, for the systems were relying on nature to supply and conserve the resources. Moreover, commercial over harvesting, inappropriate regulatory frameworks, and land-use changes also threatening the WEPs with decreasing opportunities for its collection and devaluing its potential to address household food and nutrition needs (Johns and Eyzaguirre, 2006). These plant species are undervalued and under-used [272, 273] and tend to diminish in less remote areas [26]. Moreover, indigenous and rural communities lack adequate knowledge of or access to improved processing, packaging technologies, and marketing [256]. Unfortunately, no developed system of post harvest handling of WEPs also exists.

In spite of their diverse use and an important part of world food basket, the WEPs are excluded in agricultural policies, ignored in food policy and have been neglected by research, extension programmes and national statistics [41, 103, 274]. Such unrecognized status unable any policy to include WEP in it and stick to conventional food security strategies concentrating only staple foods [275]. Although, dietary documentation of WEPs though exists all around the world from ancient times [276-281] but on nutritional composition and therapeutic properties information is very less and moreover, also currently lags behind that of cultivated species [24, 128, 282-286]. Unfortunately, due to lack of attention and authenticated data, they remain commercially under-exploited and are threatened with extinction due to unsustainable use and habitat destruction. Further, relatively little research or attention has been given to the development of indigenous trees as crops with potential for wider cultivation [180, 244, 245]. Regardless of the numerous efforts to document WEPs and associated indigenous knowledge, under-estimation of value of these WEPs can lead to neglect of ecosystems that nurture them and the indigenous knowledge systems that are related to them [113].

Although the potential role of WEPs in providing fruit security has been identified, little data is available on their identification, use and management by the communities or user's preferences. Unfortunately, it is yet to introduce and evaluate the WEPs that have potential for cultivation considering the diverse agro-climatic conditions available in the country [287]. Full utilization of introduced germplasm depends on its successful conservation, multiplication and exploitation. This needs research support for evolving superior varieties, genetic conservation, development of production technology and post harvest management [273]. Not much work has been done on this aspect. The scientific community is still unknown about its potential. Studies on factors influencing specifically the species diversity of wild fruits and their documentation are in frequent [8, 132, 288].

A major problem in food policy at international arena is that the nutritional value of many wild foods are irreplaceable by store-bought foods [41], while ignoring the importance of WEPs to dietary diversity and food security [41, 228, 289]. In addition to this, the dynamics of wild food markets are also less understood [290, 291]. Wild edibles are mostly consumed domestically and if excess, are sold only in local markets for hard cash by the collector. There are many difficulties in popularization at market and at consumer level. Major fruit and vegetable distributors do not market the wild edibles. The consumers are unaware about these plant products, their mode of usage, expectation of sensory qualities, and mode of storage and ripening. Due to lack of sustained and informative research on the wild edibles, there are constraints on its utilization and popularity. These plants which can become an alternative source of food security are ignored by people or by market because of lack of sufficient information on its production and nutritional quality.

1.5. Promotion and Conservation of Wild Edibles for Food and Nutritional Security

Tropical forests are repository of WEPs and now are recognized as key to nutritional and food security, and livelihoods for the vulnerable citizens in these countries [1, 34, 181, 292-294]. However, systematic attempts to quantify this is very less to support integrated food security policy and practice which is due to less quantified information on diverse and complex policy frameworks linking forests and WEPs [1, 41, 295]. Bridging this information gap require more empirical understanding on the potential benefits of WEPs in terms of its dietary diversity, health, and biodiversity for better support to decision-makers [181, 228, 294].

WEPs not only are sources of supplemental food, nutritionally balanced diets, medicine, fodder and fuel, but also for developed methods of resource management which may be fundamental to the conservation of some of the world's important habitats [118, 203, 296]. Since food and phyto-resources are shrinking globally with increasing population, new alternatives are required for enriching resource base of our food basket. These species need to be conserved in their natural habitats. This requires exploration, listing of these plants with their ethnobiological value. This information will be helpful to understand and analyze the human-plant relationship and potential for their use in day-to-day life which is essential for their proper management and conservation [297-300]. Efforts should, therefore, be made to develop and promote these plants to reduce the loss of biodiversity in the country for improved nutritional status.

In areas where production is difficult, the use of indigenous plants and food materials can be promoted along with staples like rice, wheat, sorghum or millet to maximize the production and use of cereals. In addition to improved status of cereals, the above will contribute to conservation of the plant genetic diversity, maintenance of cultural diversity and improvement of the general well-being of communities and the environment. These in overall will ensure the food security and improved health status of the nation [26, 114]. The wild edibles need to be conserved through law or policy and creating awareness about the importance of these plants among the people both residing in forest or rural and urban. Their importance and use value are restricted only in the knowledge and locality of tribal and rural people. Malnutrition and hidden hunger robs in developing countries robs of the dignified living of its citizen. It is fundamental to recognize the fact that nutritional security and biodiversity are strongly linked for enlisting policy support to secure wild food use and preserve their habitats [1, 34, 41, 228, 233]. In addition to this, wild food and its resources should be integrated with national food policies in more customized ways as defensive strategies to increase food security [33, 34, 41, 185, 228, 233, 275].

Further, policy institutions should link biodiversity and natural resources to reinforce conventional agricultural response strategies at arange of scales through innovative multi-sectoral and multi-actor processes [41, 182, 294, 301, 302). This requires comprehensive food composition data especially critical for most vulnerable communities to malnutrition [21, 124, 303-305]. Fortunately, now United Nation's policy interventions has initiated using biodiversity sustainably integrating environmental sustainability and gender equality for achieving food security in its human nutrition program [301, 306]. However, policy interventions integrating WEPs and food security should be more at local landscape level, which needs more research and policy exploration [228, 301, 307, 308].

Interest in wild edibles has grown significantly with the increasing awareness in linking participatory biodiversity conservation with integrated rural development at landscape levels [302, 309-311]. Since traditional knowledge on WEPs is being eroded through acculturation and loss of plant diversity along with indigenous people and their cultural background which again needs research support on WEPs so as to preserve this knowledge for benefit of future societies [10]. Therefore, it is a high time to bring out better resource management through participatory sustainable conservation and the positive attributes of these important wild edibles to understand the increased contribution, they can make to the health while using as a medicine, nutrition and also as additional source of income through value addition of the marginal population.

More recently edible wild bio-resources are being viewed as untapped or under-utilized resources that could play a significant role in rural development, poverty alleviation, livelihood and nutritional security of local communities through bio-prospecting with the applications of suitable science and technological interventions [69, 97, 309, 312, 313]. The ability of a given wild bio-resources to continue meeting both subsistence and market needs, however, largely depends upon sustainable harvesting and appropriate management practices. Initiatives of government incentives can promote small-scale village enterprises for WEPs to encourage its post harvest processing and value addition which will also reduce over-harvesting and promote conservation [257,258]. It is required to develop 'nutrient sensitive' value chains integrating traditional and scientific knowledge systems with informal and formal institutions for better public support [302, 314]. Thus, it is important to treat bio-prospecting in the context of a strong benefit sharing system among industries/firms, the ecosystem and the projected communities in such manner so that the sharing of knowledge for bio-prospecting will multiply the faunal and floral diversity of that region and enrich the cultural and moral ethics with sustained life-support systems [260].

2. Conclusion

In order for wild edible plants to be better appreciated, more work should be undertaken to determine their nutritional composition so that they can be compared with widely cultivated major fruit crops. Nutritional value of these fruits and their value added products needs further study so as to determine how they compare nutritionally with the modern diet. Use of these plants also has the potential through selective conservation and domestication which can contribute to the maintenance of plant biodiversity. Traditional conservation practices related to these species are a dimension of indigenous knowing that can be researched into. Further research should address the issue of marketing and pricing of indigenous minor or wild edible fruit trees and their products. There is a need to distinguish/recognize these plants and their value added products in the local or national or international market. It is also important to analyze the market environment for these plants compared with alternative possibilities, such as exotic or agricultural crops. It should also be noted that some markets involve higher risks and distant markets involve higher transport costs. Relating these to product prices and potential benefits to the farmer is crucial. In addition, females and young people deserve more consideration for drawing conclusion on perception of local people about values and diversified uses of these fruits. Still there is a scope to incorporate more contextual variables for explaining more variations embedded with local people's perception on values and usage of wild edibles. In general, future contemplations on wild edibles need to keep species of high consensus to forefront and make use of those informants identified as having high species competence.

3. References

1. Sunderland, T. (2011). Food security: why is biodiversity important? Int. Forest. Rev. 13(3): 265–274.

2. Lulekal, E., Asfaw, Z., Kelbessa, E. and Van Damme, P. (2011). Wild edible plants in Ethiopia: A review on their potential to combat food insecurity. *Afrika Focus* 24: 71-121.

3. Getahun, A. (1974). The role of wild plants in the native diet in Ethiopia. *Agro-ecosystems* 1: 45-56.

4. Edwards, S.B. (1991). Crops with wild relatives found in Ethiopia. Plant Genetic of Ethiopia Cambridge University Press, Cambridge. Eds. Engles J.M.M.; Hawkes, J.G. and Worede, M. pp. 42-47.

5. Asfaw, Z., and Tadesse, M. (2001). Prospects for Sustainable Use and Development of Wild Food Plants in Ethiopia. *Economic Botany* 5: 47-62.

6. Gemedo-Dalle, T., Maass, L.B. and Isselstein, J. (2005). Plant biodiversity and Ethnobotany of Borana pastoralists in southern Oromia, Ethiopia. *Economic Botany* 59: 43-65.

7. Balemie, K. and Kebebew, F. (2006). Ethnobotanical study of wild edible plants in Derashe and Kuch Districts, South Ethiopia. *Journal of Ethnobiology and Ethnomedicine* 2: 53.

8. Mengistu, F. and Hager, H. (2008). Wild Edible Fruit Species Cultural Domain, Species Competence and Preference in Three Districts of Amhara Region, Ethiopia. *Ethnobotany Research and Applications* 6: 487-502.

9. Addis, G. (2009). *Wild and semi-wild edible plants of Hamar and Xonso (South Ethiopia) with emphasis on their ethnobotany and nutritional composition of selected species*. Ph.D. Thesis Addis Ababa University.

10. Asfaw, Z. (2009). The future of wild food plants in southern Ethiopia: ecosystem conservation coupled with enhancement of the roles of key social groups. *ActaHorticulturae* (ISHS) 806: 701-708.

11. Nair, C.T.S. and Merry, F.D. (1995). Status of research on non-wood forest products through agroforestry: issues and strategies. In: *Report of expert consultation on non-wood forest products*, Yogyakarta, Indonesia, 17-27 January 1995. Non-wood forest products 3. FAO, Rome.

12. Sekeroglu, N., Ozkutlu, F., Deveci, M., Dede O. and Yilmaz, N. (2006). Evaluation of some wild plants aspect of their nutritional values used as vegetable in eastern black sea region of Turkey. *Asian Journal of Plant Science* 5: 185-189.

13. Sarma, H., C.M. Sarma and D.K. Bhattacharya, (2008). Host specificity of *Cuscutareflexa* Roxb. in the Manas Biosphere Reserve, Indo-Burma Hotspot. *International Journal of Plant Product* 2: 175-180.

14. Jeeva, S., Laloo, R.C. and Mishra, B.P. (2006a). Traditional agricultural practices in Meghalaya, north-east India. *Indian Journal of Traditional Knowledge* 5: 7–18.

15. Jeeva, S., Laloo, R.C., Venugopal, N., Kharlukhi, L. and Misra, B.P. (2006b). Traditional knowledge and biodiversity conservation in the sacred groves of Meghalaya. *Indigenous Journal of Traditional Knowledge* 5: 563-568.

16. Jeeva, S., Mishra, B.P., Venugopal, N. and Laloo, R.C. (2005). Sacred Forests: traditional ecological heritage in Meghalaya. *J. Scott. Res. Forum* 1: 93-97.

17. Anonymous, (1992). *Safeguarding the Genetic Basis of Africa's Traditional crops*. International Plant Genetic Resources Institute (IPGRI).

18. Beluhan, S. and Ranogajec, A. (2010). Chemical composition and non-volatile components of Croatian wild edible mushrooms. *Food Chemistry* 124: 1435-1452.

19. Sundriyal, M. and Sundriyal, R.C. (2003). Underutilized edible plants of the Sikkim Himalaya: Need for domestication. *Current Science* 85: 731-736.

20. Sekeroglu, N., Ozkutlu, F., Deveci, M., Dede O. and Yilmaz, N. (2006). Evaluation of some wild plants aspect of their nutritional values used as vegetable in eastern black sea region of Turkey. *Asian Journal of Plant Science* 5: 185-189.

21. Afolayan, A. and Jimoh, F. (2009). Nutritional quality of some wild leafy vegetables in South Africa. *International Journal of Food Science and Nutrition* 60: 424-431.

22. Mbuvi, D. and Boon, E. (2009). The livelihood potential of non-wood forest products: The case of Mbooni division in Makueni District, Kenya. Environment Development Sustainability 11: 989-1004.

23. Kruijssen, F. (2008). ITFnet-International Tropical Fruits Network. www.itfnet.org.

24. Nazarudeen, A. (2010). Nutritional composition of some lesser-known fruits used by ethnic communities and local folks of Kerala. *Indian Journal of Traditional Knowledge* 9: 398-402.

25. Burlingame, B. (2000). Wild nutrition. *Journal of Food Composition and Analysis* 13: 99-100.

26. Ohiokpehai, O. (2003). Promoting the nutritional goodness of traditional food products. *Pakistan Journal of Nut*rition 2: 267–270.

27. Hammer, K., Heller, J. and Engels, J. (2001). Monographs on under-utilized and neglected crops. *Genetic Resources & Crop Evolution*, 48(1): 3–5.

28. Gruère, G., Giuliani, A. and Smale, M. (2006). Marketing of under-utilized plant species for the benefit of the poor: A Conceptual framework. Environment and Production Technology Division. Retrieved from International Food Policy.

29. Farooq, S. and Azam, (2002). Food security in the new Millennium-I: The role of agricultural biodiversity. *Pakistan Journal of Biological Sciences*, 5(12): 1345-1351.

30. Lobo, V., Patil, A., Phatak, A. and Chandra, N. (2010). Free radicals, anti-oxidants and functional foods: Impact on human health. *Pharmacognosy*, 4: 118–126.

31. Huang, W.J., Zhang, X. and Chen, W.W. (2016). Role of oxidative stress in Alzheimer's disease (Review). Biomed. Rep, 4: 519–522.

32. Lock, K., Pamerleau, J., Causer, L., Altmann, D.R. and McKee, M. (2005). The global burden of disease attributable to low consumption of fruit and vegetables: Implications for the global strategy on diet. *Bull World Health Organ*, 83: 100–108.

33. Tontisirin, K., Nantel, G., Bhattacharjee, L. (2002). Food-based strategies to meet the challenges of micronutrient malnutrition in the developing world. Proc. Nutr. Soc. 61(2): 243–250.

34. Vinceti, B., Termote, C., Ickowitz, A., Powell, B., Kehlenbeck, K., Hunter, D. (2013). The contribution of forests and trees to sustainable diets. Sustainability. 5: 4797–4824.

35. Mahapatra, A.K. and Panda P.C. (2012). Wild edible fruit diversity and its significance in the livelihood of indigenous tribals: Evidence from eastern India.

36. Hunter, L.M., Patterson, L. and Twine, W. (2009). HIV/AIDS, Food Security and Role of the Natural Environment: Evidence from the Agincourt Health and Demographic Surveillance Site in Rural South Africa: Working Paper on Population Programme, Institute of Behavioural Science, University of Colorado at Boulder.

37. Mithofer, D. and Waibel, H. (2004). Seasonal Vulnerability to poverty and indigenous fruit use in Zimbabwe. Rural Poverty Reduction through Research for Development and Transformation conference. DeutscherTropentag, October 5-7, 2004. Berlin, University.

38. Bell, J. (1995). The Hidden Harvest. In Seedling, the quarterly newsletter of Genetic Resources Action International (GRAIN). www. grain.org. Accessed 9 October, 2007.

39. Anonymous, (2005). *Building on gender, agro-biodiversity and local knowledge. A training manual.* Food and Agriculture Organization, Rome.

40. Namrata; Kumar, L., Ghosh, D., Dwivedi, S.C. and Singh, B. (2011). Wild edible plants of Uttarakhand Himalaya: a potential neutraceutical source. Research Journal of Medicinal Plant 5: 670-684.

41. Bharucha, Z. and Pretty, Z. (2010). The roles and values of wild foods in Agricultural systems. *Phil. Trans. R. Soc. B.* 365: 2913-2926.

42. Anonymous, (1999). *Use and potential of wild plants in farm households.* FAO Rome, Italy.

43. Nazarudeen, A. (1999). *An Assessment of the Economically Useful Wild Fruits of Kerala.* Ph.D. thesis, University of Kerala, Thiruvananthapuram, Kerala.

44. Bhutia, K.D. (2012). *Studies on physico-chemical characteristics and utilization aspects of some indigenous minor fruits of Sikkim.* Ph.D. Thesis. Uttar Banga Krishi Vishwavidyalaya, Pundibari, West Bengal, India. Unpubl.

45. Nutritional composition of some lesser-known fruits used by ethnic communities and local folks of Kerala. *Indian Journal of Traditional Knowledge* 9: 398-402.

46. Walter, M. and Hamilton, A. (1993). *The vital wealth of plants.* Gland, Switzerland: WWF.

47. Grivetti, L.E. and Ogle, B.M. (2000). Value of traditional foods in meeting macro and micro nutrient needs: The wild plant connection. *Nutritional Research Review* 13: 31-46.

48. Anonymous, (2005). *Building on gender, agrobiodiversity and local knowledge.* A training manual. FAO, Rome.

49. Rathore, M. (2009). Nutrient content of important fruit trees from arid zone of Rajasthan. *Journal of Horticulture and Forestry* 1: 103-108.

50. Singh, H.B. and Arora, R.K. (1972). Raisan (*Digitaria* sp.) A minor millet of the Khasi hills, India. *Economic Botany* 27: 376.

51. Singh, H.B. and Arora, R.K. (1973). Soh Phlong, *Moghaniavestita*, a leguminous root crop of India. *Economic Botany* 27: 332.

52. Bhandari, M.M. (1974). Famine foods in Rajasthan Desert. *Economic Botany* 28: 73.

53. Pundir, Y.P.S. and Singh, D. (2002). Ethno-Botanical wild food plants of Jaunsar-Bawar (Western Himalaya), Uttaranchal. *The Indian forester* 128: 571.

54. Hebbar, S.S., Harsha, V.H. Shripathi, V. and Hegde, G.R. (2003). Wild fruits of Dharwad, Karnataka. *Journal of Economic and Taxonomic Botany* 27: 982.

55. Agrahar-Murugkar, D. and Subbulakshmi, G. (2005). Nutritive values of wild edible and species consumed by the Khasi tribe of India. *Ecology of Food and Nutrition* 44: 207-223.

56. Panda, T. (2014). Traditional knowledge on wild plants as livelihood food in Odisha, India. *Journal of Biology and Earth Sciences* 4: B144-B159.

57. Anjula, P., Tomer, A.K., Bhandari, D.C. and Pareek, S.K. (2007). Towards collection of wild relatives of crop plants in India. *Journal of Genetic Resources and Crop Evolution* 55: 187-202.

58. Badhani, A., Sakalani, S. and Mishra, A.P. (2011). Variation in biochemical's and anti-oxidant activity of some wild edible fruits of Uttarakhand. *Report and Opinion* 3: 1-10.

59. Samant, S.S. and Dhar, U. (1997). Diversity, endemism and economic potential of wild edible plants of Indian Himalaya. *International Journal of Sustainable Development and World Ecology* 4: 179-191.

60. Samant, S.S., Dhar, U. and Rawal, R.S. (2001). Diversity and distribution of wild edible plants of Indian Himalaya. In: *Plant Diversity of the Himalaya*, eds. Pande, P.C. Samant, S.S. Gyanodaya Prakasha, Nainital, India. pp. 421-482.

61. Mugnozza, G.T.S. (1996). Ethics of biodiversity conservation. In: *Biodiversity, science and development: towards a new partnership*, eds. Di Castri, F. and Younes, T. CAB International, United Kingdom. Pp. 622-629.

62. Verheij, E.W.M. and Coronel, R.E. (1991). *Plant resources of South-East Asia: edible fruits and nuts*. Pudoc-DLO, Wageningen the Netherlands.

63. Hooker, J.D. (1872-1997). *The Flora of British India*. Vol. I-VII. London.

64. Hedrick, U.P. (1919). *Sturtevent's Notes on Edible Plants of the World*. J.B. Lyon Co. Bombay.

65. Singh, S.C. (1960). Some wild plants of food value in Nepal. *Journal of Tribhuban*. University Kathmandu 4: 50–56.

66. Turner, N.J. (1975). *Food plants of coastal first peoples*. Vancouver: UBC Press.

67. Becker, B. (1983). The contribution of wild plants to human nutrition in the Ferlo (N Senegal). *Economic Botany* 1: 257–267.

68. Haridasan K., Bhuyan, L.R. and Deori, M.L. (1990). Wild Edible plants of Arunachal Pradesh. *ArunachalForest News* 18: 1-8.

69. Maikhuri, R.K., Semwal, R.L., Singh, A. and Nautiyal, M.C. (1994). Wild fruit as a contribution to sustainable rural development: A case study from the Garhwal Himalaya. *International Journal of Sustainable Development World Ecology* 1: 56-68.

70. Maikhuri, R.K., Nautiyal, S., Rao, K.S. and Semwal, R.L. (2000). Indigenous knowledge of medicinal plants and wild edibles among three tribal sub-communities of the central Himalayas, India. *Indigenous Knowledge and Development Monitor* 8: 7-13.

71. Maikhuri, R.K., Rao, K.S., Chauhan, K., Kandari, L.S., Prasad, P. and Rajashekharan, C. (2003). Development of marketing of medicinal plants and other forest products-can it be a pathway for effective management and conservation? *The Indian Forester* 129: 169-178.

72. Chakravarty, S., Shukla, G. and Suresh, C.P. (2009). Forest genetic resources in north east India. In: *Proc. International Conference on Quality Seed and Food Security*, eds. Rahman, G.M.M. and Wadud, M.A., Pp. 87-93. Bangladesh Agricultural University, Mymensingh Bangladesh.

73. Malla, S., Shukla, G. and Chakravarty, S. (2012). Utilization and conservation of wild plants by the tribal communities of Tripura. *Indian Forester* 138: 1002-1007.

74. Shanmugam, S., Sakthivel, B. and Rajendran, K. (2010a). Ethnobotanical studies on Paliyar tribes of Attuparai village in Theni District, Tamil Nadu. In: *International conference on Medicinal Plants and Herbal Drugs: Challenges and opportunities in cultivation, Sustainable Utilization and Conservation*. Post Graduate and Research Department of Botany, Pachaiyappa's college, Chennai, Tamil Nadu. 4th-6th March. Abs. No.74. 82p.

75. Shanmugam, S., Vinoth, M., Annadurai, B., Sakthivel, B. and Rajendran, K. (2010b). Ethnobotanical Survey on Paliyar tribes of Kannur village in Agamalai hills of Theni District, Tamil Nadu. In: *National level seminar on Emerging trends in Plant Biotechnology*. Post Graduate and Research Department of Plant Biology and Plant Biotechnology, Government Arts College, Coimbatore, Tamil Nadu. 4th-5th February. Abs. No.84. Pp 114-115.

76. Shanmugam, S., Muthuraja, G., Dhanashekharan, M., Annadurai, M. and Gobinathan, S. (2012). *Ethnobotanical study on the plants used as leafy vegetables by the Paliyar tribes of Pachalpur in Dindigul district of Tamil Nadu, India. Life Sciences Leaflets* 4: 44-47.

77. Shukla, G. and Chakravarty, S. (2012). Ethnobotanical plant use of Chilapatta Reserve Forest in West Bengal. *Indian Forester* 138: 1116-1124.

78. Shukla, G., Subba, M. and Chakravarty, S. (2013a). Ethnomedicinal plant use by locals in backward areas of Hardoi District, Uttar Pradesh (India). In: *Proceedings of the 2nd International Symposium on Minor Fruits and medicinal Plants on Minor Fruits and Medicinal Plants for better lives*, ed. Wathugala, D.L.; Subhasinghe, S.; Athapaththu, N.S.B.N.; Fonseka, K. and Priyadharshani, N.D.N., Pp. 42-49. Faculty of Agriculture, University of Ruhuna, Sri Lanka.

79. Shukla, G., Kumar, R., Subba, M., Kumari, A. and Chakravarty, S. (2013b). Wild vegetable biodiversity in Ranchi district of Jharkhand. *Life Sciences Leaflets* 11: 23-30.

80. Suresh, C.P., Bhutia, K.D., Shukla, G., Pradhan, K. and Chakravarty, S. (2013). Free listed wild edible fruit species of Sikkim Himalayas and their uses. In: *Proceedings of the 2nd International Symposium on Minor Fruits and medicinal Plants on Minor Fruits and Medicinal Plants for better lives*, ed. Wathugala, D. L.; Subhasinghe, S.; Athapaththu, N.S.B.N.; Fonseka, K. and Priyadarshani, N.D.N, Pp. 17-37. Faculty of Agriculture, University of Ruhuna, Sri Lanka.

81. Suresh, C.P., Pradheep, K., Bhutia, K.D. and Chakravarty, S. (2013). Wild relatives of fruits in Eastern Himalayan region of India. In: *Perspectives in Plant Biodiversity*, ed. Muthuchelian, K. Daya Publishing House, New Delhi. Pp. 107-118.

82. Suresh, C.P., Bhutia, K.D., Shukla, G., Pradhan, K. and Chakravarty, S. (2014). Wild edible tree fruits of Sikkim Himalayas. *Journal of Tree Sciences* 33: 43-48.

83. Bhutia, K.D., Pala, N.A., Shukla, G., Pradhan, K., Suresh, C.P. and Chakravarty, S. (2015). Informants consensus and knowledge on the use of wild edible fruits from, Sikkim. *Journal of Hill Agriculture*6: 207-212.

84. Bhutia, K.D., Suresh, C.P., Pala, N.A., Shukla, G., Chakravarty, S. (2018). Neutraceutical potential of some wild edible fruits of Sikkim Himalaya, India. *Ethnomedicine* 12(1&2): 106-112.

85. Biswakarma, S., Sarkar, B.C., Shukla, G., Pala, N.A. and Chakravarty, S. (2015). Traditional application of ethno-medicinal plants in Naxalbari area of West Bengal, India. *International Journal of Usufructus Management* 16(1): 36-42.

86. Biswakarma, S., Pala, N.A., Shukla, G. Vineeta and Chakravarty, S. (2017a). Plants for liver and jaundice treatment: A case study from forest fringe communities in North Bengal, India. *Forestry Ideas* 23: 145-151.

87. Biswakarma, S., Pala, N.A., Shukla, G., Vineeta and Chakravarty, S. (2017b). Ethnomedicinal plants used to cure stomach disorders in forest fringe communities in northern part of West Bengal. *Indian Journal of Natural Products and Resources* 8: 370-380.

88. Sarkar, B.C., Biswakarma, S., Shukla, G., Pala, N.A. and Chakravarty, S. (2015). Documentation and utilization pattern of Ethnomedicinal plants in Darjeeling Himalayas, India. *International Journal of Usufructus Management* 16(2): 3-11.

89. Chakravarty, S., Bhutia, K.D., Suresh, C.P., Shukla, G. and Pala, N. A. (2016). A review on diversity, conservation and nutrition of wild edible fruits. *Journal of Applied and Natural Science* 8: 2346-2353.

90. Raj, A.J., Biswakarma, S., Pala, N.A., Shukla, G., Vineeta; Kumar, M., Chakravarty, S., Bussman, R.W. (2018). Indigenous uses of ethnomedicinal plants among forest-dependent communities of northern Bengal, India. *Journal of Ethnobiology and Ethnomedicine*14: 8 DOI 10.1186/s13002-018-0208-9

91. Vineeta; Pala, N.A., Shukla, G. and Chakravarty, S. (2018a). Traditionally used medicinal plants for treatment of stomach disorders in West Bengal, India: a scrutiny and analysis from secondary literature. *Ethnomedicine* 12(3): 162-182.

92. Vineeta, Shukla, G., Pala, N.A., Dobhal, S. and Chakravarty, S. (2018b). Influence of seed priming on germination and seedling growth of *Adeanthera pavonina* in subhumid region of West Bengal, India. *Indian Journal of Tropical Biodiversity* 26: 87-91.

93. Shiva, B., Bhargav, V. and Nimbolkar, P.K. (2017). Under-Exploited and Nutritionally Rich Wild Fruits of Telangana, India. *International Journal of Current Microbiology and Applied Sciences* 6: 2682-2694.

94. Sharma, I.P., Kanta, C., Semwal, S.C. and Goswami, N. (2017). Wild Fruits of Uttarakhand (India): Ethnobotanical and Medicinal Uses. *International Journal of Complementary & Alternative Medicine* 8: 00260. DOI: 10.15406/ijcam.2017.08.00260

95. Majumdar, K. and Datta, B.K. (2009). Traditional wild edible fruits for the forest dwellers of Tripura, India. *East Himalayan Society for Spermatophyte Taxonomy*. Pleione 3(2): 167-178.

96. Thakur, M. (2014). Underutilized food crops: treasure for the future India. *Food Science Research Journal*, 5(2): 174-183.

97. Brahma, S., Narzary, H. and Basumatary, S. (2013). Wild edible fruits of Kokrajhar district of Assam, North-East India. *Asian Journal of Plant Science and Research*, 3(6): 95-100.

98. Paul A. (2013). Minor and uncultivated fruits of Eastern India. Paper published in the Full Paper Proceedings of the 2nd International Symposium on Minor Fruits and Medicinal Plants for better lives, 20thDec, 2013, F/Agriculture, University ofRuhuna, Mapalana, Srilanka, 54-67.

99. Ya, Li., Jiao-Jiao, Zhang., Dong-Ping, Xu., Tong, Zhou., Yue, Zhou., Sha, Li and Hua-Bin, Li. (2016). Bioactivities and Health Benefits of Wild Fruits. *International Journal of Molecular Science*, 2-27.

100. Friedman, J., Bolotin, D., Rios, M., Mendosa, P., Cohen, Y. and Balick, M.J. (1993). A novel method for identification and domestication of indigenous useful plants in Amazonian Ecuador, eds. Janick, J. and Simon J.E. New Crops Wiley, New York Pp. 167-174.

101. Bussmann, R.W., Gilbreath, G.G., Solio, J., Lutura, M., Lutuluo, R., Kunguru, K., Wood, N. and Mathenge, S.G. (2006). Plant use of the Maasai of Sekenani Valley, Maasai Mara, Kenya. *Journal of Ethnobiology and Ethnomedicine* 2: 22.

102. Medley, K.E. and Kalibo, H.W. (2007). Ethnobotanical survey of 'wild' woody plant resources at mount Kasigau, Kenya. *Journal of East African Natural History* 96: 149-186.

103. Campbell, B.M. (1987). The use of wild fruits in Zimbabwe. Economic Botany 41: 375-385.

104. Zinyama, L.M., Matiza, T. and Campbell, D.J. (1990). The use of wild foods during periods of food shortage in rural Zimbabwe. *Ecology of Food and Nutrition* 24: 251-265.

105. Johns, T. and Kokwaro, J.O. (1991). Food Plants of the Luo of Siaya District, Kenya. *Economic Botany* 45: 103-113.

106. Addis, G., Urga, K. and Dikasso, D. (2005). Ethnobotanical Study of Edible Wild Plants in Some Selected Districts of Ethiopia. *Human Ecology* 33: 83-118.

107. Carol, B. and Angela, F. (1990). Africa Ark: *Tribe and ancient culture of Ethiopia and the Horn of East Africa*. Incorporate press, New York pp. 249-315.

108. Guinand, Y. and Lemessa, D. (2000). *Wild food plants in Southern Ethiopia: Reflections on the role of famine-foods at the time of drought.* UNDP-EUE Filed Mission Report, Addis Ababa.

109. Genetu, B. (2005). Socio-Economic and Demographic Profile of Debub Omo Zone, SNNPR, Jinka, Ethiopia. Pp. 1-49.

110. Melnyk, M. (1998). *Indigenous enterprise for the domestication of trees and the commercialization of their fruits.* International Conference on Domestication and Commercialization of Non-Timber Forest Products in Agroforestry Systems. Non-Forest Products 9. FAO, Rome.

111. Shrestha, P.M. and Dhillion, S.S. (2006). Diversity and traditional knowledge concerning indigenous food species in a locally managed forest in Nepal. *Agroforestry Systems* 66: 55-63.

112. Jasmine, T.S., Jeeva, S., Febreena, G.L., Mishra, B.P. and Laloo, R.C. (2007). Wild edible plants of Meghalaya, North-east India. *Natural Product Radiance* 6: 410-426.

113. Pilgrim, S., Cullen, L., Smith, D.J. and Pretty, J. (2008). Ecological knowledge is lost in wealthier communities and countries. *Environmental Science and Technology* 42: 1004–1009.

114. Ohiokpehai, O. and Ramosweu, (1999). The role of agricultural processing and nutrition in improving food security at household level. In *Proceedings of the national post harvest conference* held October 6-8 1998, Sebele, Botswana.

115. Pauline, M. and Linus, W. (2004). Status of indigenous fruits in Kenya. In: *Review and appraisal on the status of Indigenous fruits in Eastern Africa.* A synthesis report for IPGRI-SAFORGEN, eds. Chikamai, B., Eyog-Matig, O. and Mbogga, M. http://www.bioversity international.org.

116. Tabuti, J.R.S., Lye, K.A. and Dhillion, S.S. (2004). The status of wild food plants in Bulamogi County, Uganda. *International Journal of Food Sciences and Nutrition,* 55: 485-498.

117. Maundu, P.M. (1996). Utilization and conservation status of wild food plants in Kenya. In: *The biodiversity of African plants,* eds. Maesen L.J.G. van der, Burgt, X.M. van der and J.M. van Medenbach de Rooy. Dordrecht, the Netherlands, Kluwer Academic Publishers. Pp. 678-683.

118. Edeoga, H.O., Okwu, D.E. and Mbaebie, B.O. (2003). Minerals and nutritive value of some Nigerian medicinal plants. *Journal of Medicinal and Aromatic Plant Science* 25: 1010-1015.

119. Wilkes, H.G. (1981). *New or potential crop or what to anticipate for the future.* Paper presented at the annual meeting of the American Association for the advancement of Science, January 1981, Toronto, Canada.

120. Anonymous, (1982). *Ecological Aspect of Development in the Humid Tropics.* National Academy of Sciences, Washington, DC. NRC.

121. Reddy, K.N., Pattanaik, C., Reddy, C.S. and Raju, V.S. (2007). Traditional knowledge on wild food plants in AndhraPradesh, *Indian Journal of Traditional Knowledge* 6: 223-229.

122. Mohan Ram, H.Y. (2000). *Plant Resources of Indian Himalaya.* 9th G.P. Pant Memorial Lecture, G B Pant Intititute of Himalayan Development, Gangtok, Sikkim.

123. Cooper, H.D. and Spillane, C. (1996). Promoting the identification, conservation, and use of wild plants for food and agriculture in the Mediterranean: *The FAO global plan of action,* Rome: Plant Production and Protection Division, FAO.

124. Mishra, S., Maikhuri, R.K., Kala, C.P., Rao, K.S. and Saxena, K.G. (2008). Wild leafy vegetables: A study of their subsistence dietetic support to the inhabitants of Nanda Devi Biosphere Reserve, India. *Journal of Ethnobiology and Ethnomedicine* 4: 15-23.

125. Tiwari, J.K., Ballabha, R. and Tiwari, P. (2010). Some promising wild edible plants of Srinagar and its adjacent area in Alaknanda valley of Garhwal Himalaya, India. *Journal of American Science* 6: 167-174.

126. Aberoumand, A. (2009). Nutritional evaluation of edible Portula-caoleracea as plant food. *Food Analytical Method* 2: 204-207.

127. Sundriyal, R.C., and Sharma, E. (1996). Anthropogenic pressure on tree structure and biomass in the temperate forest of Mamlay watershed in Sikkim. *Forest Ecology and Management* 81: 113–134.

128. Mahapatra, A.K., Satarupa, M., Basak, U.C. and Panda, P.C. (2012). Nutrient Analysis of Some Selected Wild Edible Fruits of Deciduous Forests of India: an Explorative Study towards Non Conventional Bio-Nutrition in *Advance Journal of Food Science and Technology* 4: 15-21.

129. Anonymous, (1970-1988). Wealth of India: raw materials, Vol. 1–12. Council of Scientific and Industrial Research, Delhi (Reprinted).

130. Duke, J.A. and Atchley, A. (1986). *Handbook of proximate analysis tables of higher plants.* CRC Press, Inc., Boca Raton, Florida. *Economic Botany* 50: 243-251.

131. Neog, M. and Mohan, N.K. (1994). Minor and less known fruits of Assam, *Indian Horticulture* 39: 28-31.

132. Getachew, A., Kelbessa, U. and Dawit, D. (2005). Ethnobotanical study of edible indigenous plants in some selected districts of Ethiopia. *Human Ecology* 33: 83-118.

133. Sundriyal, M. and Sundriyal, R.C. (2001). Wild edible plants of the Sikkim Himalaya: Nutritive values of selected species. *Economic Botany* 55: 377-390.

134. Campbell, B., Luckert, M. and Scoones, I. (1997). Local Level Valuation of Savannah Resources: A Case Study from Zimbabwe. *Economic Botany* 51: 57-77.

135. Foppes, J. and Ketphanh, S. (2004). *NTFP use and household food security in Lao PDR*, Draft paper prepared for the NAFRI/ FAO EM 1093 "Symposium on Biodiversity for Food Security" Vientainne, 14th September, 2004.

136. Akinnifesi, F.K., Kwesiga, F.R., Mhango, J., Mkonda, A., Chilanga, T. and Swai, R. (2006). Domesticating priority for Miombo indigenous fruit trees as a promising livelihood option for small holder farmers in Southern Africa. In: *Proceedings XXVI IHC Citrus Subtropical and Tropical Fruit Crops*, eds. Albrigo, L.G. and Galan, Sauco, V. *ActaHorticulturae* 632: 15-30.

137. Akinnifesi, F.K., Kwesiga, F., Mhango, J., Chilanga, T., Mkonda, A., Kadu, C.A.C., Kadzere, I., Mithofer, D., Saka, J.D.K., Sileshi, G., Ramadhani, T and Dhliwayo, P. (2006). Towards the development of Miombo fruit trees as commercial tree crops in southern Africa. *Forests, Trees and Livelihoods* 16: 103-121.

138. Mazhar, F., Buckles, D., Satheesh, P.V. and Akhter, F. (2007). *Food sovereignty and uncultivated biodiversity in South Asia.* Academic Foundation. New Delhi, India. 84p.

139. Marwat, S.K., Fazal-urRehman; Usmaan, K., Khakwani, A.A., Ghulam, S., Anwar, N., Sadiq, M. and Khan, S.K. (2011). Medico-ethnobotanical studies of wild edible fruits plants species for the flora of north-western Pakistan (D.I. Khan District) *Journal of Medical Plant Research* 5: 3679-3686.

140. Katende, A.B., Birnie, A. and Tengnas, B. (1995). *Useful trees and shrubs for Uganda. Identification, propagation and management for agricultural and pastoral communities.* Regional Soil Conservation Unit (RSCU), Technical Handbook No. 10. Swedish International Development Authority (Sida), 710p.

141. Katende, A., Segawa, P., Birnie, A., Holding, C. and Tengnas, B. (1999). *Wild food plants and mushrooms of Uganda.* Nairobi: Regional Land Management Unit (RELMA), Technical Handbook No. 19. Swedish International Development Cooperation Agency (Sida), 490p.

142. Maghembe, J.A. (1995). Achievement in the establishment of indigenous fruit trees of the Miombo woodlands of southern Africa. In: *Improvement of indigenous fruit trees of the Miombo woodlands of Southern Africa,* eds. Maghembe, J.A., Ntupanyama, Y. and Chirwa, P.W. ICRAF, Nairobi, Kenya. Pp. 39-49.

143. Bowes, G. (1999). *A colour atlas of plant propagation and Conservation.* University of Glasgow, Scotland Manson publishing. 224p.

144. Grosskinsky, B. and Gullick, C. (2000). Exploring the potential of indigenous wild food plants in Southern Sudan. *Proceedings of a workshop held in Lokichoggio, Kenya, June 3-5 1999.* Publication services provided by The Mitchell Group, Inc. (TMG) pursuant to the following USAID contract: VAFR/SD Support Services Contract Number AOT-C-00-99-00224-00. pp 1-111.

145. Kwesiga, F.R., Akinnifesi, F.K., Ramadhani, T., Kadzere, I. and Sako, J. (2000). Domestication of Indigenous Fruit trees of the Miombo in southern Africa. In: *The domestication and commercialization of indigenous fruit trees in the SADC region,* eds. Shumba, E.M., Lusepani, E. and Hangula, R.SADC Tree Seed Centre Network.

146. Nyoka, B.I. and Rukumi, D. (2000). The domestication and commercialization of indigenous fruit trees in the SADC region. In: *Proceedings of a SADC Tree Seed Centre Network technical meeting.* Windhoek, Namibia, 13 to 14 March, 2000, CIDA and FAO. Pp. 62-65.

147. Akinnifesi, F.K., Chilanga, T., Mkonda, A. and Kwesiga, F.R. (2004). Domestication of *Uapacakirkianain* southern Africa: A preliminary evaluation of provenances in Malawi and Zambia. In: *Proceedings of Regional Agroforestry Conference on agroforestry impacts on livelihoods in Southern Africa: putting research into practice,* eds. Rao, M.R. and Kwesiga, F. R. World Agroforestry Centre: Nairobi, Kenya. pp. 85-92.

148. Mithöfer, D. and Waibel, H. (2004). *Seasonal vulnerability to poverty and indigenous fruit use in Zimbabwe.* Rural Poverty Reduction through Research for Development and Transformation Conference. Deutscher Tropentag, October 5-7, 2004. University of Hannover, Berlin.

149. Jain, S.K. (1963). Wild plant-foods of the tribals of Bastar (Madhya Pradesh), *Botanical Survey of India* 30: 56-80.

150. Sinha, R. and Lakra, V. (2005). Wild tribal food plants of Orissa. *Indian Journal of Traditional Knowledge* 4: 246-252.

151. Sundriyal, M., Sundriyal, R.C., Sharma, E. and Purohit, A.N. (1998). Wild edibles and other useful plants of Sikkim Himalayas, India. *Oecologia Montana* 7: 43-54.

152. Pardo-de-Santayana, M., Tardio, J., Blanco, E., Carvalho, A.M., Lastra, J.J., San-Miguel, E. and Morales, R. (2007). Traditional knowledge of wild edible plants used in the northwest of the Iberian Peninsula (Spain and Portugal): A comparative study. *Journal of Ethnobiology and Ethnomedicine* 3: 27.

153. Campbell, A. (1986). The use of wild food plants and drought in Botswana. *Journal of Arid Environment* 11: 81-91.

154. Campbell, B.M. (1986). The importance of wild fruits for peasant households in Zimbabwe. *Food Nutrition* 12: 38-44.

155. Edwards, S.B. (1992). Traditional tree crops in Ethiopia. Historical records and economic importance. In: *Proceedings of the second Natural Resources Management for conservation and development conference.* Institute of Agricultural Research, Natural Resource Management for Conservation and development, Addis Ababa. pp. 10-13.

156. Abbink, J. (1993). Meén ritual, medicinal and other plants: A contribution to south-west *Ethiopian Ethnobotany. J. Eth. St.* 26: 1-21.

157. Scherr, S. (1995). Economic analysis of agroforestry systems: the farmers' perspective In: *Costs, Benefits and adoption of agroforestry project experience in Central America and the Carribean*, ed. Scherr S. World Bank working paper 14, CATIE–IFPR-World Bank projects. Washington, USA, pp. 28-44.

158. Eriksen, S. and Mutimba, S. (1998). *Climate change adaptation in Kenya. A review of current policies.* ACTS Discussion Paper. Global Biodiversity Forum, November 1998, Buenos Aires.

159. Nkomeshya, A.K. (1998). *The role of forests in the livelihoods of selected households in Central, Copper belt and Luapula provinces of Zambia*. PFAP, Zambia.

160. Guinand, Y. and Dechassa, L. (2000). *Indigenous food plants in southern Ethiopia Reflections on the role of 'famine foods' at the time of drought.* UNEUE Survey, Addis Ababa.

161. Mithöfer, D. and Waibel, H. (2003). Income and labour productivity of collection and use of indigenous fruit tree products in Zimbabwe. *Agroforestry Systems* 59: 295-305.

162. Akinnifesi F.K., Kwesiga F.R., Mhanga, J., Mkonda, A., Chilanga, T. and Swai, R. (2004). Domesticating priority Miombo indigenous fruit trees as a promising livelihood option for small farmers in southern Africa. *ActaHorticulturae* 632: 15-30.

163. Mojeremane, W. and Tshwenyane, S.O. (2004). *Azanzagarckeana*: a valuable edible indigenous fruit tree of Botswana. *Pakistan Journal of Nutrition* 3: 264–267.

164. Tiisekwa, B.P.M., Ndabikunze, B.K., Samson, G. and Juma, M. (2004). Suitability of some indigenous tree fruits for manufacturing juices and jams in Tanzania. In: *Agroforestry impacts on livelihoods in southern Africa: Putting research into practice*, eds. Rao, M.R.; Kwesiga, F.R. Proceedings of the Regional Agroforestry Conference held in Warmbaths, South Africa 20-24 May, 2002. World Agroforestry Centre (ICRAF), Nairobi, Kenya. Pp 331-335.

165. Kebu, B. and Fassil, K. (2006). Ethnobotanical study of wild edible plants in Derashe and Kucha Districts. South Ethiopia. *Journal of Ethnobiology and Ethnomedicine* 2: 53.

166. Redzic, S.J. (2006). Wild edible plants and their traditional use in the human nutrition in Bosnia- Herzegovina. *Ecology of Food and Nutrition* 45: 189-232.

167. Bhattarai, S.; Chaudhary, R.P. and Taylor, R.S.L. (2009). Wild edible plants used by the people of Manang District, Central Nepal. *Ecology of Food and Nutrition* 48: 1-20.

168. Piya, L., Maharajan, K.L. and Joshi, N.P. (2011). Forest and food security of indigenous people: A case of chepangs in Nepal. *Journal of International Development and Cooperation* 17: 113-135.

169. Arum, G. (1989). Resources. KENGO.

170. Karachi, M., Ruffo C.K., Lema, N.K. and Minae, S. (1991). *Use of multipurpose trees in Western Tanzania: results of an Ethnobotanical survey.* AFRENA Report No. 49. Nairobi: ICRAF.

171. Maghembe, J.A. and Seyani, J.H. (1991). *Multipurpose trees used by small scale farmers in Malawi: results of an Ethnobotanical survey*. AFRENA Report No. 42. Nairobi: ICRAF.

172. Poulton, C. and Poole, N. (2001). *Poverty and Fruit Tree Research: Issues and Options Paper*, www. nrinternational.co.uk/forms2/frpzf0141b.pdf, DFID Forestry Research programme, Wye College, Ashford, England, 70p.

173. Boyer, J.S. (1982). Plant productivity and environment. *Science* 218: 443-448.

174. Ludlow, M. and Muchow, R.C. (1990). A critical evaluation of the traits for improving crop yield in water limited environments. *Advance Agronomy* 43: 107-153.

175. Babu, S.C. (2000). Rural nutrition interventions with indigenous plant foods: a case study of vitamin A deficiency in Malawi. *Biotechnol. Agron. Soc. Environ.* 4: 169-179.

176. Harris, F.M.A. and Mohammed, S. (2003). Relying on nature: wild foods in Northern Nigeria. *Ambio* 32: 24-29.

177. Anonymous, (2009). *The state of food insecurity in the world*. FAO, Rome.

178. Anonymous, (1999). Agricultural Research Master Plan. Volume III-Part 5. Horticultural crops research technical report. Bahir Dar. Bureau of Agriculture (BOA).

179. Debarata, D. (2002). Wild Food Plants of Medinipur, West Bengal Used During Drought and Flood. In: *Ethnobotany and Medicinal Plants of Indian Subcontinent*, ed. Maheshwari, J.K. Scientific Publishers, Jodhpur, India.

180. Akinnifesi, F.K., Mhango, J., Sileshi, G. and Chilanga, T. (2008). Early growth and survival of three Miombo indigenous fruit tree species under fertilizer, manure and dry-season irrigation in southern Malawi. *Forest Ecology and Management* 255: 556-557.

181. Arnold, M., Powell, B., Shanley, P., Sunderland, T.C.H. (2011). Editorial: forests, biodiversity and food security. Int. Forest. Rev. 13(3): 259–264.

182. Shumsky, S., Hickey, G.M., Johns, T., Pelletier, B., Galaty, J. (2014). Institutional factors affecting wild edible plant harvest and consumption in semi-arid Kenya. Land Use Policy 38: 48–69.

183. Shumsky, S., Hickey, G.M., Pelletier, B., Johns, T. (2015). Understanding the contribution of wild edible plants to rural socio-ecological resilience in semi-arid Kenya. Ecol. Soc. 19(4): 34.

184. Wunder, S., Börner, J., Shively, G., Wyman, M. (2014). Safety nets, gap filling and forests: a global-comparative perspective. World Dev. 64: s29–s42.

185. Hickey, G.M., Pouliot, M., Smith-Hall, C., Wunder, S. and Nielsen, M.R. (2016). Quantifying the economic contribution of wild food harvests to rural livelihoods: A global-comparative analysis. *Food Policy* 62: 122-132.

186. Ercisli, S. and Sagbas, H.I. (2017). Wild edible fruits: a rich source of biodiversity. *ANADOLU, Journal of AARI*27: 116-122.

187. Pierce A.R., Emery, M.R. (2005). The use of forests in times of crisis: ecological literacy as a safety net. *Forests, Trees and Livelihoods* 15: 249-252.

188. Eromosele, I.C., Eromosele, C.O. and Kuzhkzha, D.M. (1991). Evaluation of mineral elements and ascorbic acid contents in fruits of some wild plants. Plant Food. *Human Nutrition* 41: 151-154.

189. Nordeide, M.B., Hatloy, A., Folling, M., Lied, E. and Oshoug, A. (1996). Nutrient composition and nutritional importance of green leaves and wild foods in an agricultural district, Koutiala, in Southern Mali. *International Journal of Food Science and Nutrition* 47: 455-468.

190. Ondachi, P. (2001). Nutritional composition of some indigenous fruits in dry lands. In: *Proceedings of the Regional Social Forestry Extension Seminar for Semi-arid and Arid areas.* KEFRI, Muguga, Kenya.

191. Orech, F.O., Aagaard-Hansen, J. and Friis, H. (2007). Ethno ecology of traditional leafy vegetables of the Luo people of Bondo district, Western Kenya. *International Journal of Food Science and Nutrition* 58: 522-530.

192. Bhatt, I.D., Rawat, S., Badhani, M. and Rawal, R.S. (2017). Nutraceutical potential of selected wild edible fruits of the Indian Himalayan region. *Food Chemistry* 215: 84-91.

193. Jardin, C. (1967). Fruits. In: *List of foods used in Africa.* FAO, Rome. Pp. 118-170.

194. Schmidt, D.R. (1971). Comparative yields and composition of eight tropical leafy vegetables grown at two fertility levels. *Agronomy Journal* 63: 546-550.

195. Williamson, J. (1975). *Useful plants of Malawi.* Revised ed. University of Malawi, Zomba. 336p.

196. Anonymous, (1982). *Fruit bearing forest species.* Technical notes. Food and Agricultural Organization, Forestry Paper No. 34. Rome.

197. Mwanjumwa, L.B. (1982). *Survey of production and utilization of fruits and vegetables of Machakos District and their nutritional evaluation with emphasis on indigenous species.* M.Sc. Thesis, University of Nairobi.

198. Okafor, J.C. (1982). Promising trees of agroforestry in southern Nigeria. In *Agroforestry in the African Humid Tropics,* ed. MacDonald, L.H. Proceeding of a Workshop held in Ibadan,Nigeria, 27th April - 1st May, 1981. United Nations University Press.Tokyo, Japan.

199. Abdoellah, O.S. and Marten, G.C. (1986). The Complementary notes of home gardens, upland fields and rice fields for meeting nutritional needs in West Java. Traditional Agriculture in Southeast Asia, ed. Marten, G.C. West View Press, Boulder, Colorad. Pp. 293-325.

200. Tucker, J.B. (1986). Amaranth - The one and future crop. Bio Science, 36: 9-13.

201. Sims, L.S. and Peterkin, B.B. (1987). Contributions of fruits and vegetables to human nutrition. In: *Horticulture and human health*, eds. Quebedeaux, B. and Bliss, F. Prentice- Hall, Englewood Cliffs, New Jersey. Pp. 9-167.

202. Kapur, S.K., and Y.K. Sarin, (1990). *Flora of Trikuta Hills (Shri Vaishno Devi Shrine)*. Bishen Singh Mahendra Pal Singh, Dehradun.

203. Maikhuri, R.K. (1991). Nutritional value of some lesser known wild food plants and their role in tribal nutrition: A case study in North-East India. *Journal of Tropical Science* 31: 397-405.

204. Dhyani, P.P. and Khali, M.P. (1992). Himalayan *Ficus* diversity holds promise. *Hima-Paryavaran* 4: 6-7.

205. Packham, J. (1993). *The value of indigenous fruit-bearing trees in Miombo woodland areas of south-central Africa*. RDFN paper 15. Pp. 10-15.

206. Chwenya, J. (1994). Potential for agronomic improvement of indigenous plant Germplasm in African Agriculture: A case study of indigenous vegetables in Kenya. In: *Proceedings of a CTA/IPGRI/ UNEP seminar*, ed. Putter A., 59 October 1992, Nairobi, Kenya. CTA, The Netherlands/IPGRI, Rome, Italy.

207. Singh, V. (1995). Lesser known wild edibles of Sikkim Himalaya. *Journal of Economic and Taxonomic Botany* 19: 385–390.

208. Kalenga, Saka, J.D. and Msonthi, J.D. (1994). Flowering plants of the World. Nutritional value of edible fruits of indigenous wild trees in Malawi. *Forest Ecology and Management* 64: 245-248.

209. Maghembe, J.A., Kwesiga, F.R., Ngulube, M., Prins, H. and Malaya, F.M. (1994). Domestication potential of indigenous fruit trees of Miombo woodlands of Southern Africa. In: *Tropical trees potential for domesticating and rebuilding of forest resources HMSO*, eds. Leakey, R.R.B. and Newton. London. pp. 220-229.

210. Saka, J., Msothi, J.D. and Maghembe, J.A. (1994). The nutritional value of edible fruits of indigenous wild trees of Malawi. *Forest Ecology and Management* 64: 245-248.

211. Malaisse, F. (1997). *Se nourrirenforet Claire africaine: Approcheecologique et nutritionelle*. Les presses gronomoques de Gembloux/CTA.

212. Maghembe, J.A., Simons, A.J., Kwesiga, F.R. and Rarieya, (1998). *Selecting indigenous trees for domestication in Southern Africa: Priority setting with farmers in Malawi, Tanzania, Zambia, Zambia and Zimbabwe*. ICRAF, Nairobi.

213. Leakey, R.R.B. (1999). Potential for novel food products from agroforestry trees *Food Chemistry* 64: 1-14.

214. Sundriyal, M. (1999). *Distribution, Propagation and Nutritive value of some wild edible plants in Sikkim Himalaya*. Ph. D. thesis. H.N.B. Garhwal University, Srinagar, Uttaranchal, India.

215. Agte, V.V., Tarwadi, K.V., Mengale, S., Chiplonkar, S.A. (2000). Potential of traditionally cooked green leafy vegetables as natural sources for supplementation of eight micronutrients in vegetarian diets. *Journal of Food Composition and Analysis* 13: 885-892.

216. Lockett, C.T., Calvert, C.C. and Grivetti, L.E. (2000). Energy and micronutrient composition of dietary and medicinal wild plants consumed during drought. *International Journal of Food Science and Nutrition* 51: 195-208.

217. Loura, B.B., Mapongmetsem, P.M., Nkongmeneck, B.A. Foko, J., Kapchie, N.V. and Denedoumba, D. (2000). Characterization physico-chimique de deuxMyrtaceaesalimentairesendemiques des sava nessoudano-guineennes (Adamaoua, Cameroun). *Cameroon Journal of Biochemical Sciences* 10: 31-39.

218. Ogle Britta, M. (2001). *Wild vegetables and Micronutrient Nutrition-Studies on the Significance of Wild vegetables in Women's Diets in Vietnam*, Comprehensive summaries of Uppsala, Dissertations from the Faculty of Medicine.

219. Ogle Britta, M., Dung, N.N.X., Thanh, D.T. and Hambraeus, L. (2001). The contribution of Wild Vegetables to micronutrient intakes among women: An example from the Mekong Delta, Vietnam *Ecology and Food Nutrition* 40: 159-184.

220. Ogle Britta, M., Tuyet, M.T., Duyet, H.N. and Dung, N.N.X. (2003). Food, Feed or Medicine: The multiple functions of edible wild plants in Vietnam. *Economic Botany* 57: 103-117.

221. Ogle, B.M., Dung, N.N.X., Do, T.T. and Hambraeus, L. (2004). The contribution of wild vegetables to micronutrient intake among women. An example from the Mekong Delta, Vietnam. *Ecology of Food Nutrition* 40: 159-184.

222. Parvathi, S. and Kumar, V.S.F. (2002). Studies on chemical composition and utilization of the wild edible vegetable Athalakkai (*Momordicatuberose*). Plant Foods. *Human Nutrition* 57: 215-222.

223. Gockowski, J., Mbazo'o J., Mbah, G. and Moulende T.F. (2003). African traditional leafy vegetables and the urban and peri-urban poor. *Food Pol.*, 28: 221-235.

224. Shackleton, C.M. (2003). The prevalence of use and value of wild edible herbs in South Africa. *South African Journal of Science* 99: 23-25.

225. Agrahar-Murugkar, D. and Pal, P.P. (2004). Intake of nutrients and food sources of nutrients among the Khasi tribal woman in India. *Nutrition* 20: 268-273.

226. El, S.N. and Karakaya, S. (2004). Radical scavenging and iron-chelating activities of some greens used as traditional dishes in Mediterranean diet. *International Journal of Food Science and Nutrition* 55: 67-74.

227. Ansari, N.M., Houlihan, L., Hussain, B. and Pieroni, B. (2005). Anti-oxidants activity of five vegetables traditionally consumed by South-Asian migrants in Bradford, Yorkshire, UK. *Phytotherapy Research* 19: 907-911.

228. Johns, T. and Eyzaguirre, P.B. (2006). Linking biodiversity, diet and health in policy and practice. Proc. Nutr. Soc. 65: 182–189.

229. Delang, C.O. (2006). Indigenous systems of forest classification: Understanding land-use patterns and the role of NNTP's in shifting cultivators' subsistence economies. *Environmental Management* 37: 470-486.

230. Kuhnlein, H., Erasmus, B., Creed-Kanashiro, H., Englberger, L., Okeke, C., Turner, N., Allen, L. and Bhattacharjee, L. (2006). Indigenous peoples' food systems for health: Finding interventions that work. *Public Health Nutrition* 9: 1013-1019.

231. Kuhnlein, H., Erasmus, B. and Spigelski, D. (2009). Indigenous peoples' food systems. Centre for Indigenous People's Nutrition and Environment. FAO, Rome.

232. Singh, V. and Garg, A.N. (2006). Availability of essential trace elements in Indian cereals, vegetables and species using INAA and the contribution of species to daily dietary intake. *Food Chemistry* 94: 81-89.

233. Toledo, I. and Burlingame, B. (2006). Biodiversity and nutrition: a common path toward global food security and sustainable development. *J. Food Compos. Anal.* 19: 477–483.

234. Fentahun, M.T. and Hager, H. (2009). Exploiting locally available resources for food and nutritional security enhancement: wild fruits diversity, potential and state of exploitation in Amhara region of Ethiopia. *Food security* 1: 207-219.

235. Kumar, C.S., Gaddala, N., Vanamala, S., Naresh, V. and Elumalai. (2011b). A short review on therapeutic uses of *couroupitaguianensis* aubl. *International Research Journal of Pharmaceutical and Applied Sciences* 1: 105-108.

236. Kumar, R., Chauhan, P.K., Bhardwaj, V.S., Kumar, A. and Kumar, M. (2011a). *In vitro*-investigations of anti-oxidants & phytochemical activities of aqueous extracts of *Terminalia belerica&Terminalia chebula.International Journal of Institutional Pharmacy and Life Sciences* 1: 18-29.

237. Seal, T. 2011. Evaluation of some wild edible plants from nutritional aspect used as vegetable in Meghalaya state of India. *World Applied Sciences Journal*12: 1282-1287.

238. Martin, G.J. (1995). *Ethnobotany, A people and plants conservation manual.* Chapman and Hall London, New York, Tokyo. pp. 23-29.

239. Leakey, R.R.B. and Simons, A.J. (1998). The domestication and commercialization of indigenous trees in agroforestry for the alleviation of poverty. *Agroforestry Systems* 38: 165-176.

240]. Shackleton, S.E., Shackleton, C.M., Dzerefos, C.M. and Mathabela, F.R. (1998). Use and trading of wild edible herbs in central the Lowveld savanna region, South Africa. *Economic Botany* 52:251-259.

241. Nesamvuni, C. Steyn, N.P. and Potgieter, M.J. (2001). Nutritional value of wild, leafy plants consumed by the Vhavenda. *South African Journal of Science* 97: 51-54.

242. Leakey, R.R.B.; Schreckenberg, K. and Tchoundjeu, Z. (2003). The participatory domestication of West African indigenous fruits. *International Forestry Reviews* 5: 338-347.

243. Kadu, C.A.C., Imbuga, M., Jamnadass, R. and Dawson, I.K. (2006). Genetic management of indigenous fruit trees in southern Africa. A case study of *Sclerocarya birrea* based on nuclear and chloroplast variation. *South African Journal of Botany* 72: 421-427.

244. Akinnifesi, F.K., Ajayi, O.C., Sileshi, G., Kadzere, I. and Akinnifesi, A.I. (2007). Domesticating and commercializing indigenous fruit and nut tree crops for food security and income generation in Sub-Saharan Africa. In: *Proceedings New Crops International Symposium*, 3-4 September, Southampton, United Kingdom, 45: 82-87.

245. Akinnifesi, F.K., Sileshi, G., Ajayi, O.C., Chirwa, P.W. and Harawa, R. (2008). Contributions of Agroforestry research and development to livelihood of smallholder farmers in Southern Africa. 2. Fruit, medicinal, fuelwood and fodder tree systems. *Agriculture Journal* 3: 76-88.

246. Hoe, V.B. and Siong, K.H. (1999). The nutritional value of indigenous fruits and vegetables in Sarawak. *Asia Pacific Journal of Clinical Nutrition* 8: 24-31.

247. Teklehaimanot, Z. (2004). Exploiting the potential of agroforestry tree: *Parkiabiglobosa*and *Vitellariaparodoxa*in Sub-Saharan Africa. *Agroforestry System* 61: 207-220.

248. Ham, C. (2003). *Commercial Products from the Wild Consortium.* Background report on indigenous fruit commercialization activities in selected SADC countries. Department of Forest Science, University of Stellenbosch, South Africa.

249. Kadzere, T., Hove, L., Gatsi, T., Masarirambi, M.T., Tapfumaneyi, L., Maforimbo, E. and Magumise, I. (2004). Current status of post harvest handling and traditional processing of indigenous fruits in Zimbabwe. In: *Agroforestry impacts on livelihoods in Southern Africa: Putting research into practice*, eds. Rao, M.R. and Kwesiga, F.R. Proceedings of the regional agroforestry conference held on 20-24 May, 2002. Warm baths, South Africa. World Agroforestry Centre (ICRAF), Nairobi, Kenya. Pp. 353-363.

250. Ramadhani, T. (2002). *Marketing of indigenous fruits in Zimbabwe.* Socio-economic Studies on Rural Development, Vol. 129. Wissens chaftsverlag Vauk. Kiel, Germany.

251. Saka, J.D.K., Swai, R., Mkonda, A., Schomburg A., Kwesiga F. and Akinnifesi F.K. (2004). Processing and utilization of indigenous fruits of the Miombo in southern Africa. Agroforestry Impacts on Livelihoods in Southern Africa: Putting Research into Practice. In: *Proceedings of Regional Agroforestry Conference on Agroforestry Impacts on Livelihoods in Southern Africa: Putting Research into Practice*, eds. Rao, M.R. and Kwesiga, F.R. World Agroforestry Centre: Nairobi, Kenya. Pp. 343-352

252. Saka, J.D.K., Rapp, I., Ndolo, V., Mhango, J. and Akinnifesi, F.K. (2006). A comparative study of the physicochemical and organoleptic characteristics of *Uapacakirkiana, Strychnoscocculoides, Adansoniadigitata* and *Mangifera indica* products. *International Journal ofFood Science and Technology* 42: 836-841.

253. Akinnifesi, F.K., Ajayi, O.C., Sileshi, G., Kadzere, I. and Akinnifesi, A.I. (2005). *Domesticating and commercializing indigenous fruit and nut tree crops for food security and income generation in sub-Saharan Africa.* Nairobi. World.

254. Akinnifesi, F.K., Jordaan, D., Ham, C. (2005). Building opportunities for small to commoditize indigenous fruit trees and products in

southern Africa: processing, markets and rural livelihoods. In: *The global food and product chain - dynamics, innovation, conflicts, strategies*, eds. Tielkes, E.; Hülsebusch, C.; Häuser, I. *et al*. Deutscher Tropentag, Stuttgart-Hohenheim. University of Hohenheim, Book of abstracts.

255. Mawula, R. (2009). *Regeneration of threatened indigenous species in Uganda, the caseof Parinari curatellifolia*. M.Sc. Thesis. International Master Programme at the Swedish Biodiversity Centre. No. 67.

256. Tewari, D. and Campbell, J.Y. (1995). Developing and sustaining non-timber forest products: some policy issues and concerns with special reference to India. *Journal of Sustainable Forestry* 3: 53–79.

257. Hegde, R., Suryaprakash, S., Achpth, L. and Bawa, K.S. (1996). Extraction of non-timber forest products in the forest of Biligiri Rangan Hills, India. I: Contribution to rural income. *Economic Botany* 50: 243–251.

258. Uma Shankar; Murali, K.S., Uma Shanker, R., Ganeshaiah, K.N. and Bawa, K.S. (1996). Extraction of non timber forest products in the forest of Biligiri Rangan Hills, India. 3. Productivity, extraction and prospects of sustainable harvest of Amla (*Phyllanthusemblica*). *Economic Botany* 50: 270–279.

259. Gairola, Y. and Biswas, S. (2008). Bioprospecting in Garhwal Himalayas, Uttarakhand. *Current Science*, 90(9): 1139–1143.

260. Singh, D., Wangshu, L. and Prahlad, V.C. (2008). *Processing and marketing feasibility of underutilized fruit species of Rajasthan*. IAMO Forum, 1-12.

261. Chandra, D.S. and Prakash, J. (2009). Minor fruits: A livelihood opportunity for the tribal peoples of Tripura. IInd International Symposium on pomegranate and minor including Mediterranean fruits, ISPMMF, 2009.

262. Pareek, O.P. (2001). *Ber*. International Centre for Crops. Southampton (U.K.). Research Institute; www.ifpri.org; 1-54.

263. Mahato, S.K. and Chhetri, B. (2015). Important aspects of some underutilized fruits Crops and their wild relatives for food and nutritional Security in Darjeeling Himalayas. *International Journal of Agricultural Science and Research (IJASR)* 5: 29-42.

264. Benz, B., Cevallos, J., Santana, F., Rosales, J. and Graf, S. (2000). Losing knowledge about plant use in the Sierra de Manantlan Biosphere Reserve, Mexico. *Economic Botany* 54: 183-191.

265. Byg, A. & Balslev, H. (2001). Traditional knowledge of *Dypsis fibrosa* (Arecaceae) in Eastern Madagascar. *Economic Botany* 55: 263-275.

266. Ladio, A.H. and Lozada, M. (2003). Comparison of wild edible plant diversity and foraging strategies in two aboriginal communities of North-western Patagonia. *Biodiversity and Conservation* 12: 937-951.

267. Anonymous, (1995). *Non wood forest products and nutrition.* Report of the International Expert Consultation on Non-Wood Forest Products. Non-Wood Forest Products 3. Rome. FAO.

268. Mao, A.A., Hynniewta, T.M. and Sanjappa, M. (2009). Plant wealth of North-east India with reference to ethnobotany. *Indian Journal of Traditional Knowledge* 8: 96-103.

269. Locay, L. (1989). From hunting and gathering to agriculture. *Economic Development and Cultural Change* 37: 737-756.

270. Ondachi, P.W. (1999). *Nutritional studies of indigenous fruit trees in support of conservation.* Annual Research Report, part 1. African Academy of Sciences.

271. Bagra, G., De, L.C. and Singh, V.B. (2006). Collection and evaluation of some indigenous lesser known fruits of Arunachal Pradesh. In: *Horticulture for Sustainable Income and Environmental Protection.* Vol. 1, Pp. 224-231.

272. Byron, E. (2003). *Market integration and health: morbidity, diet and nutritional status in the Tsimane' Ameridians of Lowland Bolivia.* Ph. D. thesis, University of Florida, Gainesville.

273. Rathore, D.S. (2001). Overcoming malnutrition through non-traditional fruits. *Indian Horticulture* Jan-Mar: 8-12.

274. Roe, D. and Elliott, J. (2004). Poverty reduction and biodiversity conservation: rebuilding the bridges. *Oryx* 38: 137-139.

275. Johns, T., Eyzaguirre, P.B. (2007). Biofortification, biodiversity and diet: a search for complementary applications against poverty and malnutrition. Food Policy 32(1): 1–24.

276. Athenaeus, (1927). The *Deipnosophists.* (Gulick, C.B. translator), G.P. Putnams Sons.

277. Apicius, (1958). *The Roman Cookery Book.* A critical Translation of the Art of Cooking, for use in the study and the Kitchen (Flower, B. and Resenbaun, E. translator), George G. Harrap. London.

278. Arano, L.C. (1976). *Tacuinum Sanitatis. The Medieval Health Handbook.* New York. N.Y.

279. Darby, W.J., Ghalioungui, P. and Grivetti, L.E. (1977). *Food: The Gift of Osiris.* Academic Press, London.

280. Sharma, P. (Ed.) (1981). *Caraka Samhita. Agnivesa's Treatise Refined and Annonated by Craka and Redected by Dr Dhabala, vol.1 Sitrasthana to Indryasthana.* Chaukambha Orientalis, Delhi.

281. Simoons, F.J. (1991). *Food in China. A Cultural and Historical Inquiry.* CRC Press, Boca Raton.

282. Anonymous, (1984). Food and Fruit Bearing Forest Species. Examples of South-Eastern Asian Forestry Paper no. 44/2 FAO, Rome.

283. Aloskar, L.V., Kakkar, K.K. and Chakra, O.J. (1992). *Second Supplement to Glossary of Indian Medicinal Plants with Active Principles,* Part-I (A-K), 1965- 1981, NISC, CSIR, New Delhi, pp: 265-266.

284. Gopalan, C., Sastri, B.V.R. and Balasubramanian, S.C. (2004). *Nutritive Value of Indian Foods.* National Institute of Nutrition, Indian Council of Medical Research, Hyderabad, India, pp. 2-58.

285. Vincetti, B., Eyzaguirre, P. and Johns, T. (2008). The nutritional role of forest plant foods for rural communities. In: *Health and Forests: A global overview of issues practice and policy,* ed. Coler, C.J.P. London, UK. Pp. 63-96.

286. Gruere, G.L., Nagarajan, O., King, O.D.I. (2009). The role of collective action in the marketing of under-utilized plant species: Lessons from a case study on minor millets in South India. *Food Policy* 34: 39-45.

287. Chadha, K.L. and Pareek, O.P. (1988). Genetic Resources of Fruit Crops: Achievements and Gaps. *Indian Journal of Plant Genetic Resources* 1: 43-48.

288. Demel, T. and Abeje, E. (2004). Status of indigenous fruits in Ethiopia. In: *Review and Appraisal on the status of Indigenous fruits in Eastern Africa.* A synthesis report for IPGRI - SAFORGEN, eds. Chikamai, B., Eyog-Matig, O., Mbogga, M. http://www. bioversity international.org/Networks/saforgen/publications.htm.

289. Anonymous, (2014). *Promotion of Under-utilized Indigenous Food Resources for Food Security and Nutrition in Asia and the Pacific.* Rome,m FAO. http://www.fao.org/3/a-i3685e.pdf.

290. Humphrey, J. (2007). The supermarket revolution in developing countries: tidal wave or tough competitive struggle? *J. Econ. Geogr.* 7: 433-450.

291. Reardon, T. and Hopkins, R. (2007). The supermarket revolution in developing countries: policies to address emerging tensions among supermarkets, suppliers and traditional retailers. *Eur. J. Dev. Res.* 18: 522-545.

292. Murray, C.H. (1991). Caring for forests in a changing world. *Food Policy* 16: 213-218.

293. Vinceti, B., Eyzaguirre, P. and Johns, T. (2008). The nutritional role of forest plant foods for rural communities. Human Health Forests. A Glob Overview Issues Pract. Policy. People Plants Int. Conserv. Ser. 63–96.

294. Sunderland, T., Powell, B., Ickowitz, A., Foli, S., Pinedo-Vasquez, M., Nasi, R. and Padoch, C. (2013). *Food Security and Nutrition*. Center for International Forestry Research (CIFOR), Bogor, Indonesia.

295. Powell, B., Hall, J. and Johns, T. (2011). Forest cover, use and dietary intake in the East Usambara Mountains, Tanzania. *Int. Forest. Rev.* 13: 305-317.

296. Cotton, C.M. (1996). *Ethnobotany: Principles and Applications.* Chichester: John Wiley and Sons.

297. Alcorn, J.B. (1981). Huastec non-crop resource management. *Human Ecology* 9: 395-417.

298. Alcorn, J.B. (1981). Factors influencing botanical resource perception among the Huastec: suggestions for future ethnobotanical inquiry. *Journal of Ethnobiology* 1: 221-230.

299. Bye, R.A. (1979). Incipient domestication of mustards in north-west Mexico, *Kiva* 44: 237-256.

300. Gangwar, K.K., Deepali, Gangwar, R.S. (2010). Ethnomedicinal plant diversity in Kumaun Himalaya of Uttarakhand, India. *Nat. Sci.* 8: 66-78.

301. Johns, T., Powell, B., Maundu, P., Eyzaguirre, P.B. (2013). Agricultural biodiversity as a link between traditional food systems and contemporary development, social integrity and ecological health. *J. Sci. Food Agric.* 93: 3433-3442.

302. Vira, B., Jamnadass, R., Kleinschmit, D., McMullin, S., Mansourian, S., Nuefeldt, H., Parotta, J.A., Sunderland, T.C.H. and Wildburger, C. (2015). Conclusions. In: Vira, B., Wildburger, C. and Mansourian, S. (Eds.), *Forests, Trees and Landscapes for Food Security and Nutrition: A Global Assessment Report*. IUFRO World Series Volume 33. International Union of Forest Research Organizations (IUFRO). Pp. 153-157.

303. McBurney, R.P.H.; Griffin, C., Paul, A.A., David, D.C. and Greenberg, C. (2004). The nutritional composition of African wild food plants. *Journal of Food Composition Analysis* 17: 277–289.

304. Flyman, M.V. and Afolayan, A.J. (2006). The suitability of wild vegetables for alleviating human dietary deficiencies. *South African Journal of Botany* 72: 492–497.

305. Frison, E.A., Smith, I.F., Johns, T., Cherfas, J. and Eyzaguirre, P.B. (2006). Agricultural biodiversity, nutrition and health. *Food Nutrition Bulletin* 27: 167-179.

306. Anonymous, (2015). *Sustainable Development Goals*. UN, New York. Available at http://www.un.org/sustainabledevelopment/sustainable-development-goals/

307. Phalan, B., Balmford, A., Green, R.E., Scharlemann, J.P.W. (2011). Minimizing the harm to biodiversity of producing more food globally. *Food Policy* 36: S62-S71.

308. Burlingame, B. and Dernini, S. (Eds.) (2012). *Sustainable Diets and Biodiversity*. FAO/Biodiversity International, Rome.

309. Maikhuri, R.K., Rao, K.S. and Saxena, K.G. (2004). Bioprospecting of wild edibles for rural development in the central Himalayan mountain of India. *Mt Res Dev*. 24: 110-113.

310. Negi, V.S., Maikhuri, R.K. and Rawat, L.S. (2011). Non-timber forest products (NTFPs): a viable option for biodiversity conservation and livelihood enhancement in central Himalaya. *Biodiversity Conservation* 20: 545-559.

311. Maikhuri, R.K. Rawat, L.S., Negi, V. and Purohit, V.K. (2007). *Eco-friendly appropriate technologies for sustainable development of rural ecosystems in Central Himalaya*. G.B. Pant Institute of Himalayan Environment and Development, Almora.

312. Maikhuri, R.K., Negi, V., Rawat, L.S. and Purohit, V.K. (2007). Promoting value addition in potential wild edibles of central Himalaya for sustainable livelihood and small scale enterprise development. G.B. Pant Institute of Himalayan Environment and Development, Almora.

313. Maikhuri, R.K. and Ramakrishnan, P.S. (1992). Ethnobiology of some tribal societies of Arunachal Pardesh, north-eastern India. *Journal of Economic Taxonomic Botany* 10: 61–78.

314. Lowitt, K., Hickey, G.M., Ganpat, W. and Phillip, L. (2015). Linking communities of practice with value chain development in smallholder farming systems. World Dev. 74: 363-373.

□□□

Chapter 13

Evolution of Plant Polysaccharides: Towards Functional Foods for Human health Improvement

Kiran Thakur[1] and Zhao-Jun Wei*[2]

[1] Associate Professor, [2]Professor, School of Food Science and Engineering, Hefei University of Technology, Hefei-230009, People's Republic of China

Kiran Thakur: kiran.thakur2@gmail.com, kumarikiran@hfut.edu.cn

*Corresponding Author: Zhao-Jun Wei: zjwei@hfut.edu.cn

ABSTRACT

Due to the increasing consumer's demand for novel foods coupled with growing population, it is imperative to look for unconventional and corresponding sources of natural food ingredients. Plant-derived polysaccharides have long been considered useful in food and feed, medicine and pharmaceutics. Owing to their multifaceted health benefits, it is essential to review the currently available information and make valid recommendations for the use of plant-derived polysaccharides as functional foods. Dietary intake of these polysaccharides offers a highly effective approach to influence the gut microbiota structure and composition by re-storing the gut dysbiosis for improved human health. Nevertheless, particular carbohydrate exerts their prebiotic effect in a strain or species-specific manner. There is an enormous potential to gain insights into different factors which affect their prebiotic effect. Thus, this chapter aims to update the information available on their sources, biological activities, and their intestinal fermentation and gut microbiota modulation.

Keywords: Plant polysaccharides; functional foods; prebiotic; gut modulation

1. Introduction

Past decades have witnessed the evolution of functional foods which led to significant improvement in human health. This has advanced the demand for natural products with higher functional values for physical well-being. For this reason, it becomes essential to shift the industrial and academic research focus for the development of novel potential compounds with improved functional properties [1].Various phytochemicals particularly plant-derived polysaccharides are being used as food ingredients, medicines and pharmaceutical products. The recent interests in the development of polysaccharides research owe to their therapeutic nature [2]. Moreover, natural phytochemicals extracted from the medicinal herbs offer the most attractive strategies for the treatments of various diseases [3]. The latest understanding due to mechanistic approach has highlighted the emerging role of these bioactive molecules for their novel industrial applications [4]. On the basis of their chemical structure, they constitute organic macromolecules which are formed by the polymerization of simple sugar units with a key function of energy storage and structural components formation. For example, at the cellular level, starch represents an energy storing polysaccharide in the cytoplasm; while cellulose acts as a structural component of the plant cell wall [5].

Recent studies revealed the pharmacological and biochemical significance of polysaccharides isolated from natural source due to their beneficial activities. The plant polysaccharides in both simple and complex glycoconjugated forms can support multi-faceted functions, such as antioxidant, immuno-modulation, anti-cancerous anti-tumour, radioprotection, anti-diabetes, hepatoprotection, anti-bacterial, anti-virus, and anti-fatigue functions [1].These bioactive molecules are also known to improve the food quality as thickeners, to enhance the water retention and also to stabilize the emulsions and dispersions [6]. Polymers derived from the plant (polysaccharides) and animal (proteins) kingdoms serve as potent materials for drug delivery and re-generative medicine due to their biochemical similarity with human extracellular matrix components, these polymers can be readily recognized and accepted by the body. Besides the above mentioned attributes, these bio-active molecules are naturally abundant, easily isolated, which can also be chemically modified to meet varying techno-functional requirements. In addition, enzymatic and/or hydrolytic degradations of these compounds in biologic environments release body friendly degradation by products which can be helpful for health promotion [7]. The polysaccharides from microbes (bacteria, fungi, and yeasts), algae, plants, and animals are not present in purified forms; they are always entrapped with proteins, polynucleotides, lipids and mineral substances. Therefore, to study their individual effect *in vivo*, purification is a crucial step to boldly claim their functionality. Previous research has shown that different polysaccharides obtained from micro-organisms, plants, and animals have diverse properties. Natural polysaccharides are known to possess ideal bio-activities which are not always satisfying the need of consumers. There are many factors which affect the physiological functions of plant derived polysaccharides such as extraction method, diverse structures, compositions, and confirmations [8,9,4, 2]. Few studies on structure–activity relationships of polysaccharides confirmed that not all the natural polysaccharides, are conducive to have bio-activities due to their structure and physico-chemical properties [10,11]. The studies revealed that the bio-active polysaccharides with high molecular weight may not be suitable for penetrating into organisms to exert pharmacological effects. The bio-activities of polysaccharides are directly or indirectly controlled by the original structure, which give rise to idea of structural modification as a fundamental way to achieve desirable and ideal functions with pharmacological activities [10, 11]. In the present time, various degenerative and terminal diseases are emerging with increased morbidity and mortality. In this regard, global attention has been shifted from conventional treatments to functional foods with little or no side effects, such as prebiotics due to their potential to modulate the gut microbiota, strengthen the immune system, and prevent degenerative diseases [12].

From past decade, various researchers have emphasized that diet modulates the composition of our gut microbiota [13]. As known, diet is a complex mixture; however, studies have proved that when dietary carbohydrates transit through the gastro-intestinal tract, they can influence the composition and stability of the gut microbiota [14,15]. Consequently, "prebiotic" compounds which are non-digestible by the host but can enhance the metabolism of beneficial gut bacteria[16]. Prebiotic compounds, such as the human milk-, galacto-, fructo-, xylo-, and pectin oligosaccharides and lactulose [17,18], exopolysaccharides of microbial origin[19] and plant polysaccharides, such as inulin and fructo-oligosaccharides or enzymatically synthesized, such as galacto-oligosaccharides are currently available prebiotics in the market [20]. The only known phenomenon by which prebiotics lead to healthy gut is their break down into smaller units which makes them available as a substrate during fermentation. The production of SCFAs as a result of fermentation, thus provides an acidic environment in the colon which indirectly suppresses the growth of pathogens and thereby enhance the immune function, improve digestion and eliminate the faeces as well as reduce the potential of getting irresistible bowel syndrome [21].

This chapter will focus mainly on plant-derived polysaccharides and their role in managing metabolic disorders through beneficial gut microbial manipulation. It should also be noted that plant polysaccharides are widely studied till date. Therefore, it is not feasible to compile the whole information in this chapter; however, we will discuss selected plant derived polysaccharides, their sources and biological activities and their fermentation by intestinal bacteria.

1.1. Types of Bio-active Polysaccharides

Polysaccharides are classified into diverse types due to different structures, chemical constituents, sources, solubility, and functions. Considering their chemical composition, the polysaccharides are divided into homo-polysaccharides or hetero-polysaccharides [22]. On the basis of their origin, plant bioactive molecules are known as dietary fibers, herbs and wood, plants, algae and lichen, and other bio-active polysaccharides which are derived from animals and acquire related structural characteristics (e.g. sulphate) [4].

1.2. Selective Stimulation of Beneficial Bacteria in the Gut

Recently, there are many studies published which claim the pre-biotic role of plant derived bio-active molecules. However, we have to understand the fact that exhaustive studies need to be carried out to claim the role of these pre-biotic compounds because not all the dietary carbohydrates or non-digestible plant materials can exert prebiotic effects [23, 24]. Moreover, diverse composition of different substrates can influence the community diversity. In this section, few studies are highlighted on the basis of the remarkable outcomes. The pre-biotic prospective of polysaccharides extracted from seaweeds has been complied in one review with a special focus on improved growth of colonic bifidobacteria. The authors recommended the use of seaweed polysaccharides as functional foods to enhance the intestinal immune system [25].

Another study claimed the pre-biotic use of rapeseed polysaccharides on the basis of improved proliferation of bifidobacteria and lactobacilli as well as enhanced acid production [26]. Intriguingly, the growth of probiotic bacteria was augmented as the sugar content was increased. Another factor, molecular weight was noticed to influence the accessibility of a carbohydrate to bacteria. As known previously, non-digestible carbohydrates with a lower molecular weight (short-chain fructo-oligosaccharides and inulin) can be better pre-biotics [27]. The ideal monosaccharide composition (fructose, glucose, galactose and xylose) is recognized to exert better growth effects on pro-biotics [28]. Branched polysaccharides with higher degree of polymerization and water solubility were the another important factors which were noteworthy to effect the prebiotic ability of any bio-active compound.

The traditional Chinese medicines polysaccharides, such as ginseng polysaccharides and ginsenosides in *Du-Shen-Tang* were confirmed to restore the homeostasis of gut microbiota particularly enhancing the growth of *Lactobacillus* spp. and *Bacteroides* spp. [29].

One research finding published in BMC biology very clearly mentioned that specific plant bio-active molecules can enormously stimulate the selected gut microbiota, but these modifications are strains and species dependent along with other factors especially pH, which plays a key role in determining the pre-biotic effect of any non-digestible food. Not all the probiotics would get stimulated by a given prebiotic approach. Therefore, it is crucial to design the tailor-made combination for *in vitro* efficacies before the animal trials can be approved [30].

Another review discussed the role of xylooligosaccharides from agricultural by-products with various claims, such as improved beneficial gut microflora, reduced blood glucose and cholesterol levels and pro-carcinogenic enzymes along with enhanced mineral absorption from large intestine and strengthened immune system [31]. Interestingly, apple pectin resulted in extremely precise improvement of specific bacterial operational taxonomic units (*Bacteroides*) during human trials. Among the *Firmicutes, Eubacterium eligens* was largely promoted by pectin. Going further, the authors demonstrated how the different complexities of different substrates can influence the community diversity which was higher in the pectin-fed than in the inulin-fed fermentors [32]. Another study demonstrated that fructans from *Aloe Vera* induced bacterial growth better than inulin. Through expression analysis, it was revealed that fructans could stimulate the population of *Bifidobacterium* spp. which led to greater amounts of SCFAs; without affecting the branched-chain fatty acids [33]. The high-throughput pyrosequencing technique revealed the prebiotic potential of *Lentinula* edodes-derived polysaccharides in nice model by reducing the diversity (decreased *Bacteroidetes* and increased *Proteobacteria*) and evenness of gut microbiota along the intestine, especially in the cecum and colon [34].

1.3. Stimulation of Host Immunity

There is a dearth of information about the relationship between host immunity and plant based polysaccharides. The "post-biotics " i.e. the fermentation products or prebiotics itself might participate in the immuno-modulation process. Mushrooms polysaccharides, glucans, and heteroglycans are reported with immuno-modulating and anti-tumour activity [35].

Another study confirmed the immune stimulating effects of polysaccharides extracted from medicinal plant *Tinospora cordifolia*. Being non-cytotoxic and non-proliferating to normal lymphocytes as well as tumor cells, this bioactive compound could trigger natural killer cells, T cells, and B cells as well as elicited the synthesis of specific interleukins [36]. There is one interesting review published which covers the immuno-modulatory functions of plant cell wall polysaccharides [37]. The review highlighted the importance of plant polysaccharides to promote the immune system for considerable boost to fight and eliminate intruders during any infection. Particularly, polysaccharides from higher plants are recognized as ideal biological response modifiers for complement activation, lymphocytes proliferation and macrophages stimulation without any side effect [20]. Various types such as pectic (acid and neutral), hemicellulosic, and sulphated polysaccharides are known to have immuno-modulatory and immuno-stimulatory effects. Among the hemicellulosic polysaccharides, *Aloe Vera* (Acemannan), *Tamarindus indica* (Xyloglucan) and *Echinacea purpurea* (Xyloglucan) were reported with immuno-modulatory activity and latter could activate macrophages and trigger the production of interleukine-1, respectively [37].

1.4. Prevention of Pathogen Adherence to the Epithelial Membrane of the Gut/ *in vitro* Anti-bacterial Ability

In addition to some previously reported oligosaccharides (galacto-oligosaccharide) which block the binding of pathogenic bacteria to the same sugar on the epithelial membrane of the gut, there are some plant derived polysaccharides which exhibit appreciable anti-bacterial activity. In our previous studies, sequentially extracted polysaccharides from onion, *Polygonatum* and peony seed dreg have shown remarkable anti-bacterial action against Gram-positive (*Bacillus subtilis* and *Staphylococcus aureus*) as well as Gram-negative (*Escherichia coli* and *Salmonella typhimurium*) *in vitro* [38-40]. Previous studies showed that sulfated polysaccharides from seaweeds exerted an anti-microbial effect by inhibiting the adhesion of pathogens (such as *Helicobacter pylori*) to the mucous membrane [41,25]. The polysaccharides from olive leaves showed remarkable anti-bacterial activity against *Micrococcus luteus* (ATCC 4698) and *Salmonella enterica* (ATCC 43972) [6].

From the previous research, it was revealed that the permeability of bacterial cell walls and cell membranes can be affected by polysaccharides [40] which can increase the water-soluble intracellular proteins or inhibit the bacterial growth. Previous reports indicated that polysaccharides may serve as a barrier to inhibit bacterial growth by blocking the input of nutrients. Our recent study revealed that sulphation modification can be considered as an effective method for improvement of the anti-microbial potential of plant based polysaccharides [40] Ultimately, because of the stimulation of probiotic growth and fermentation, these polysaccharides might contribute to host health, particularly with respect to defense against pathogenic bacteria by decreasing the pH in the colon.

a) Anti-obesity

In one recent study, apple polysaccharides could re-store the gut dysbiosis in high-fat diet fed rats. The administration of apple polysaccharides led to higher abundance of *Bacteroidetes* and *Lactobacillus*, while lowered the *Firmicutes* and *Fusobacteium* which resulted into significantly increased total SCFAs level [32]. In another study, *Fuzhuan* brick tea polysaccharides were shown to attenuate the metabolic syndrome in high-fat diet induced mice in association with modulation in the gut microbiota [42].

b) Anti-inflammatory

Seaweeds polysaccharides (fucoidan from *Fucus evanescens*) could decrease the intensity of the inflammatory process in the liver of mice with induced chronic toxic hepatitis and to promote the normalization of morphometric parameters and liver function parameters [25].

c) Anti-oxidant properties

Our previous work reported the anti-oxidative potential of polysaccharides extracted from peony seed dreg, onion, *Polygonatum*, Mulberry and *Vaccinium* [43-47, 2]. It was evident from our results that different compositions and structures of polysaccharide extracted from different plant niches varied significantly for scavenging the free radicals and in their reducing power. Similar to anti-bacterial results, sulphation modification resulted in improved anti-oxidant ability in peony seed dreg polysaccharides. We also confirmed that in addition to sugar, nucleic acid and protein content may also affect the anti-oxidant properties to some extent which need to be investigated further. In another studies, polysaccharides extracted from olive leaves and potato peel waste showed strong DPPH-radical scavenging, reducing power and â-carotene bleaching inhibition activities [6, 9].

d) Anti-cancerous potential

The anti-cancerous activities of various plant polysaccharides have been reported due to their non-toxic nature [48]. In our previous study, the molecular mechanism behind the anti-cancerous effects of peony seed dreg polysaccharides was studied in prostate cancer cells, colon cancer cells, human breast cancer cells, and cervical cancer cells. Different fractions extracted by separate physical methods showed variation in their anti-cancerous potential (Cell growth inhibition, cell cycle arrest and apoptosis and mRNA expression of cell cycle related genes) [3]. Nevertheless, new preparation and modification methods can largely impact the anti-cancerous polysaccharides which can be an active area of future research.

1.5. Plant derived Polysaccharides as prebiotics: Fermentation by Intestinal Bacteria

The digestion of plant polysaccharides starts right away in the mouth from mechanical stress and with the release of salivary enzymes, such as amylase. The kind and magnitude of the process of acid hydrolysis in the stomach and enzyme digestion in the intestine depend on the constitution of the monosaccharide units, the array of covalent bonds and their position within the polymer, anomeric forms and substitutions on the sugar molecules of the polysaccharide. On the basis of hydrolysis, these polysaccharides are classified into two groups: fermentable and non-fermentable. Non-fermentable polysaccharides pass to the large intestine and are eventually excreted out as waste/feces. However, the indigestible but fermentable polysaccharides are consumed (i.e., metabolized) by the microflora of the large intestine and are fermented to produce diverse products or metabolites that act as an additional energy source for the host. SCFAs, the major microbial metabolites found in the human gut, are produced mainly as a result of this microbial fermentation of such polysaccharides [49].

In vitro methods of fermentation use batch and continuous fermentation systems to investigate anaerobic fermentation. Fecal samples are a major source of mixed bacteria population, although pure culture of selected bacteria could also be used. Fermentation is measured *in vitro* through pH determination, enumeration of bacterial populations, SCFA quantification, and disappearance of candidate prebiotics in fermented solution. All these indices have been widely used in studying the prebiotic activity of various polysaccharides [50, 51]. *In vivo* methods use fecal samples and content of gastro-intestinal tracts collected from anaesthetized experimental animals after oral administration of proposed prebiotic diets. Heteroxenic rats, which harbour human fecal flora, are good experimental models for evaluating prebiotic activity *in vivo*[52]. Fecal samples obtained are screened for fermentation products, such as gases and SCFA. In a study, GIT and digest samples from the caecum and proximal colon of pigs were aseptically removed and screened for colonic pH, bacterial population, and SCFA [12].

The various methods for studying microbial population can be classified as either culture-dependent or culture independent [53]. Culture-dependent techniques are based on the principle of culturing selected bacterial species on agar media, which support their growth, followed by morphological, biochemical, and physiological assays [53, 12, 49]. This method is able to examine the physiological function of living bacteria strains, detect specific intestinal pathogens, and determine genotypes of cultured isolates. Culture-independent techniques were derived as a result of the increased interest in prebiotic research and the need for more efficient and reliable methods for quantifying microbial populations [52]. These methods operate on the principle of quantifying genetic elements or gene expression and do not require culturing of bacteria. They qualitatively and quantitatively study a wide range of bacterial species. Culture independent techniques include fluorescent *in situ* hybridization (FISH), quantitative polymerase chain reaction, DNA profiling, flow cytometry, DNA microarray and DNA sequencing targeting the 16S ribosomal RNA[49,52]. A detailed review on these methodologies for intestinal bacterial population determination is available [12]. Advanced high throughput techniques and *in silico* analysis depict the relationship between diet and human health with a special reference to GIT microbiota composition and function. SCFAs influence gastro-intestinal epithelial cell integrity, glucose homeostasis, lipid metabolism, appetite regulation, and immune function [49].

2. Conclusions and future outlook

The collected information and above discussion would be useful in the formulation and designing of novel foods containing these functional ingredients from natural sources with human health-promoting characteristics. Therefore, dietary manipulation of gut-associated microflora should be a top priority for the management of various metabolic disorders. Further research should be carried out to identify novel polysaccharides, determine their effect individually and in combination and identify their role in modulating gut microbes and their interaction with host genes targeting specific lifestyle disorders. In case of existing plant based polysaccharides, their underlying mechanisms should be studied for the reported functions.

Currently, most of the recognized pre-biotics offer rapid fermentation in the proximal part of the colon. Recently, this drawback has been highlighted due to occurrence of major colonic diseases in the distal part of the GIT. Therefore, it is imperative to design target specific pre-biotic with improved functional properties which can be used in the desired site of host body. The extraction and purification of plant bio-active molecules is the major route of development of agricultural waste which can be further improved through technological advances in molecular separations and identifications. Among all the carbohydrates consumed, the cell wall polysaccharides constitute a major part of dietary fiber. Therefore, more attention should be drawn towards dietary fiber and resistant forms of starch to positively lower down the risk factors for various chronic diseases. The major and recent evidences about plant polysaccharides present in various edible plants, and how these polysaccharides could be exploited to improve host metabolic health and ameliorate the pathophysiology of metabolic diseases should be the focus of researchers to take this particular area to next level.

Despite the abundance of polysaccharides and their increasing utilization as food materials; however, a relatively low number of polysaccharides are approved for use as food ingredients due to safety issues. Among them, plant-based polysaccharides are dominant which can be used as promising food ingredients. The commercialization of these polysaccharide-based plant extracts after their safety evaluation can pave the way for utilization of various unexplored agricultural waste by-products. Before making any claim regarding the use of plant based polysaccharides as novel food ingredients, there safety must be evaluated in terms of their origin, isolation method, composition, purity, stability, physico-chemical properties, toxicology and allergenicity [54]. Scientists have explored natural sources (plants and micro-organisms) of food grade polysaccharides for their ability to modulate gut microbiota. While several reports have emerged recently claiming the health benefits of such polysaccharides, the mechanisms underlying their function are not studied so far which is a great limitation to scale up the production process of potent functionalized food polysaccharides. All these above challenges should be dealt well to develop strategies to combat the rising diet-related influences which are occurring in both developed countries and rapidly expanding economies, such as China and India.

3. References

1. Gullon, B. Gullon, P. Tavaria, F. Alonso, J.L. and Pintado, M. (2015). *In vitro* assessment of the prebiotic potential of *aloe vera* mucilage and its impact on the human microbiota. *Food & Function*. 6(2): 525-531.
2. Li, L. Liao, B.Y. Thakur, K. Zhang, J.G. Wei, Z.J. (2018). The rheological behaviour of polysaccharides sequential extracted from *Polygonatum cyrtonema* Hua. *International Journal of Biological Macromolecules*. 109: 761–771.
3. Zhang, F. Shi, J.J. Thakur, K. Hu, F. Zhang, J.G. Wei, Z.J. (2017). Anti-cancerous potential of polysaccharide fractions extracted from

peony seed dreg on various human cancer cell lines via cell cycle arrest and apoptosis. *Frontiers in Pharmacology*. 8: 102.

4. Liu, J. Willfor, S. Xu, C. (2015). A review of bio-active plant polysaccharides: Biological activities, fictionalization, and biomedical applications. *Bioactive Carbohydrates & Dietary Fibre*. 5(1): 31-61.

5. Ahmadi, S. Mainali, R. Nagpal, R. and Sheikh-Zeinoddin, M. *et al.* (2017). Dietary Polysaccharides in the Amelioration of Gut Page Microbiome Dysbiosis and Metabolic Diseases. *Obesity & Control Therapies*. 4(3): 1-15.

6. Khemakhem, I. Abdelhedi, O. Trigui, I. Ayadi, M.A. and Bouaziz, M. (2018). Structural, anti-oxidant and anti-microbial activities of polysaccharides extracted from olive leaves. *International Journal of Biological Macromolecules*. 106: 425-432.

7. Shelke, N.B. James, R. Laurencin, C.T. and Kumbar, S.G. (2014). Polysaccharide biomaterials for drug delivery and re-generative engineering. *Polymers for Advanced Technologies*. 25(5): 448–460.

8. Hu, Z. Hong, P. Cheng, Y. Liao, M. and Li, S. (2018). Polysaccharides from *Enteromorpha tubulosa*: optimization of extraction and cytotoxicity. *Journal of Food Processing & Preservation* . *42* (1). https://doi.org/10.1111/jfpp.13373.

9. Jeddou, K.B. Chaari, F. Maktouf, S. Nouri-Ellouz, O. Helbert, C.B. and Ghorbel, R.E. (2016). Structural, functional, and anti-oxidant properties of water-soluble polysaccharides from potatoes peels. *Food Chemistry*. 205: 97-105.

10. Li, S. Xiong, Q. Lai, X. Li, X. Wan, M. and Zhang, J. *et al.* (2016). Molecular modification of polysaccharides and resulting bio-activities. *Comprehensive Reviews in Food Science & Food Safety*. 15(2): 237-250.

11. Li, X.L. Tu, X.F. Thakur, K. Zhang, Y.S. Zhu, D.Y. Zhang, J.G. and Wei, Z.J. (2018). Effects of different chemical modifications on the anti-oxidant activities of polysaccharides seasonal extracted from peony seed dreg . *International's Journal of Biological Macromolecules*. 112: 675-685.

12. Okolie, C.L. Rajendran, S.R.C.K. Udenigwe, C.C. Aryee, A.N.A. and Mason, B. (2017). Prospects of brown seaweed polysaccharides (BSP) as prebiotics and potential immunomodulators. *Journal of Food Biochemistry*. 41(5): 12392.

13. Cockburn, D.W. and Koropatkin, N.M. (2016). Polysaccharide degradation by the intestinal microbiota and its influence on human health and disease. *Journal of Molecular* Biology. 428: 3230–3252.

14. Fava, F. Gitau, R. Griffin, B.A. Gibson, G.R. Tuohy, K.M. and Lovegrove, J.A. (2013). The type and quantity of dietary fat and carbohydrate alter faecal microbiome and short-chain fatty acid excretion in a metabolic syndrome 'at-risk' population. *International Journal of Obesity*. 37: 216–223.

15. Xu, Z.J. and Knight, R. (2015). Dietary effects on human gut microbiome diversity. *Brazilian Journal of Nutrition*. 113: S1–S5.

16. Rastall, R.A. and Gibson, G.R. (2015). Recent developments in prebiotics to selectively impact beneficial microbes and promote intestinal health. *Current Opinion in Biotechnology*. 32: 42–46.

17. Wilson, B., and Whelan, K. (2017). Pre-biotic insulin-type fructans and galacto oligosaccharides: definition, specificity, function, and application in gastro-intestinal disorders. *Journal of Gastroenterology and Hepatology*. 32(Suppl. 1): 64–68.

18. Ferrario, C. Statello, R. Carnevali, L. Mancabelli, L. and Milani, C. (2017). How to Feed the Mammalian Gut Microbiota: Bacterial and Metabolic Modulation by Dietary Fibers. *Frontiers in Microbiology*. 8: 1749-1760.

19. Patel, A.K. Michaud, P. Singhania, R.R. Soccol, C.R. and Pandey, A. (2010). Polysaccharides from Pro-biotics as Food Additives. *Food Technology and Biotechnology*. 48(4): 451–463.

20. Thakur, K. Zhang, J.G. Wei, Z.J. Kumar, N. Tomar, S.K. and Pophaly, S.D. (2018). Cross-Talk Between Functional Foods and Gut Health . *Nutraceuticals and Innovative Food Products for Healthy Living and Preventive Care* . DOI: 10.4018/978-1-5225-2970-5.ch009.

21. Azmi, A.F. Mustafa, S. Hashim, D.M. and Manap, Y.A. (2012). Pre-biotic activity of polysaccharides extracted from *Gigantochloa levis* (buluh beting) shoots. *Molecules*. 17(2): 1635-1651.

22. Xiao, Z. Tappen, B.R. Ly, M. Zhao, W. Canova, L.P. Guan, H. and Linhardt, R.J. (2011). Heparin mapping using heparin lyases and the generation of a novel low molecular weight heparin. *Journal of Medicinal Chemistry*. 54(2): 603–610.

23. Gibson, G.R. (2004).From pro-biotics to pre-biotics and healthy digestive system. *Journal of Food Science*. 69(5): 141-143.

24. Singdevsachan, S.K. Auroshree, P. Mishra, J. Baliyarsingh, B. Tayung, K. and Thatoi, H. (2016). Mushroom polysaccharides as potential pre-biotics with their anti-tumour and immuno-modulating properties: a review. *Bio-active Carbohydrates & Dietary Fibre.* 7(1): 1-14.

25. Zaporozhets, T.S. Besednova, N.N. Kuznetsova, T.A. Zvyagintseva, T.N. Makarenkova, I.D. Kryzhanovsky, S.P. and Melnikov, V.G. (2014). The pre-biotic potential of polysaccharides and extracts of seaweeds. *Russian Journal of Marine Biology.* 40(1): 1-9.

26. Wang, X. Huang, M. Yang, F. Sun, H. Zhou, X. and Guo, Y. Wang, X. and Zhang, M. (2015). Rapeseed polysaccharides as pre-biotics on growth and acidifying activity of pro-biotics in *in vitro. Carbohydrate* Polymers. 125(1): 232-240.

27. Li, P. Qi, L.W. Liu, E.H. Zhou, J.L. and Wen, X.D. (2008). Analysis of Chinese herbal medicines with holistic approaches and integrated evaluation models. TrAC Trends in Analytical Chemistry. 27: 66–77.

28. Manning, T.S. and Gibson, G.R. (2004). Microbial-gut interactions in health and disease. Pre-biotics. *Best Practice & Research: Clinical Gastroenterology.* 18(2): 287-98.

29. Zhou, S. S. Xu, J. Zhu, H. Wu, J. Xu, J. D. and Yan, R., *et al.* (2016). Gut microbiota-involved mechanisms in enhancing systemic exposure of ginsenosides by co-existing polysaccharides In ginseng decoction:. *Scientific Reports.* 6: 22474.

30. Chung, W.S.F. Walker, A.W. Louis, P. Parkhill, J. Vermeiren, J. and Bosscher, D. *et al.* (2016). Modulation of the human gut microbiota by dietary fibres occurs at the species level. *BMC Biology.* 14(1): 3.

31. Samanta, A.K. Jayapal, N. Jayaram, C. Roy, S. Kolte, A.P. Senani, S. and Sridhar, M. (2015). Xylooligosaccharides as pre-biotics from agricultural by-products: production and applications. *Bioactive Carbohydrates & Dietary Fibre.* 5(1): 62-71.

32. Wang, S. Li, Q. Zang, Y. Zhao, Y. Liu, N. and Wang, Y., *et al.* (2017). Apple polysaccharide inhibits microbial dysbiosis and chronic inflammation and modulates gut permeability. In HFD-fed rats. *International Journal of Biological Macromolecules.* 99: 282-292.

33. Quezada, M.P., Salinas, C., Gotteland, M. & Cardemil, L. (2017). Acemannan and fructans from *Aloe vera* (aloe barbadensis miller) plants as novel pre-biotics. *Journal of Agriculture and Food Chemistry.* 65(46): 10029-10039.

34. Xu, X. and Zhang, X. (2015). *Lentinula edodes*-Derived Polysaccharide Alters the Spatial Structure of Gut Microbiota in Mice. *PLoS ONE* 10(1): e0115037.

35. Shin, J.Y. Song, J.Y. Yun, Y.S. Yang, H.O. Rhee, D.K. and Pyo, S.(2002). Immuno-stimulating effects of acidic polysaccharides extract of *Panax ginseng* on macrophage function. *Immunopharmacology and Immunotoxicology*. 24: 469–82.

36. Nair, P.K. Rodriguez, S. Ramachandran, R. Alamo, A. Melnick, S.J. and Escalon, E. *et al.* (2004). Immune stimulating properties of a novel polysaccharide from the medicinal plant *Tinospora cordifolia*. *International Immuno-pharmacology,* 4(13): 1645.

37. Abougheangone, S. Nguemaona, E. Boudjeko, T. and Driouich, A. (2011). Plant cell wall polysaccharides: immuno-modulators of the immune system and source of natural fibers. *Current Topics in Phytochemistry*. 10: 1-16.

38. Ma, Y.L. Zhu, D.Y. Thakur, K. Wang, C.H. Wang, H. Ren, Y.F. Zhang, J.G. Wei Z.J. (2018). Anti-oxidant and anti-bacterial evaluation of polysaccharides sequentially extracted from onion (*Allium cepa* L.). *International Journal of Biological Macromolecules*. 111: 92–101.

39. Li, L. Thakur, K. Liao, B.Y. Zhang, J.G. and Wei, Z.J. (2018). Anti-oxidant and anti-microbial potential of polysaccharides seasonal extracted from *Polygonatum cyrtonema* hua. *International Journal of Biological Macromolecules*. 114: 317-323.

40. Li. X.L. Thakur, K. Zhang, Y.Y. Tu, X.F. Zhang, Y.S. Zhu, D.Y. Zhang, J.G. and Wei, Z.J. (2018). Effects of different chemical modifications on the anti-bacterial activities of polysaccharides sequentially extracted from peony seed dreg. *International Journal of Biological Macromolecule*. https://doi.org/10.1016/j.ijbiomac. 2018.05.082.

41. Smith A.G. O'Doherty, J.V. and Reilly, P. *et al.* (2011). The effects of laminarin derived from *Laminaria digitata* on measurements of gut health: selected bacterial populations, intestinal fermentation, mucin gene expression and cytokine gene expression in the pigs, *Brazilian Journal of Nutrition*. 105: 669–677.

42. Chen, G. Xie, M. Wan, P. Chen, D. Dai, Z. and Ye, H. *et al.* (2018). Fuzhuan brick tea polysaccharides attenuate metabolic syndrome in high-fat diet induced mice in association with modulation in the gut microbiota. *Journal of Agricultural & Food Chemistry*. DOI: 10.1021/acs.jafc.8b00296.

43. Li. X.L. Tu, X.F. Thakur, K. Zhang, Y.S. Zhu, D.Y. Zhang, J.G. and Wei, Z.J. (2018). Effects of different chemical modifications on the anti-oxidant activities of polysaccharides seasonal extracted from peony seed dreg. *International Journal of Biological Macromolecules*. 112: 675–685.

44. Shi, J.J. Zhang, J.G. Sun, Y.H. Qu, J. Li, L. Prasad, C. and Wei, Z.J. (2016). Physico-chemical properties and anti-oxidant activities of polysaccharides sequentially extracted from peony seed dreg. *International Journal of Biological Macromolecules*. 91: 23–30.

45. Zhu, D.Y. Ma, Y.L. Wang, C.H. Wang, H. Ren, Y.F. Zhang, J.G. and Wei, Z.J. (2017). Insights into physico-chemical and functional properties of polysaccharides periodically extracted from onion (*Allium cepa* L.). *International Journal of Biological Macromolecules*. 105(Pt 1): 1192-1201.

46. Liao, B.Y. Zhu, D.Y. Thakur, K. Li, L. Zhang, J.G. and Wei Z.J. (2017). Thermal and anti-oxidant properties of polysaccharides sequentially extracted from mulberry leaves (*Morus alba* L.). Molecules 22: 2271.

47. Xu, Q.X. Shi, J.J. Zhang, J.G. Li, L. Jiang, L. and Wei Z.J. (2016). Thermal, emulsifying and rheological properties of polysaccharides sequentially extracted from *Vaccinium bracteatum* Thumb leaves. *International Journal of Biological Macromolecules*. 93: 1240–1252.

48. Meng, X. Liang, H. and Luo, L. (2016). Anti-tumour polysaccharides from mushrooms: a review on the structural characteristics, anti-tumour mechanisms and immuno-modulating activities. *Carbohydrate Research*. 424: 30–41.

49. Holscher, H.D. (2017). Dietary fiber and pre-biotics and the gastro-intestinal microbiota. *Gut Microbes*. 8(2): 172–184.

50. Charoensiddhi, S. Conlon, M.A. and Vuaran, M.S. (2016). Impact of extraction processes on prebiotic potential of the brown seaweed *Ecklonia radiata* by *in vitro* human gut bacteria fermentation. *Journal of Functional Foods*. 24: 221–230.

51. Kong, Q. Dong, S. Gao, J. and Jiang, C. (2016). *In vitro* fermentation of sulfated polysaccharides from *E. prolifera* and *L. japonica* by human fecal microbiota. *International Journal of Biological Macromolecules*. 91: 867–871.

52. Roberfroid, M.B. (2008). Pre-biotics: Concept, definition, criteria, methodologies, and products. In G.R. Gibson & M.B. Roberfroid

(Eds.), Handbook of pre-biotics (pp. 39–68). Boca Raton, FL: CRC Press, Taylor & Francis Group.

53. Gong, J. and Yang, C. (2012). Advances in the methods for studying gut microbiota and their relevance to the research of dietary fiber functions. FRIN. 48: 916–929.

54. Yoo, H.D. Kim, D. Paek, S.H. and Oh, S.E. (2012). Plant Cell Wall Polysaccharides as Potential Resources for the Development of Novel Pre-biotics. *Biomolecules and Therapeutics*. 20(4): 371-379.

❑❑❑

Chapter 14

Recent Trends in Nanotechnology in Food Industry-Applications, Opportunities and Challenges

M. R. Manikantan[1], R. Pandiselvam[*1], and N. Varadharaju[2]

[1]Physiology, Biochemistry and Post Harvest Technology Division, [1]ICAR-Central Plantation Crops Research Institute, Kasargod – 671 124, Kerala, India, [2]Agricultural Engineering College & Research Institute, [2]Tamil Nadu Agricultural University, Coimbatore-641 003, India, [*]Corresponding author : anbupandi1989 @yahoo.co.in

ABSTRACT

Application of nanotechnology in the food processing industry will provide new tools to enhance safety and the nutritional value of food products. This chapter will review the recent trends in nanotechnology in various fields of food science and technology including nutraceuticals and functional food development, food packaging (nanocomposite, nanobiodegradable packaging, nanoantimicrobial packaging, nanointelligent packaging), nanofiltration, detection of food borne pathogens, and shelf-life extension of food products. Also, it describes new current food laws on safety issues and regulatory concerns on nanoprocessed food products.

1. Introduction

Nanotechnology is a rapidly growing science in use today, for producing and utilizing nano sized particles including atoms, macromolecules or molecular clusters with the size of approximately 1–100 nm to create and use materials that have novel properties [1]. Ultra small scale materials have inimitable properties due to high surface to volume ratio and other novel and improved physical, thermal, optical and electrical properties like solubility, colour, diffusivity, toxicity, strength, magnetic, thermodynamic, optical, etc.

[2]. Hence, it is apparent that the new nanosized particles replace the old because of their enhanced function has brought new industrial revolution [3].

Nanotechnology offers a range of opportunities for the design, production, application and development of materials/devices or structures with improved properties in various areas like biochemistry, food packaging, food safety, wastewater, water treatment, agriculture, medicine, and construction materials for machines, floors, and electronics, etc., increase the magnitude of its transformative potential [4]. Even agricultural production and food industries have witnessed the use of nanotechnology so that can be integrated into number of food systems and food packaging products.

Ultra small scale level of foods can affect the efficiency, safety, nutritional properties, molecular synthesis, and bioavailability of new products [5]. The development of new functional materials, microscale and nanoscale processing, product development, and design of methods and instrumentation in food production may benefit from nanotechnology. The major goals of nanotechnology to the food industry are improving food safety and functionality of foods, cost-effective storage and distribution, pathogen detection, and protection of the environment. This book chapter will describe recent scientific results that address the uses of nanotechnology in food industry and also discussed some negative facts associated with nanotechnology. However, the full potential of nanotechnology in the agricultural and food industry has still not been realized.

1.1. Applications in Food Industry

The applications of nanotechnology in food industry can be re-capitulated in two groups, such as food nanostructured ingredients and food nanosensing [6]. The applications of this technology in food sector are summarized in Table 1. The food nanostructured ingredients can be used as carriers for smart delivery of nutrients, food additives, anti-caking agents, fillers for improving mechanical strength and durability of the packaging material, anti-microbial agents, etc., whereas food nanosensing can be applied to enhance the food quality and safety [7].

1.1.1. Nano-nutraceuticals and Nano-functional foods

Agri-Food Canada defines nutraceuticals and functional foods as "food components that provide demonstrated physiological benefits or reduce the risk of chronic disease, above and beyond their basic nutritional functions. A functional food is similar to a conventional food, while a nutraceutical is isolated from a food and sold in dosage form, in both cases the active components occur naturally in the food.

Table 1: Application of nanotechnology in food sector

Food sector	Purpose/Material used	Approaches
Food Processing	Nanofiltration	Passage of materials on the basis of shape and size
	Nanocapsules for modification of absorption	Nanoceramic pan to reduce time of roasting and amount of consumed oil, reduction of trans fatty acids due to usage of plant oil instead of hydrogenated oil and finally resulted in safe nano food development of nanocapsules that can be incorporated into food to deliver nutrients to enable increased absorption of nut
	Nanofabrication for heat and mass transfer	Enhanced heat resistance of packages
Food Packaging	Nanocomposites application-increasing the barrier, mechanical and heat-resistance properties	Nanocomposites application as barriers, coating, release device, and novel packaging modifying the permeation behavior of foils, increasing barrier properties (mechanical, thermal, chemical, and microbial), developing active antimicrobial surfaces, sensing as well as signaling microbiological and biochemical changes, developing dirt repellent coatings for packages
	Antimicrobial Packaging	Polypropylene flims coated with TiO_2 nanoparticles to inhibit the growth of E.coli
		Carbohydrate-phytoglycogen exhibited sustained antimicrobial activity against plated L. monocytogenes with efficacy that lasts several times longer than free nisin
	Edible coatings- To preserve the quality of fresh foods during extended storage	Gelatin-based edible coatings containing cellulose nanocrystal
Quality control and food safety	Nanosensors and Nanobiosensors	Detection of chemical contaminants
		Food born pathogen identification by measurement of nucleic acid, protein or any other indicator metabolite of micro-organism
		Amperometric detection of coliforms based on multiwall carbon nanotubes/ Nafion composite film modified glass carbon electrode
		TiO2 nanoparticles throughout EVOH films observed photo-activated biocidal properties against nine food poisoning and spoilage bacteria and yeasts
Food Nutrition	Target delivery	Nanocapsulation for controlled release of nutrients, proteins, anti-oxidants, and flavours
		Nanomycells for targeted delivery of nutrients (nutrition nanotherapy)
	Nano-selenium-enriched Nanotea	Enhance uptake and bioavailability of selenium
	Nutri-Nano CoQ-10	Increase absorption of the fat-soluble Coenzyme Q-10
	Nanoceuticals Slim Shake Chocolate	Enhanced taste and flavour
	Vifit Calcimel	Ca+2 enriched fermented dairy drink to enhance bioavailability
Nanobiocatalyst	Industrial glucose production	Glucoamylase-based nanobiocatalyst (immobilizing the enzyme covalently with glutaraldehyde and retained by electrostatic interactions to graphite nanosheets or to carboxyl-functioned MNPs

(Source: Rashidi and Khosravi-Darani [8], *Singh et al.* [6]; *Patel et al.* [9])

1.1.2. Nanocochleates

Nanocochleates are novel drug delivery system based upon entrapment of desired drug molecule into the multi-layered structure containing solid-lipid bilayer in the form of sheet rolled up in spiral shape [10]. The nanocochleate structure provides protection to encapsulated molecule form surrounding harsh environment. In food and beverage industry, attempts have been made to add micronutrients and anti-oxidants to food substances. But these anti-oxidants degrade during manufacturing and food storage. Nanocochleates delivery system protects these substances from degradation. Ex. Polyphenols and Resveratrol are the substances present in most foods and wine, respectively. They gets degraded and oxidized, when exposed to air. Nanocochleates solve early oxidation by indivdually cpaturing and wrapping then in a phospholipid wrap, and maintaining the internal nutrients secure from water and oxygen.

BioDelivery Sciences International have developed nanocochleates which are 50 nm coiled nanoparticles and can be used to deliver nutrients, such as vitamins, lycopene and omega 3 fatty acids more efficiently to cells, without affecting the colour or taste of food. The delivery vehicle is made of soyphophatidylserine which is 100% safe. It provides a protective coat for range of nutrient additives (www.biodeliverysciences.com).

1.1.3. Nanoclusters™

R&D Scientists at Royla Body Care also developed NanoClusters™, a nanosize powder that combines with nutritional supplements. When consumed, it reduces the surface tension of foods and supplements to increase wetness and absorption of nutrients. It replaces our previous microclusters ingredient. Bioral nanocochleates is a licenced and patented delivery technology, where bioral omega 3 nanocochleates is introduced. The common sources of omega-3-fatty acid are cold water fish which have an unpleasant fishy odor and taste.

But the bioral omega 3 formulation has some advantages, like:

- It feels like flour
- No fish odour or taste
- Tested in food produts
- Can subjected to baking and heating
- Ability to protect omega 3 fatty acid from degradation

1.2. Food Packaging

In the food packaging arena, nano materials are being developed with enhanced mechanical and thermal properties to ensure better protection of food from exterior mechanical, thermal, chemical or microbiological effects. This would endow packaged food with an additional level of safety and functionality. It would also offer advantages along the supply chain and potentially increase the shelf-life of food. Some of the potential uses of nanotechnology in food packaging include modifying the permeation behaviour of films, increasing barrier properties, improving mechanical and heat resistance properties, developing active anti-microbial and anti-fungal surfaces, sensing and signalling microbiological and biochemical changes [11].

Consumers demands food to be fresh for long time, and the packaging materials should be ease for handling, safe and healthy to human and also to the environment. A major problem in food science is determining and developing an effective packaging material. Using nanoparticle technology, Bayer has developed an even more airtight plastic packaging that will keep food fresher and longer than their previous plastics and the plastics of their competitors. Researchers at Bayer Polymers refer to this new plastic as a "hybrid system", as it is enriched with an enormous number of silicate nanoparticles. When this plastic is processed into a thin film and wrapped over food, it does a better job than previous plastics of preventing food from going bad on the shelf and it helps prevent odours from one food mixing with another.

The most problematic for food packaging engineers is oxygen because it spoils the fat in meat and cheese and turns them pale. Due to the nature of the nanoparticles in Durethan®, Bayer's new plastic material, air cannot penetrate it like other conventional plastics. The embedded particles have a maze like arrangement in the plastic, acting like barriers, which makes it difficult for gases, like oxygen, to pass through the packaging. They actually increase the distance the gas molecules have to travel by causing those molecules to zigzag around the silicate plates in effect increasing the amount of time it will take for the molecules to completely penetrate.

1.2.1. Polymer Nanocomposites

Nanocomposite technology has been described as the next great frontier of material science in packaging. This technology was developed to improve barrier performance pertaining to gases, such as oxygen and carbon dioxide [12]. It also enhances the barrier performance to ultraviolet rays, as well as adding strength, stiffness, dimensional stability, and heat resistance. New plastics created with this technology demonstrate an increased shelf-life and are less likely to shatter. Once perfected, these plastics will offer these improved characteristics at competitive prices. It will also make them attractive for use in food and beverage packaging and pharmaceutical packaging applications.

"Polymer nanocomposites consist of resins (either thermoset or thermoplastics) that have fillers added with a least one dimension measured in nanometers. These fillers fall into three categories: nanotubes, nanoscale oxides and metals, and nanoclays. Nanotube-based nanocomposites are used for electrostatic dissipation applications; nanoscale oxides and metals are used for abrasion-resistant films; and nanoclay-based nanocomposites are used for barrier packaging applications. "Nanocomposites have received considerable attention over the past decade because of their potential to achieve property enhancements significantly greater than that attainable using conventional fillers or polymer blends.

Conventional fillers, such as minerals, metals, and fibers have been added to polymers for decades to produce composites with improved mechanical and thermal properties. High-aspect-ratio nano-fillers also provide improvements in barrier performance and clarity without a significant increase in density, which is not feasible with conventional composites.

Nanocomposite studies at Soldier Science Center (SSC) have incorporated montmorillonite, a layered silicate commonly used in polymeric nanocomposite systems. Montmorillonite layered silicates (MLS) have high cation exchange, high surface area (around $750m^2/g$), and large aspect ratio with a platelet thickness of 10 Å. The closely stacked structure of the silicate layers and chemical incompatibility between the hydrophilic clay and hydrophobic polymer are two significant challenges, when developing an exfoliated nanocomposite with improved properties.

Organically modifying the MLS by replacing inorganic exchange cations with organic onium ions on the gallery surfaces has been shown to compatibilize the MLS and polymer, and also leads to increased basel spacing between the silicate layers due to the bulky organic ions.

Greatest property improvements have been seen in exfoliated systems, where silicate layers have delaminated and are homogeneously dispersed into the polymer matrix. Polymer nanocomposites studied at Natick have been shown to be anywhere from intercalated to highly intercalated. SSC has studied both biodegradable and re-cyclable polymer nanocomposites for applications in Meal-Ready-to-Eat (MRE) packaging. Biodegradable polymers include polylactic acid (PLA) and polyhydroyalkanoates (PHA). Recyclable polymers include LDPE, ethylene co-vinyl alcohol (EVOH), and polyethylene terephthalate (PET). These polymers were chosen based on their barrier properties and potential use as food packaging.

Properties which have been shown to undergo substantial improvements include:

- Mechanical properties e.g. strength, modulus and dimensional stability
- Decreased permeability to gases, water and hydrocarbons
- Thermal stability and heat distortion temperature
- Flame retardancy and reduced smoke emissions
- Chemical resistance
- Surface appearance
- Electrical conductivity
- Optical clarity in comparison to conventionally filled polymers

Triton Systems and the US Army are conducting further work on barrier performance in a joint investigation. The requirement here is for a non-refrigerated packaging system capable of maintaining food freshness for three years. Nanoclay polymer composites are currently showing considerable promise for this application. USFDA has approved the use of nanocomposite in contact with foods.

1.2.2. Nano Degradable Packaging

The use of nanomaterials to strengthen bioplastics (plant-based plastics) may enable bio based nanocomposite plastics to be used instead of fossil-fuel based plastics for food packaging and carry bags. Bio based nanocomposites are a new class of materials in food packaging industry with improved barrier and mechanical properties as compared to those of neat biopolymers. They are biodegradable and they are also produced from renewable resources. So, these make them environment friendly. It should be noted that barrier properties and especially mechanical properties of bio based nanocomposite films are stronger than edible films and synthetic polymeric films. Unlike edible films, they could not have been consumed as a part of food.

Bio based nanocomposites are produced from incorporation of nanoclay into biopolymers (or Edible films). Advantages of bio based nanocomposites are numerous and possibilities for application in the packaging industry are endless. Bio based nanocomposites can be used to extend the shelf-life of the fresh products, such as fruits and vegetables by controlling respiratory exchange. Also, it can improve the quality of fresh, frozen, and processed meat, poultry, and seafood products by retarding moisture loss, reducing lipid oxidation and discolouration, enhancing product appearance, and reducing oil uptake by battered and breaded products during frying.

Bio based nanocomposites are composed of bio-polymer, nanoclay and usually compatibilizing agents. Major component of bio based nanocomposites is biopolymers. Bio-polymers have great commercial potential for bio plastic and edible films, but some of the properties, such as brittleness, low heat distortion temperature; high gas permeability and low melt viscosity for further processing restrict their use in a wide range of applications [13]. As mentioned before, modification of bio-polymers with nanotechnology is an effective way to improve their properties. Bio-polymers derived from renewable resources are broadly classified according to method of production. This gives the following three main categories [14]:

i) Bio-polymers directly extracted/removed from natural materials (mainly plants), such as hydrocolloids (polysaccharides and proteins). The most frequently utilized polysaccharides were cellulose and starch (and their derivatives), chitosan, seaweed extracts (carrageenans and alginates), exudates (arabic gum), seed (guar gum), xanthan and gellan gum and pectin. Proteins include collagen, gelatin, casein, whey proteins, corn zein, wheat gluten and soy proteins.

ii) Bio-polymers produced by classical chemical synthesis from renewable bio derived monomers like polylactate.

iii) Bio-polymers produced by micro-organisms or genetically transformed bacteria like Polyhydroxyalkanoates.

Hence, bio-polymers which can be used in bio based nano-composites formulation are numerous.

The utilization of special compatibilizing agents (modifier) between the two basic materials (bio-polymer and nanoclay) for the preparation of bio based nanocomposite is necessary. Layered silicates are characterized by a periodic stacking of mineral sheets with a weak interaction between the layers and a strong interaction within the layer. The space in-between the layers is occupied by cations. By cation exchange reactions between the clay and organic cations (such as alkyl ammonium salts), the layered silicate can be transformed into organically modified clay. The inter-layer distance will increase by using voluminous modifiers. If this modifier is compatible with bio-polymer as well, a homogeneously and nanoscaled distribution (exfoliation) of the clay sheets can be effected in the polymer matrix. The pure clay shows an inter-layer distance of 1.26 nm. It has been proven by XRD analysis that most of the layers are indeed *swollen* after the modification reaction. The inter-layer distance changes to 2.34 nm, an increase of nearly 100% compared to the pure clay.

Bio-polymers like starch present some drawbacks, such as the strong hydrophilic behaviour (poor moisture barrier) and poorer mechanical properties than the conventional non-biodegradable plastic films used in the food packaging industries [15-17]. So, incorporation of nanoclay in bio-polymers like starch can improve its properties, such as barrier and mechanical properties [18]. The most commonly used nanoclays include montmorillonite, a 2:1 phyllosilicate [19].

1.2.3. Anti-microbial nanopackaging materials

The incorporation of anti-microbial compounds into food packaging materials has received considerable attention. Films with anti-microbial activity could help control the growth of pathogenic and spoilage micro-organisms. An anti-microbial nanocomposite film is particularly desirable due to its acceptable structural integrity and barrier properties imparted by the nanocomposite matrix, and the anti-microbial properties contributed by the natural anti-microbial agents impregnated within. Materials in the nanoscale range have a higher surface-to-volume ratio, when compared with their microscale counterparts. This allows nanomaterials to be able to attach more copies of biological molecules, which confers greater efficiency.

1.2.4. Polymer composites with nano-metals or metal oxides

Polymer nanocomposites incorporating metal or metal oxide nanoparticles are utilized mainly for their anti-microbial action, abrasion resistance, UV absorption, and strength. Some nano materials have been used to develop active packaging that can absorb oxygen and therefore, keep food fresh. Other nano materials have been incorporated as UV absorbers to prevent UV degradation in plastics, such as PS, PE and PVC. The commercially important nanomaterials in this respect include nanosilver and nanozinc oxide for anti-microbial action, nanotitanium dioxide for UV protection in transparent plastics, nanotitanium nitride for mechanical strength and as a processing aid, and nano silica for surface coating.

The most common nanocomposites used as anti-microbial films for food packaging are based on silver, which is well known for its strong toxicity to a wide range of micro-organisms, with high temperature stability and low volatility. Some mechanisms have been proposed for the anti-microbial property of silver nanoparticles (Ag-NPs): adhesion to the cell surface, degrading lipopolysaccharides and forming "pits" in the membranes, largely increasing permeability; penetration inside bacterial cell, damaging DNA; and releasing anti-microbial Ag^+ ions by Ag-NPs dissolution. The latter mechanism is consistent with findings by Kumar and Munstedt [20], who affirmed that the anti-microbial activity of silver-based polymers depends on releasing of Ag^+, which binds to electron donor groups in biological molecules containing sulphur,oxygen or nitrogen.

It is important to note that the surface biocides, such as nano silver, in packaging materials are not intended to have a preservative effect on the food. Instead, the biocidal agent is intended to help maintain the hygienic condition of the surface by preventing or reducing microbial growth. Where the use of a nano material gives a preservative effect in the packaged product, there would be a requirement for additional regulatory authorization as a direct food additive in most countries. Based on the anti-microbial action of nano silver, a number of "active" FCMs have been developed that are claimed to preserve the food materials by inhibiting the growth of micro-organisms. Examples include food storage containers and plastic storage bags. Nano silver has also been incorporated into the inner surface of some domestic refrigerators to prevent microbial growth and maintain a clean and hygienic environment in the fridge. The discovery of anti-microbial properties of nano zinc oxide and nano magnesium oxide at the University of Leeds may provide more affordable materials for such applications in food packaging [21]. A plastic wrap containing nano zinc oxide is also available, which is claimed to sterilize under indoor lighting.

Nanoscale chitosan also has antibacterial activity. One possible anti-microbial mechanism by nanoscale chitosan involves interactions between positively charged chitosan and negatively charged cell membranes, increasing membrane permeability and eventually causing rupture and leakage of intracellular material. Another two anti-microbial mechanisms suggest: chelation of trace metals by chitosan, inhibiting enzyme activities; and, in fungal cells, penetration through the cell wall and membranes to bind DNA and inhibit RNA synthesis. Carbon Nano Tubes (CNTs) have been also reported to have anti-bacterial properties. Direct contact with aggregates of CNTs was demonstrated to be fatal for *E. coli*, possibly because the long and thin CNTs puncture microbial cells, causing irreversible damages.

1.2.5. Coatings containing nanoparticles

Coatings that contain nanoparticles are used to create anti-microbial, scratch resistant, anti-reflective, or corrosion-resistant surfaces. This involves the coating of nanoparticulate form of a metal, metal oxide or a film resin substance with nanoparticles. Examples of FCMs with nanocoating include anti-bacterial kitchenware, cutting boards and teapots. High-barrier nanocoatings have also been developed that contain numerous nanodispersed platelets per micron of coating thickness to increase the barrier properties of PET; this enhances the oxygen barrier, when used in food and drink applications, ensuring longer shelf life. The coatings have been reported to be very efficient at keeping out oxygen and retaining carbon dioxide and can rival traditional active packaging technologies, such as oxygen scavengers [22]. Examples include a nanocoating which is an aqueous-based nanocomposite barrier coating that provides an oxygen barrier with a 1–2 micron coating for food packaging use, and plasma arc deposition of amorphous carbon inside PET bottles as a gas barrier.

Waxy coating is used widely for some foods, such as apples and cheeses. Recently, nanotechnology has enabled the development of nanoscale edible coatings as thin as 5 nm thick, which are invisible to the human eye. These edible nano-coatings could be used on meats, cheese, fruits, vegetables, confectionery, bakery goods and fast foods. They could provide a barrier to moisture and gas exchange, act as a vehicle to deliver colours, flavours, anti-oxidants, enzymes and anti-browning agents and could also increase the shelf-life of manufactured foods, even after the packaging is opened. The U.S. Company Sono-Tec Corporation announced in early 2007 that it has developed an edible anti-bacterial nano-coating, which can be applied directly to bakery goods. Another trend in this respect is the chemical release nano-packaging. This technique enables food packaging to interact with the food. The exchange can be processed in both directions. Packaging can release nanoscale antimicrobials, anti-oxidants, flavours, fragrances or nutraceuticals into the food or beverages to extend its shelf-life or to improve its taste or smell.

In many instances chemical release packaging also incorporates surveillance elements, where the release of nano-chemicals will occur in response to a particular trigger event. Conversely, nano-packaging using carbon nanotubes are being developed with the ability to pump out carbon dioxide that would otherwise result in food or beverages deterioration. Nano-packaging that absorbs undesirable flavours is also in development stage.

The chemical release packaging is also designed to release biocides in response to the growth of a microbial population, humidity or other changing conditions. Other packaging and food contact materials incorporate antimicrobial nanomaterials, that are designed not to be released, so that the packaging itself acts as an anti-microbial agent. These products commonly use nanoparticles of silver, although some use nanosize zinc oxide or chlorine dioxide and other materials. Nano magnesium oxide, nano copper oxide, nano titanium dioxie and carbon nanotubes are also predicted for future use in anti-microbial food packaging.

1.2.6. Nanofiltration

Nanofiltration water treatment is a liquid separation membrane technology. Nanofiltration is selected, when Reverse Osmosis and Ultrafiltration are not the correct choice for separation. Nanofiltration can perform separation applications, such as demineralization, colour removal, and desalination. In concentration of organic solutes, suspended solids, and polyvalent ions, the permeate contains monovalent ions and low-molecular-weight organic solutions like alcohol.

Nanoporous membranes are suitable for a mechanical filtration with extremely small pores smaller than 10 nm. Nanofiltration is mainly used for the removal of ions or the separation of different fluids. Magnetic nanoparticles offer an effective and reliable method to remove heavy metal contaminants from waste water by making use of magnetic separation techniques. Using nanoscale particles increases the efficiency to absorb the contaminants and is comparatively inexpensive compared to traditional precipitation and filtration methods (www.geafiltration.com).

Nanofiltration systems offer cost-effective alternative to commercial reverse osmosis units, as Nanofiltration systems can be operated at low pressures. The level of dissolved solids to be removed is less than what is typically encountered in brackish water or seawater. Sometimes, referred to as membrane softening or membrane water softeners, nanofiltration is an alternative to salt based water softening. Like reverse Osmosis, the mass transfer mechanism in Nanofiltration is diffusion. Though generally quite similar in terms of membrane chemistry, the Nanofiltration membrane allows the diffusion of certain ionic solutes (such as sodium and chloride), predominantly monovalent ions, as well as water. Larger ionic species, including divalent and multivalent ions, and more complex molecules are highly retained.

Since monovalent ions are diffusing through the Nanofiltration membrane along with the water, the osmotic pressure difference between the solutions on each side of the membrane is not as great and this typically results in somewhat lower operating pressure with Nanofiltration compared with Reverse Osmosis. Micro Filtration removes particulate material ranging in size from 0.1 to 1.0 microns (1,000 to 10,000 angstroms) and larger. Nanofiltration is used for separating materials less than 0.001 microns (10 angstroms) in size.

Some typical applications for Nanofiltration are:

- Desalination of food, dairy and beverage products or by-products
- Partial De-salination of whey, UF permeate or retentate as required
- De-salination of dyes and optical brighteners
- Purification of spent clean-in-place (CIP) chemicals
- Color reduction or manipulation of food products
- Concentration of food, dairy and beverage products or by-products
- Fermentation byproduct concentration.

1.2.7. Nano-emulsions

Nano-emulsions are nano-sized emulsions, which are manufactured for improving the delivery of active pharmaceutical ingredients. These are the thermodynamically stable isotropic system in which two immiscible liquids are mixed to form a single phase by means of an emulsifying agent, i.e., surfactant and co-surfactant. The droplet size of nano-emulsion falls typically in the range 20–200 nm. The main difference between emulsion and nano-emulsion lies in the size and shape of particles dispersed in the continuous phase. These nano-emulsions find upsurge of interest not only in the pharmaceutical, medical, cosmetic, and engineering industry but also revolutionized the food industry. In the food industry, nano-emulsions are being explored to encapsulate, stabilize, and deliver lipophilic constituents like flavours, omega-3 fatty acids, vitamins, preservatives, nutraceuticals. They have a number of potential advantages over conventional emulsions like incorporation into optically transparent products, may enhance the texture, stability, and bioavailability of products. Food proteins like soybean protein isolate, whey protein isolates are used as safer stabilizers for nano-emulsions. Formation of nano-emulsions requires processes like high-energy emulsification with the help of high-pressure homogenizers, sonication, and some low-energy emulsification.

1.3. Intelligent Packaging based on Nanosensors

Nanotechnology has also enabled the development of nanosensors that can be applied as labels or coatings to add an intelligent function to food packaging in terms of ensuring the integrity of the package through detection of leaks (for foodstuffs packed under vacuum or inert atmosphere), indications of time–temperature variations (e.g. freeze–thaw–refreezing), or microbial safety (deterioration of foodstuffs). Examples include an indicator that turns from transparent to blue, informing the consumer that air has entered the modified atmosphere of the packaged materials. For this type of application, nanotechnology-derived printable inks have been developed. One example is an oxygen detecting ink containing light-sensitive titanium oxide (TiO_2) nanoparticles, which only detect oxygen when they are "switched on" with UV light. Other conductive inks for ink jet printing based on copper nanoparticles have also been developed [23].

Food safety also requires confirmation of the authenticity of products. This is where application of nanobarcodes incorporated into printing inks or coatings has shown the potential for use in tracing the authenticity of the packaged product [24]. Food quality indicators have also been developed that provide visual indication to the consumer, when a packaged foodstuff starts to deteriorate. An example of such food quality indicators is a label based on detection of hydrogen sulphide, which is designed for use on fresh poultry products. The indicator is based on a reaction between hydrogen sulphide and a nanolayer of silver [25]. The nanosilver layer is opaque light brown, but when processed meat starts to deteriorate silver sulphide is formed and the layer becomes transparent, indicating that the food may be unsafe to consume.

Other materials developed for potential food packaging applications are based on nanostructured silicon with nanopores. The potential applications include detection of pathogens in food and variations of temperature during food storage. Another relevant development is aimed at providing a basis for intelligent preservative packaging technology that will release a preservative only when a packaged food begins to spoil (ETC Group, 2004 [26]).

Based on applied studies of the surface properties of materials, several types of gas sensors have been developed, which translates chemical interactions between particles on the surfaces into a response signal. Metal oxide gas sensors are one of the most popular types of sensors because of their high sensitivity and stability. Conducting polymers (CPs) or electro active conjugated polymers, which can be synthesized either by chemical or electro-chemical oxidation, are very important because of their electrical, electronic, magnetic and optical properties, which are related to their conjugated ð electron backbones. Polyene and polyaromatic CPs such as polyaniline (PANI), polyacetylene, polypyrrole (PPy) has been widely studied. Electrochemically polymerized CPs have a remarkable ability to switch between conducting oxidized (doped) and insulating reduced (undoped) state, which is the basis of many applications. Food spoilage is caused by micro-organisms, whose metabolism produces gases which can be detected by conducting polymer nanocomposites (CPC) or metal oxides, which can be used for quantification and/or identification of micro-organisms based on their gas emissions. Sensors based on CPC consist on conducting particles embedded into an insulating polymer matrix. The resistance changes of the sensors produce a pattern that corresponds to the gas under investigation.

Scientists at Kraft, as well as at Rutgers University and the University of Connecticut, are working on nano-particle films and other packaging with embedded sensors that will detect food pathogens. Referred as "electronic tongue" technology, the sensors can detect substances in parts per trillion and would trigger a colour change in the packaging to alert the consumer if a food has become contaminated or if it has begun to spoil.

Researchers in the Netherlands are going one further to develop intelligent packaging that will release a preservative, if the food within begins to spoil. This "release on command" preservative packaging is operated by means of a bio-switch developed through nanotechnology.

Georgia Tech in the United States used modified carbon nanotube as bio-sensor to detect micro-organisms, toxic substances and spoilage of foods or beverages. Another example, Opel, which makes Opalfilm incorporating 50nm carbon black nano particles, was used as bio-sensor that can change colour in response to food spoilage.

Nanotechnology is also enabling sensor packaging to incorporate cheap Radio Frequency Identification (RFID) tags. The nano-enabled RFID tags are much smaller, flexible and can be printed on thin labels. This increases the tags versatility and thus, enables much cheaper production. Other nano-based track and trace packaging technologies are also being improved. For instance, a United States company Oxonica Inc, has developed nano-barcodes to be used for individual items or pellets, which must be read with a modified microscope. These have been developed primarily for anti-counterfeiting purposes.

Food safety

We come into contact with millions of bacteria in our daily life. It is found in the air we breath, in the food we eat and in most of the things we touch. Organic dyes are the most commonly used biolabels to stain bacteria for detection. Organic dyes are expensive and their flouroscence degrades with time. So there is a need for better alternative. Recent advances in field of luminescence nanocrystals have led to new area of research in fluorescent labeling by quantum dots. QDs have several advantages over conventional organic flurophores.

Detection of *E-coli* 0157:H7 using quantum dots made successful. Nanobioluminescence detection spray, can also be used which binds to the surface of the microbes and emits a visible glow. More intense the glow, higher is the bacterial contamination. This helps in easy detection of contaminated food and beverage.

Improved biosensor technology may be used to detect gases present in packaged foods as a measure of integrity of the packaging material, compounds released during food spoilage or deterioration, and the presence of pathogens or toxins in foods. Such sensors could be incorporated into packaging to alert consumers, producers, and distributors as to the safety status of foods or could be used to detect pathogens in processing plants.

The gas sensors of electronic nose (e-nose) are composed of nanoparticles (e.g. Zinc oxide nanowires) whose resistance changes, when a certain gas is made to pass over it. This change in resistance generates a change in electrical signal that forms the fingerprint for gas detection. This finger print pattern derived from the sensor is used to determining the type, quality and quantity of the odour being detected. The advantage of using nanoparticles is that they have improved surface area for better gas adsorption.

1.4. Opportunities in Nanotechnology based Food Design

Understanding the nature of nanostructures in foods allows a better selection of raw materials and control of processing and packaging in order to optimize product quality and shelf-life. A large number of nanoscience tools have been developed over the last 10-20 years, such as the atomic force microscope (AFM), which has enabled researchers to gain atomic and molecular scale insight. Additionally, a variety of synthesis techniques have resulted in design of new functional nanoparticles.

Some food processing methods utilize enzymes to alter food components to improve flavor, nutritional value or other characteristics. Immobilization of these enzymes on nanoparticles may aid in dispersion through food matrices and enhance their activity. Nano-silicon dioxide particles with reactive aldehyde groups were constructed and found to covalently bind to a porcine triacylglycerol lipase. These particles effectively hydrolyzed olive oil and were determined to have good stability, adaptability and reusability.

Nanofood refers to the use of nanobiotechnology and nanotools, applied during the cultivation, production, processing, or packaging of foods. Nanotechnology can be used to create tiny sensors and diagnostic machines, which prevent spoiled and contaminated foods from leaving the processing facilities and ending up at the consumers. Furthermore, nanotechnology may bring about food products, which are designed from scratch starting with simple molecular structures. The resultant products may have tailored functionality and be produced at lower cost using sustainable production methods.

It seems almost certain that most major food companies are monitoring or researching the potential benefits of nanoscience in food. Some companies are more willing to discuss this aspect of their research than others, so it is difficult to assess the precise level of interest in these topics. Kraft Foods started the first nanotechnology laboratory in 1999 and its 'Nanotek' consortium, involving 15 universities worldwide and national research laboratories was established in 2000. The food department, at Rutgers University in New York, has appointed what is believed to be the first professor of food nanotechnology. Both Unilever and Nestle have research topics involving potential uses of nanotechnology in food.

Nanoparticles of carotenoids can be dispersed in water, allowing them to be added to fruit drinks and providing improved bioavailability, and it is possible that this approach could be more widely exploited as an alternative to encapsulation and controlled release of functional ingredients. A synthetic lycopene produced by BASF has undergone toxicity tests and has been GRAS-affirmed ("generally recognised as safe") under US FDA procedures. The GRAS status would imply that the absorption of synthetic lycopene nanoparticles is considered to be the same as the absorption of the carotenoids from the consumption of natural vegetables. It should be noted, however, that the GRAS affirmation procedure permits a product to be marketed on the basis of a notification by the applicant company without the substance being specifically subject to pre-market review and formal approval by FDA. It is the responsibility of the applicant company to ensure that the food ingredients that it markets are safe and comply with all applicable legal and regulatory requirements. In the particular case of nanoparticle lycopene, for example, its use as a food colour is not within its GRAS status and will require pre-market review and approval by FDA since food colours are subject to their own specific approval procedures.

The Institute of Food Science and Technology (IFST) have approved the use of nanoparticle in foods. EU food labelling legislation specifically requires the names of ingredients to indicate details of their physical condition or any treatment, which they have undergone, where omission of such information could mislead a purchaser. Whilst the interpretation of this requirement is necessarily subjective, there is a strong argument that, where an ingredient is in nanoparticle form, its name in the ingredient list should be qualified accordingly. Where the nanoparticles are food-approved additives, their use could be indicated via a modification of the E-number system, for example by qualifying the existing E-number and name with a subscript "n".

Kraft, Nestlé, Unilever and others are employing nanotech to change the structure of food-creating "interactive" drinks containing nanocapsules that can change colour and flavour (Kraft) and spreads and ice creams with nanoparticle emulsions (Unilever, Nestlé) to improve texture. Others are inventing small nanocapsules that will smuggle nutrients and flavours into the body.

1.5. Challenges on Nano-particles and Human health

The economic growth in the field of nanotechnologies will lead to an increased variety and increased volumes of engineered nanoparticles that are produced. Even if exposure assessments and data are still lacking, it is foreseeable that some degree of exposure to engineered nanoparticles-for various segments of the population and for the environment-will occur to an increasingly extend over the coming years. Keeping in mind that these "free nanoparticles" can enter the human body over various pathways (inhalation, ingestion or via the skin) or disperse into the environment, it is important to understand the implications for human health and the ecosystems. To assess the risks of nanoparticles, established methods of chemical safety assessments have to be modified to address the special characteristics of nanoparticles. The main difference to the assessment of bulk materials is the fact that additional parameters like size, shape or surface properties will come into play. The same reason that makes nanoparticles technologically interesting leads to the fact that they represent a new category of (potentially) toxic substances. The interaction with the human body and their health effects are expected be different from molecules as well as from bulk materials of the same composition.

Inhaled particles can have two major effects on the human body

- Their primary toxic effect is to induce inflammation in the respiratory tract, causing tissue damage and subsequent systemic effects. The property that drives the inflammogenicity of nanoparticles is unknown but is expected relate to particle surface area and number of particles.
- Nanoparticles can impair the ability of macrophages to phagocytose and clear particles, and this may contribute to inflammatory reactions. Transport through the blood stream to other vital organs or tissues of the body. This may result in cardiovascular and other extra pulmonary effects.

The study on the toxic effects of manufactured nanoparticles on aquatic organisms, fish (largemouth bass) was exposed to uncoated fullerene carbon 60 (C_{60}) nanoparticles. The fullerenes are one type of manufactured nanoparticle that is being produced by tons each year. Significant lipid peroxidation (oxidation of fats) was found in the brain of the animals after exposure to 0.5 ppm uncoated nC_{60}. The study demonstrates that manufactured nanomaterials can have adverse effects on aquatic and possibly other organisms.

1.6. Conclusion

Nanotechnology is attractive to the food industry, as it promises to yield new solutions to key challenges. As nanotechnology steps forward, material or devices based on this technology become cheaper, smaller, safer, and more sensitive. Its applicability in the areas of food safety and food packaging are well known. Research and development underway includes the development of functional food, nutrient delivery systems and methods for optimizing food appearance, such as flavour, colour, and consistency. Source of increasing the speed of adaption of this new technology is linked to population growth and cost efficiency. Nanotechnology has the potential for revolutionizing the ways in which materials and products are created and the range and nature of functionalities that can be accessed. It is already having a significant commercial impact, which will assuredly increase in the future; but there still endure many challenges and opportunities to improve the current method and issues related to the consequences of this technology that must be addressed in order to assuage the consumer concerns. Also, the safety and environmental issues should be the priority while development of nanotechnology based food products. Nanotechnology can improve the public awareness but before that the regulatory authorities of different nations should come together to reveal the risks associated with the application of nanotechnology and accordingly frame regulations before incorporation of this novel technology into food industry.

References

1. Roco, M.C., Williams R.S., and Alivisatos P. (Eds.), (2000). Nanotechnology Research Directions. Springer, Kluwer Academic Publishers. 316 p. Available at: http://www.wtec.org/loyola/nano/IWGN.Research.Directions.

2. Gupta, A., Eral, H. B., Hatton, T. A., and Doyle, P. S. (2016). Nano-emulsions: formation, properties and applications. Soft-Matter 12, 2826–2841. doi: 10.1039/c5sm02958a.

3. Warad, H.C. and Dutta, J. (2005). Nanotechnology for Agriculture and Food Systems: A View. Asia Institute of Technology, Thailand. 14 p. Available at: http://www.nano.ait.ac.th.

4. Doyle, M.E. (2006). Nanotechnology:ABrief Literature Review. Food Research Institute Briefings, University ofWisconsin-Madison. 10 pages. Available at: http: //www.wisc.edu/fri

5. Aguilera, J.M. (2005). Why food microstructure? J. Food Eng. 67(1-2): 3–11.

6. Singh, T., Shukla, S., Kumar, P., Wahla, V., Bajpai, V.K. & Rather, I. A. (2017). Application of nanotechnology in food science: perception and overview. Frontiers in microbiology, 8: 1501.

7. Ezhilarasi, P.N., Karthik, P., Chhanwal, N., and Anandharama-krishnan, C. (2013). Nanoencapsulation techniques for food bio-active components: a review. Food Bioprocess Technol. 6, 628–647. doi: 10.1007/s11947-012-0944-0

8. Rashidi, L. & Khosravi-Darani, K. (2011). The applications of nanotechnology in food industry. Critical reviews in food science and nutrition, 51(8): 723-730.

9. Patel, A., Patra, F., Shah, N. & Khedkar, C. (2018). Application of Nanotechnology in the Food Industry: Present Status and Future Prospects. In Impact of Nanoscience in the Food Industry, pp. 1-27.

10. Bhosale, R.R., Ghodake, P.P., Nandkumar, A., & Mane, A.A.G. (2013). Nanocochleates: A novel carrier for drug transfer. Journal of Scientific and Innovative Research, 2(5): 964-969.

11. Manikantan, M.R., Arumuganathan, T., Rani, C.I., Kasturi, R. & Varadharaju, N. (2017). Storage stability of sugarcane juice in polypropylene-based nanocomposite packaging films. Sugar Tech, 19(4): 438-445.

12. Manikantan, M.R. & Varadharaju, N. (2011). Preparation and properties of polypropylene based nanocomposite films for food packaging. Packaging Technology and Science, 24(4): 191-209.

13. Ray, S.S. and Bousmina, M. (2005). Biodegradable polymers and their layered silicate nanocomposites: In greening the 21st century materials world. Progress in Materials Science, 50: 962–1079.

14. Petersen, K., Nielsen, V., Bertelsen, G., Lawther, M., Olsen, M.B., Nilsson, N.H. and Mortensen, G. (1999).Potential of biobased materials for food packaging. Trends in Food Science & Technology, 10: 52-68.

15. McGlashan, S.A. & Halley, P.J. (2003). Preparation and characterization of biodegradable starch based nanocomposite materials. Polymer International, 52(11): 1767-1773.

16. Park, H.M., Lee, W.K., Park, C.Y., Cho, W.J. and Ha, C.S. (2003). Environmentally friendly polymer hybrids Part 1. Mechanical, thermal, and barrier properties of thermoplastic starch/clay nanocomposites. Journal of Materials Science, 38: 909–915.

17. Park, H.M., Li, X., Jin, C.Z., Park, C.Y., Cho, W.J. and Ha, C.S. (2002). Preparation and properties of biodegradable thermoplastic starch/clay hybrids. Macromolecular Materials and Engineering, 287: 553–558.

18. Vaia, R.A., In Pinnavaia, T.J. and Beall, G.W. (2000).Polymer–clay nanocomposites (Eds.), New York, Wiley, 229–266.

19. Chiou, B.S., Yee, E., Glenn, G.M. and Orts, W.J. (2005). Rheology of starch–clay nanocomposites. Carbohydrate Polymers, 59: 467–475.

20. Kumar, R. and Munstedt, H. (2005). Silver ion release from anti-microbial polyamide/silver composites. Biomaterials, 26: 2081–2088.

21. Zhang, L., Jiang, Y., Ding, Y., Povey, M. and York, D. (2007). Investigation into the anti-bacterial behaviour of suspensions of ZnO nanoparticles (ZnO nanofluids). J. Nanoparticle Res., 9(3): 479–489.

22. Garland, A. (2004). Nanotechnology in plastics packaging. Commercial applications in nanotechnology (ed). Pira International Limited UK, 14-63.

23. Park, B.K., Kim, D., Jeong, S., Moon, J. and Kim, J.S. (2007). Thin solid films, 2007, 7/16. Vol. 515 (19): 7706–7711.

24. Han, M., Gao, X., Su, J.Z. and Nie, S. (2001). Quantum-dot-tagged microbeads for multiplexed optical coding of biomolecules. Nat. Biotechnol. 19: 631–635.

25. Smolander, M., Hurme, E., Koivisto, M. and Kivinen, S. (2004). PCT International Patent Application, WO 2004/102185 A1.

26. www.etcgroup.org

❑❑❑

Chapter 15

The phytochemical Composition and Health-Benefits of Various Cucurbita Species

H. Chikh-Rouhou[1*], R. Ilahy[2] and R. Sta-Baba[1]

[1] Centre Régional des Recherches en Horticulture et Agriculture Biologique (CRRHAB/IRESA). BP 57, 4042 Chott-Mariem, Sousse. Tunisia., [2] Institut National de Recherche Agronomique de Tunis (INRAT/IRESA). Rue Hédikarray 2049 Ariana, Tunisia.

* Corresponding author: hela.chikh.rouhou@gmail.com

ABSTRACT

Cucurbits are one of the most important crops and are widely grown in almost all the countries of the world.It has been identified as a food of great interest given its nutritional and bio-active components and its high consumption rate all year round. Screening of phytochemicals has been of interest in Cucurbitsgenotypes, as there is emerging evidence from epidemiological and clinical studies that consumption of phytochemical-rich vegetable cultivars may provide health benefits. We report here the benefits for Cucurbits consumption especially *Cucurbita* spp. (such as *Cucurbita moschata, C. maxima* and *C. pepo*). The nutrient composition differsdepending on the cultivation environments, species, and fruit parts.

Keywords: Cucurbitaceae, Cucurbita species, Phytochemicals, Health benefits.

1. Introduction

The Curcurbitaceaefamily, also referred as cucurbits form a very large group with approximately 130 genera and 800 species and can be cultivated in warmer region of worldwide and make popular food crop plants. Some of these species include squashes (*Cucurbita moschata*), zucchini (*Cucurbita pepo*), pumpkins (*Cucurbita maxima*), melons (*Cucumis melo*), cucumber (*Cucumis sativus*), watermelons (*Citrullus lanatus*), luffa (*Luffa aegyptica*) and gourds (*Lagenaria siceraria*).

Cucurbits are among the largest and most diverse plant families, cultivated worldwide in a variety of environmental conditions. Although cultivated cucurbits are very similar in above ground development and root habit, they have a large range of fruit characteristics [1]. Fruits are eaten, when immature or mature. Fruits can be baked, pickled, candied, or consumed fresh in salads or dessert. The fruits of cucurbits are very useful in terms of human health, i.e., purification of blood, removal of constipation and good for digestion and give energy.

Cucurbita spp. are widely cultivated and used as food since ancient times. Some species, such as *Cucurbita moschata, Cucurbita maxima* and *Cucurbita pepo,* revealed high levels of carotenoids [2-3]. Also, seeds, flowers and roots are consumed by humans. Seeds or fruit parts of some cucurbits are reported to possess purgatives, emetics and antihelmintics properties due to the secondary metabolite cucurbitacin content [4-5].

In recent years the cucurbits has been receiving more attention by consumers, who have shown an increased interest toward the nutraceuticals (functional compounds) that provide health benefits including prevention of diseases. Foods with a high flavonoid content, in addition to their nutritive value, can have a protective effect on human health. The selection of food plants with a high content of nutraceuticals is one of the research fields directed to disease prevention rather than cure [6-8]. Many studies were performed to validate the medicinal properties assigned to cucurbits (regarding to diabetes, hypertension, tumours and cholesterol, besides anti-bacterial, anti-helmintic and anti-inflammatory actions) [9-11].

We report here a review of some cucurbits consumption benefits especially of *Cucurbita* spp. their phytochemical composition that differs depending on the species and fruit part (peel, flesh or seed) and their anti-oxidative properties associated with medicinal bio-activities.

1.1. Phytochemistry of *Cucurbita* spp.

Cucurbita spp has been considered as beneficial to health because it contains various biologically active components, such as polysaccharides, para-aminobenzoic acid, fixed oils, sterols, proteins and peptides [12,13].

Cucurbita spp provide vitamins of the B complex, calcium and phosphorus [14]. They are important sources of carotenoids, especially â-carotene, a pre-cursor of vitamin A. Several studies [3, 15] have been developed in many countries, highlighting the importance of some pumpkins (*Cucurbita maxima*) as a source of pro-vitamin A to human diet. Anti-oxidants are commonly found in fruits and vegetables and, therefore, its consumption has been associated with protection against several non-communicable diseases [3].

Cucurbita sppflesh and seeds are rich in proteins, anti-oxidant vitamins, such as carotenoids and tocopherols [16], and minerals, but low in fat and calories. β-carotene reduces skin damage from the sun and acts as an anti-inflammatory agent. α-carotene is thought to slow the aging process, reduce the risk of developing cataracts, and prevent tumour growth. Vitamin E (tocopherols) protects the cell from oxidative damage by preventing the oxidation of unsaturated fatty acids in cell membrane [16].

Cucurbita spp. fruits are characterized by a low content of fat (2.3%, pumpkin pulp is not a rich source of oil), carbohydrates (66%), proteins (3%) [13], and by a high-carotenoids content with values of 171.9 to 461.9 $\mu g.g^{-1}$[17].

Cucurbita spp. seeds are valued for their high protein contentand useful amounts of the essential fattyacid, linoleic acid [16]. Pumpkin has received considerable attention in recent years because of the nutritional and health protective value of the seeds that are rich in health beneficial compounds like polysaccharides, carotene, mineral salts, vitamins, and others [18-19]. Pumpkin seeds, often eaten as a snack, after salting and roasting, are a good source of zinc, polyunsaturated fatty acids, and phytosterols (e.g. β-sitosterol), which can prevent chronic diseases [17].

Pumpkin seed and seed oil are a rich natural source of phytosterols [20], proteins, polyunsaturated fatty acids [21], anti-oxidant vitamins, carotenoids and tocopherols [16] and various elements [22] due to these components are attributed providing many health benefits.

Pumpkin seed oil is accepted as edible oil and as a nutraceutical. Seeds of pumpkin are consumed either roasted or raw and used in cooking and baking as an ingredient of cereals, bread, cakes and salads. Pumpkin seeds are excellent sources of protein (25-37%) and oil (37-45%), itcontain remarkably high proportions of essential amino acids[16]. Previous reports showed that C. *pepo* seeds had significantly more protein than *C. maxima* seeds. The flesh and peel of *C. maxima* had more amino acids than the other two species. *C. pepo* seeds generally had more amino acids than *C. moschata*and *C. maxima* [23].

Several previous reports have described the nutritive value of proteins, fatty acids and vitamins from pumpkin seed oils [16, 24]. Pumpkin seed oil is accepted as edible oil and as a nutraceutical. Seeds contain oil with the main fatty acid components being palmitic, palmitoleic, Linolenic, Gadoleic, total saturated fatty acids (19.35%), and total unsaturated fatty acids (80.65%) [16]. Also, it contain various components, such as *p*-aminobenzoic acid, γ-aminobutic acid, polysaccharides, peptides, proteins, carotenoids as lutein, lutein epoxide, α-carotene, β-carotene [25].

From *Cucurbita* spp. leaves and germinated seeds, several phytochemicals, such as polysaccharides, phenolic glycosides,and proteins have been isolated [26]. Various hypoglycaemic polysaccharides have been characterized from fruit pulps of Cucurbita spp. Plants [27]. Various antibiotic components including anti-fungalcomponents have been characterized from various parts of *Cucurbita* spp. plants.

Kim *et al.* [23] demonstrated that *C. maxima* had significantly more carbohydrates in the flesh than *C. pepo* and *C. moschata*. This high carbohydrate concentration contributed to the sweet taste of *C. maxima*. The *C. maxima* flesh and peel had significantly more protein than *C. pepo* or *C. moschata*.

1.2. Medicinal Bio-activities of *Cucurbita* spp.

A large number of compounds have been isolated from *Cucurbita* spp. [6], only some of them have biological activities and medicinal properties,which are described in the following sections:

1.2.1. *Anti-diabetic activity*

Various studies have been recently conducted to recognize the anti-diabetic potential of herbal formulations; pumpkin is one of them, their fruits are used for human consumption in diabetic conditions[10]. The ingestion of crude aqueous extract of pumpkin fruits is recommended for the treatment of type 2 diabetes or non-insulin-dependent diabetes mellitus. A pumpkin-rich diet could reduce blood glucose [9-10] as well as mixtures of flax and pumpkin seeds supplemented in diet of diabetic rats revealed to be helpful in preventing diabetes and its complications [11].

1.2.2. *Anti-microbial activity*

Pumpkin oil inhibits *Acinetobacter baumanii, Candida albicans, Enterococcus faecalis, Escherichia coli, Pseudomonas aeruginosa, Serratiamarcescens* and *Staphylococcus aureus*[28]. A peptide from pumpkin seeds was proved to inhibit *Botrytis cinerea, Fusariumo-xysporum* and *Mycosphaerellaarachidicola*[29].

1.2.3. *Anti-oxidant and free radical scavenger*

Cellular damage or oxidative injury arising from free radicals or reactive oxygen species (ROS) now appears to be the fundamental mechanism underlying a number of human disorders including diabetes, digestive system disorders, viral infections, neurodegenerative disorders, autoimmune pathologies and inflammation [30].

Studies have attributed that anti-oxidant properties of *Cucurbita* spp. are due to the presence of phenols and flavonoids [12,13]. The presence of phytoconstituents, such as phenols, flavonoids and tannin in plants, indicates the possibility of anti-oxidant activity and this activity will helps in preventing a number of diseases through free radical scavenging activity. Tocopherols are non-glycoside compounds of vegetable oil, and are natural anti-oxidants.

Tocopherols and carotenoids have been suggested to be fat-soluble anti-oxidants. Anti-oxidants play an important role in decreasing DNA damage, diminishing lipid peroxidation, maintaining immune function, and inhibiting malignant transformation or proliferation *in vitro*, which are thought to prevent some diseases [31]. *C. maxima* had more á-tocopherol in the flesh and peel that the other species. ã-Tocopherol was only present in *C. moschata* flesh, *C. pepo* and *C. maxima* peels, and the seeds of all 3 species [23].

In addition to fat-soluble anti-oxidants (tocopherols and carotenoids), *C. maxima* had 16 mg vitamin C per 100 g raw pumpkin [32]. Vitamin C is a strong water-soluble anti-oxidant that protects cells and cellular components from free radicals by donating electrons, and re-generating other antioxidants, such as vitamin E (tocopherols) [33].

1.2.4. *Anti-cancer/anti-tumour effects*

There is a strong interest in the use of plants as a promising source of more efficient anti-cancerdrugs. Recent studies have reported that pumpkin can benefit the treatment of benign prostate hyperplasia, because of its high β-sitosterol content [8, 34]. which has been indicated to reduce blood cholesterol and to decrease risks of certain types of cancers. Also, potent anti-cancer activities were exhibited by *Cucurbita andreana* [39].

Several cucurbitane and hexanorcucurbitane glycosides and other types of triterpenoids have been isolated from the fruits of *C. pepo* [37]. Anti-ulcercucurbitane type triterpenoid has been isolated from the Seeds of *Cucurbita pepo* [38].

Phytochemical analysis of *C. andreana* showed the presence of cucurbitacins B, D, E and I. These cucurbitacins were evaluated for their anti-inflammatory and inhibitory effects on the growth of human colon, breast and lung cancer cell lines [39]. *Cucurbita andreana* exhibited potent anticancer activities.

The cucurbitacins are of great interest because of the wide range of biological activities they exhibit in plants and animals. Currently more than 40 cucurbitacin-derived compounds and cucurbitacins have been isolated from the Cucurbitaceae family and from other species [35]. Cucurbitacins constitute a group of diverse triterpenoid substances which are well known for their bitterness and toxicity. Despite their toxicity, plant species in which they are found have been used for centuries in various pharmacopoeias. A number of compounds of this group have been investigated for their cytotoxic, hepatoprotective, anti-inflammatory and cardiovascular effects [36].

2. References

1. Kocyan, A., Zhang, L.B., Schaefer, H., and Renner, S.S. (2007). A multi-locus chloroplast phylogeny for the Cucurbitaceae and its implications for character evolution and classification. Mol. Phylogenet. Evol., 44: 553-557.

2. Boiteux, L.S., Nascimento, W.M., Fonseca, M.E.N., Lana, M.M., Reis, A., Mendonca, J.L., Lopes, J.F. and Reifschneider, F.J.B. (2007). 'Brasileirinha': Cultivar de abóbora (*Cucurbitamoschata*) de frutos bicolores comvalorornamental e aptidão para consumoverde. HorticulturaBrasileira 25(1): 103-106.

3. Rodriguez-Amaya, D.B., Kimura, M., Godoy, H.T. and Amaya Farfan, J. (2008). Updated Brazilian database on food carotenoids: factors affecting Carotenoid composition. Journal of Food Composition and Analysis 21(6): 445-463.

4. Bisognin, D.A. (2002). Origin and evolution of cultivated cucurbits.Ciencia Rural 32: 715-723.

5. Rahman, A.H.M.M., Anisuzzaman, M., Ahmed, F., Rafiul Islam, A.K.M. and Naderuzzaman, A.T.M. (2008). Study of nutritive value and medicinal uses of cultivated cucurbits. J. Applied Sci. Res 4: 555-558.

6. Caili, F., Huan, S. and Quanhong, L. (2006). A review on pharmacological activities and utilization technologies of pumpkin. Plant Foods Hum Nutr 61: 73-80.

7. Chaturvedi, P. (2012). Anti-diabetic potentials of *Momordicacharantia*: multiple mechanisms behind the effects. J Med Food 15: 101-107.

8. Dhiman, K., Gupta, A., Sharma, D.K., Gill, N.S. and Goyal, A. (2012). A review on the medicinally important plants of the family Cucurbitaceae. Asian J ClinNutr 4: 16-26.

9. Li Q.H., Tian Z., Cai T.Y. (2001). Study on the hypoglycemic action of pumpkin extract in diabetic rat. Acta Nutrimenta Sinica 25: 34-36.

10. Zhang, Y. and Yao, H. (2002). Study on effect of hypoglycemia of different type pumpkin. J Chin Food Sci 23: 118–120.

11. Makni, M., Sefi, M., Fetoui, H., Garoui, E.M., Gargouri, N.K., Boudawara, T. and Zeghal, N. (2010). Flax and Pumpkin seeds mixture ameliorates diabetic nephropathy in rats. Food ChemToxicol 48: 2407–2412.

12. Heo, S.J., Kim, J.H., Kim, J.K. and Moon, K.D. (1998). The comparison of food constituents in pumpkin and sweet-pumpkin. Korean J Food Cult 13: 91-6.

13. Achu, M.B., Fokou, E., Tchiégang, C., Fotso, M. and Tchouanguep, F.M. (2005). Nutritive value of some *Cucurbitaceae* oil seeds from different regions in Cameroon. Afr J Biotechnol 4: 1329-34.

14. Ramos, S.R.R. and Queiroz, M.A. (2005). Recursosgenéticosde abóbora no Nordeste brasileiro. In M.C. Lima, *Recursosgenéticos de hortaliças: riquezasnaturais*(pp. 99-116). São Luís: Instituto Interamericano de Cooperação para aAgricultura.

15. Gwanama, C., Lungu, D., Nichterlein, K., and Simawachi, W. (2002). Variation of fruit Beta-carotene content of tropical pumpkin (*Cucurbita moschata*Duch.) landraces in Zambia.Noticiário de RecursosFitogenéticos 129: 44-46.

16. Stevenson, D.G., Eller, F.J., Wang, L., Jane, J.L. and Wang, T. (2007). Oil and tocopherol content and composition of pumpkin seed oil in 12 cultivars. J Agric Food Chem 55: 4005-4013.

17. Applequist, W.L., Avula, B., Schaneberg, B.T., Wang, Y.H. and Khan, I.A. (2006). Comparative fatty acid content of seeds of four *Cucurbita* species grown in a common (shared) garden. J Food Compost Anal 19: 606-11.

18. Fu, C.L., Shin, H. and Li, Q.H. (2006). A review on pharmacological activities and utilization technologies of pumpkin. Plant Foods Hum Nutr 61: 70–77.

19. Fu, C.L., Tian, H.J., Cai, T.Y., Liu, Y. and Li, Q.H. (2007). Some properties of an acidic protein-bound polysaccharide from the fruit of pumpkin. Food Chem 100: 944–947.

20. Sabudak, T. (2007). Fatty acid composition of seed and leaf oils of pumpkin.Walnut almond, maize, sunflower and melon. Chem Nat Compounds 43: 465- 467.

21. Matus, Z., Molnar, P. and Szabo, L.G. (1993). Main carotenoids in pressed seeds (*Cucurbitaesemen*) of oil pumpkin (*Cucurbita pepo*.convar. pepo var. styriaca). Acta Pharm Hung 63: 247-256.

22. Glew, R.H., Glew, R.S., Chuang, L.T., Huang, Y.S. and Millson, M. (2006). Amino acid, mineral and fatty acid content of pumpkin seeds (*Cucurbita*spp) and Cyperu: esculentlls nuts in the Republic of Niger. Plant Foods Hum Nut 61: 51-56.

23. Kim, Y.M., Kim, E.J., Kim, Y.N., Choi, C. and Lee, B.H. (2012). Comparison of the chemical compositions and nutritive values of various pumpkin (*Cucurbitaceae*) species and parts.Nutr Res Pract 6(1): 21-27.

24. Murkovic, M., Mulleder, U. and Neunteufl, H. (2002). Carotenoid content in different varieties of pumpkins. Journal of Food Composition and Analysis 15(6): 633-638.

25. Rabrenovic, C.C., Dimic, E.B., Novakovic, M.M., Tesevic, V.V. and Basic, Z.N. (2014). The most important bio-active components of cold pressed oil from different pumpkin (*Cucurbitapepo* L.) seeds. Food Sci Technol 55: 521-527.

26. Zhang, Y., Chen, P., Zhang, Y., Jin, H. and Zhu, L. (2013). Effects of polysaccharide from pumpkin on biochemical indicator and pancreatic tissue of the diabetic rabbits.Int J BiolMacromol 62: 574-581.

27. Jun, H.I., Lee, C.H. and Song, G.S. (2006). Characterization of the pectic polysaccharides from pumpkin peel. Food Sci Tech 39: 554–561.

28. Hammer, K.A., Carson, C.F. and Riley, T.V. (1999). Anti-microbial activity of essential oils and other plant extracts. J Appl Microbiol 86: 985–990.

29. Vassiliou, A.G., Neumann, G.M. and Condron, R. (1998). Purification and mass spectrometry-assisted sequencing of basic anti-fungal proteins from seeds of pumpkin (*Cucurbitamaxima*). Plant Sci 134: 141–162.

30. Attarde, D.L., Kadu, S.S., Chaudhari, B.J., Kale, S.S. and Bhamber, R.S. (2010). *In vitro* Anti-oxidant activity of Pericarp of *Cucurbita maxima* Duch. ex Lam, International Journal of PharmTech Research 2(2): 1533-1538.

31. Gropper, S.S., Smith, J.L. and Groff, J.L. (2005). Advanced Nutrition and Human Metabolism, 4th ed. Belmont: Thomson Wadsworth Publishing Co. p. 381-405.

32. Dubey, S.D. (2012). Overview on *Cucurbita maxima*, International Journal of Phytopharmacy, 2(3): 68-71.

33. Keith, R.E. (2006). Ascorbic acid. In: Driskell JA, Wolinsky I, editors. Sports Nutrition: Vitamins and Trace Elements. Boca Raton, FL: CRC Press. Chapter 2.

34. Tsai, Y.S., Tong, Y.C., Cheng, J.T., Lee, C.H., Yang, F.S. and Lee, H.Y. (2006). Pumpkin seed oil and phytosterol-F can block testosterone/prazosin-induced prostate growth in rats. UrolInt 77: 269-74.

35. Guha, J. and Sen, S.P. (1975). The cucurbitacins-a review. Plant Biochem Journal 2: 12-28.

36. Miro, M. (1995). Cucurbitacins and their pharmacological effects. Phytother Res 9: 159-168.

37. Ge, S., Wang, D.C., Wu, L.J. and Deng, X.M. (2006). Triterpenes from the fruits of *Cucurbita pepo*cv Dayangua. J Shengyang Pharm 23: 55-59.

38. Gill, N.S., Kaur, S., Arora R. and Bali, M. (2011). Screening of anti-oxidant and anti-ulcer potential of *Citrulluscolocynthis* methanolic seed extract. Res J Phytochem 2: 98-106.

39. Jayaprakasam, B., Seeramn, N.P. and Nair, M.G. (2003). Anti-cancer and anti-inflammatory activities of cucurbitacins from *Cucurbita andreana*. Cancer Lett 10: 11-16.

❑❑❑

Chapter 16

Bioactive Potential of Microalgae as Future Food

Neha Mishra, Devinder Kaur and Neetu Mishra*

Centre of Food Technology, University of Allahabad, Allahabad-211002,
*Corresponding Author: Dr. Neetu Mishra, neetum1976@gmail.com

ABSTRACT

The expanding global population raised the need to search for novel food resources. Microalgae is an oldest form of life and have balanced composition of all important biomolecules, such as protein, vitamins and minerals, however, they was only traditionally consumed by Japan and some south-east countries. In recent years, they have been recognized to have potential as a source of sustainable novel food for expanding population. They have several advantages over terrestrial plants including high biomass productivity, high photosynthetic efficiency and avoid competing with arable lands and potable water. In addition, they also have potential as functional food due to the presence of various bio-active compounds, which includes polyunsaturated fatty acids (PUFA), pigments, polysaccharides and phenolic compounds. They exhibits wide variety of biological activities, such as anti-microbial, anti-tumour, anti-coagulant, antioxidant, and anti-inflammatory capacities, with the potential for the reduction and prevention of diseases. They biosynthesize and accumulates different bio-active compounds to survive in adverse environmental conditions. Therefore culture conditions, such as light, temperature, medium composition and salinity are the key factors that change its chemical composition and influence the synthesis of specific compounds of interest. In spite of this, they are untapped resource of value added biomolecules. Microalgae are the future foods with several pharmaceutical and nutraceutical applications.

Keywords: Microalgae, novel food, bio-active compounds, biological activity.

1. Introduction

Algae are diverse group of photosynthetic organism that ranges from microscopic unicellular to vary large multicellular. Algae include the microalgae (microscopic) and Macro algae (macroscopic form) the later consist of mainly seaweeds. Microalgae are diverse group of photoautotrophic micro-organism that fix carbon dioxide into valuable compounds, such as proteins, lipids, carbohydrates, carotenoids or vitamins for health, food and feed additives. There are more than 50,000 different microalgae that are present in wide habitat, ranging from marine to fresh water and from hot boiling springs to snow. They are one of the oldest form of life that originated 3 million year ago, although have been used as food by humans dates back 2000 years (1,2). Algae was discovered and traditionally consumed as food by Japan, later it was widely accepted and cultivated by countries, such as China, Thailand, Malaysia and United States. However, over the last few decades microalgae have attracted the interest of researchers due to its high nutritional value and vast variety of novel metabolites that can be used in, animal feeds, food industry, pharmaceutical industry, bioremediation and biofuel (3,4,5). Recently microalgae representa novel food resource to meet the needs of supplementary food for the expanding population. Algae have several advantages over terrestrial agriculture as high biomass productivity, high photosynthetic efficiency and no competition with portable water and arable lands (6,7). The chemical composition can be varied by manipulation of culture conditions, such as nutrients, salinity, light, temperature, and pH (8). The conditions for microalgae cultivation are important factors that influence the metabolism of these micro-organisms, thus directing the synthesis of specific compounds of interest. Despite of the fact of its good nutritional value, high photosynthesis efficiency and less demand of environmental resources, such as land, potable water, they represent a large, untapped reservoir of novel compounds. The use of microalgae as a food source is not commondue to the limited knowledge related to their biochemical composition and cultivation techniques.

1.1. Algae as a Human Food

The population of the world is growing rapidly; there is an increment in population by 90 millions in every year. The demand of food for growing population will be double in order to meet the minimum requirements. Recently microalgae have emerged as novel food resources with high nutritional value. They represent an untapped source of valuable metabolites, such as protein, polyunsaturated fatty acids (PUFA), anti-oxidant and vitamins with high therapeutic values. Microalgae are a hugely diverse group containing approximately 200,000 species. Although only handful species are currently exploited for a variety of biotechnological purposes, including, human food, animal feed, pharmaceuticals, bioremediation, and biofuel. Species of microalgae, such as *Dunaliella, Spirulina, Chlorella* are majorly used as a human food in south-east countries, such as Japan, China, Indonesia and Malaysia. They are initially explored due to their high protein content with good amino acid profile. *Nostoc* is the first microalgae used as food over 2000 years ago but later *Chlorella* and *Spirulina* species were the most exploited microalgae as human food. They have high nutritional value with high protein content, vitamin, minerals, chlorophyll and other bio-active compounds. Microalgae *Chlorella* are spherical members of the phylum Chlorophyta (green algae) and *Arthrospiraplatensis* is a filamentous Cyanobacterium that also called Sprullina. In recent years *Spirulina* has gained worldwide popularity as a food supplement (9) and World Health Organization has declared *Spirulina* as a" Super Food". In last few years, health promoting effect of microalgae have been documented and promoted its consumption. Now-a-days, wide range of microalgae are available in market in form of tablets, capsules, powder as food supplements and also incorporated into different food products, such as bakery, dairy, cereal breakfast and beverages.

1.2. Value Added Compounds of Microalgae

Algae increasingly are being marketed as "functional foods". Functional food contains many bio-active compounds, or phyto-chemicals that provide health benefits beyond providing the basic nutrition (10). Bio-active compounds are secondary metabolite that accumulates under stress to cope the adverse growth conditions (11). Microalgae have unique ability to survive in adverse environmental condition and in order to adapt new environmental surroundings, they biosynthesize and accumulates different primary and secondary bio-active compounds that have potential pharmaceutical and therapeutic values (12), such as anti-microbial, anti-tumor, anti-coagulant, anti-oxidant, and anti-inflammatory capacities, with the potential for the reduction and prevention of diseases (13,14). In most microalgae, the bio-active compounds are accumulated in the biomass;

however, in some cases, these metabolites are excreted into the medium; these are known as exo-metabolites. They produce a wide range of bioactive compounds that depends on culture conditions, such as salinity, light intensity, and medium.Some of the highly valuable bioactive products isolated from microalgae are discussed below.

1.2.1. Polyunsaturated fatty acids (PUFAs)

Polyunsaturated Fatty Acids (PUFAs) are essential fatty acids that are not synthesized by higher plants and animals human due to the absence of requisite enzyme and needed to supply it by diet. There are two types of polyunsaturated fatty acid (PUFA)-omega 3 fatty acid (n-3FA) and omega 6 fatty acid (n-6FA). Most common type of n-6 FA are Linoleic acid (LA; C18:2), Arachidonic acid (AA; C20:4) and n-3 FA are α-linolenic acid (ALA, 18:3), eicosapentaenoic (EPA; C20:5) and docosahexaenoic (DHA; C22:6) acids.Marine fatty fish, such as Salmon, Mackerel are the principle source of PUFA in human consumption (15). However, its undesirable taste, typical smell, cultural constraint, decline resources, increasing risk of contamination and prices, has created the demand for an alternative sources. Currently, microalgae explored for commercial source of PUFA and have the potential to replace fish oil (16). Most common microalgae used for the production PUFA esp. DHA and EPA are from the members of the families Thraustochytriaceae and Crypthecodiniaceae. EPA is an n-3 PUFA that plays an important role in the regulation of biological functions and prevention and treatment of a number of human diseases, such as cardiovascular disease (CVD), inflammatory disease, neuro-degenerative disease, cancer and ageing (17). Antimicrobial activity of microalgae against human pathogens, such as *Escherichia coli, Pseudomonas aeruginosa, Staphylococcus aureus,* and *Staphylococcus epidermidis,* has been attributed to its fatty acid content and its derivatives, such as γ-linolenic acid, eicosapentaenoicacid, docosahexaenoic acid, oleic acid, lactic acid, and arachidonic acid (18,19) Microalgae produces and accumulates high content of lipid which vary with its different strain and cultural condition. They release free fatty acid in media through enzymatic activity as a defensive mechanism against pathogenic bacteria, virus and other co-existing algae (20,21). These fatty acid and their derivates have anti-bacterial activity due to its ability to bring bacterial cell lyses. DHA is the major constituents, in the retina of the eye, in the grey matter of the brain and heart tissue. Several study demonstrated that adequate supply of

DHA improves brain, visual acuity and infant cognitive development. Microalgae most commonly, such as *Crypthecodiniumcohnii, Nannochloropsis sp. Isochrysisgalbana,* diatoms *Phaeodactylum-tricornutum, Nitzchia* offer a promising resource for LC-PUFA production as an alternative to fish oil because of its high DHA/EPA productivity (22,23). Now, many microalgal oil rich in EPA/DHA are commercially available e.g. For the infant formula market DHASCO™ oil from micro-algae *Crypthecodiniumcohnii* that contain 40–45% w/w of DHA content, Life's DHATM oil contain 35% or 40% DHA for food processing industry was manufactured from micro-algae *Schizochytrium.*

1.2.2. Polysaccharides

Polysaccharides *are polymeric carbohydrate structures that are formed of monosaccharide via different glycosidic bonds* (24). It represent a highly diverse class of high value-added components due to its structure and conformation which provides it wide range of applications in food processing, nutraceutical, and pharmaceutical industry (25). Most of the polysaccharides are produced and released into the medium called as exo-polysaccharides. Generally, exo-polysaccharides are sulphated-polysaccharides (contain sulphate esters), referred as Exosulphated polysaccharides (EPS). EPS hasacidic characteristics and possess several biological activities like as anti-microbial, anti-tumour, anti-coagulant and immune-modulator (26). These biological activities of polysaccharide depends on the sugar composition, position, and ratio of $1 \rightarrow 3$ and $1 \rightarrow 6$ linkage (degree of branching) and chain length (27,28). The therapeutic action due to its ability to stimulate and modulate the immune system through the activation of macrophage functions. The anti-microbial and anti-viral activity of EPS is well documented both *in vivo* and *in vitro* (29). The anti-viral activity of polysaccharides is due to the positive charges on its cell surface, which prevents the penetration of the virus into host cells (25). Microalgae, such as *Chlorella* sp. and *Dunaliella*sp., produce and secrete polysaccharides at relatively high levels (30).

1.2.3. Carotenoids

Carotenoids are lipophilic pigments belongs to class terpenoids. All are derivatives of tetraterpenes, contain 8 isoprene molecules with conjugated double bonds, are responsible for giving characteristic colour, such as yellow, orange, or red (31). They are the most common pigments produced by photosynthetic organisms and also by some non photosynthetic bacteria and fungi. They serve as accessory light harvesting pigments and play key roles in photosynthesis due to their ability to absorb photons. In addition, they also have photo-protective ability by scavenging and deactivating free radicals and reactive oxygen species. Structurally, naturally occurring carotenoids classified into: carotenes, non oxygenated carotenoids (such as β-carotene, α-carotene); and xanthophylls, which are oxygenated carotenoids. Most of xanthophylls produced by higher plants, but lutein astaxanthin, canthaxanthin and fucoxanthin are common xanthophylls synthesized by green algae. On the basis of their function, carotenoids can be divided into primary and secondary carotenoids. Primary carotenoids which are tightly linked with structural and functional components in the cellular photosynthetic apparatus and the main function are to absorb photons in visible wave lightranging from 400–550 nanometers (violet to green light) and transfer absorbed energy to chlorophylls (33). Secondary carotenoids present in lipid bodies outside the chloroplast (34) and main function is to protect chlorophyll from photo-damage and other environmental stress. They have the anti-oxidant property by which it protects the cell from reactive oxygen and nitrogen species,, and thereby promote the stability and functionality of the photosynthetic apparatus (35). In this support several studies documented the health benefits of carotenoids, which are usually related to their anti-oxidant activities and found carotenoids provide a protective role for humans as similar to microalgae (36, 37). Their ability to prevent cancer, aging, atherosclerosis, coronary heart disease, and degenerative diseases have been widely documented. Due to their anti-oxidant properties and pigmentation ability these compounds have application in the food and pharmaceutical industries. Among the 400 known carotenoid compounds, β-carotene and astaxanthin are prominent carotenoids that have been used commercially and make-up almost half of the carotenoid market globally (38).

1.2.4. β-carotene

β-Carotene has higher pro-vitamin A activity that is, they can be converted into vitamin A, which is essential for vision and the correct functioning of the immune system (39).They are most commonly used as natural food colorants and food supplementsin food industry, as additive in cosmetics, and medicine (40). Natural β-carotene contains both *cis*- and *trans*-isomers and has higher anti-oxidant activity than synthetic α-carotene as the latter consists of all- *trans*-isomer (8).

*Dunaliellasalina*is a green halophilicmicroalga recognized as a major biological source of β-carotene pigment, accounting upto 14% of its dry biomass (41). Now-a-days, *Dunaliella* powder and dried *Dunaliella are available in market for human and animal* feed use. *Protective role of* β-carotene against atherosclerosis and hyperchlolesterimia have been documented (42). Previous studies reported that supplementation *Dunaliella* rich in β-carotene inhibits the oxidation of low density lipoprotein (LDL), lowers the serum triglycerides and cholesterol levels in mouse and humans (43).

1.2.5. Astaxanthin

It is next commercially available pigments and strongest anti-oxidant in carotenoids, exhibits 100 times stronger anti-oxidant activity than α-tocopherol (44). Astaxanthin have ability to quench free radicals by donating the electrons, and prevent the oxidative damage of lipid membranes and cells. Several previous studies documented theprotective action of astaxanthin against oxidative damage, cellular necrosis, atheroscvlerosis, mascular degeneration, and an increased immune response (29). *Haematococcuspluvialis*is a green microalgae commercial produces astaxanthin upto 1-8% of the dry weight (45). Over late few years, interest for production of carotenoids from microalgae has been raised.Astaxanthin is the only carotenoids approved by FDA for direct consumtion as nutraceutical agent and to sold in capsules.

1.2.6. Phycobiliproteins

Phycobiliproteins are water-soluble assessory pigment most commonly present in microalgae belongs to cyanobacteria, rhodophyta, and cryptophyte. There are three major types of phycobiliproteins, namely phycocyanin, allophycocyanin and phycoerythrin,. Phycocyaninis a blue photosynthetic pigment that belongs to the group of phycobiliproteins found in large quantities in the *Spirulinaplatensis* may vary from 0.11 to 12.7% dry weight (46). They are major components of the photosynthetic light-harvesting antenna pigments (phycobilisomes) and formed by a protein backbone co-valently linked to tetrapyrrolechromophoric prosthetic groups(phycobilins). The cyanobacterium *Arthrospira* and the rhodophyte *Porphyridium* are the main commercial producers of phycobiliproteins (phycoerythrin and phycocyanin). Initially, they were explored as colouring agent and commonly used as natural colorants in food, such as like candies, chewing gums, jellies, ice creams, soft drinks, dairy products and also in cosmetics, such as lipsticks, eyeliners and eye shadows. Recently there health-promoting potential have been recognized and are widely used as fluorescent markers in immunology laboratories. Several studies showed the protective effect of phycocyaninaganist atherosclerosis, arthritis, tumours, human erythrocytes and cell lysis. (46,47). These beneficial effects of the pigment are attributed to its anti-oxidant potential which is 16 times than trolox (vitamin E analog) and 20 times more effective than vitamin C (48).

1.2.7. Phenolic and polyphenols compounds

These are the secondary metabolites accumulates in stress in order to cope adverse condition. Polyphenolic are a large and diverse group of chemical compounds includes more than 8000 compounds. Depending on the basic structure, they can be divided into 10 different classes. Phenolic compounds is the major class of polyphenolic compounds and posses one benzene ring substituented by atleast one hydroxyl group (49). It can act as anti-oxidants by scavenging free radicals, chelating metal ions, and improving the endogenous anti-oxidant system. Other common polyphenolsare flavonoids (i.e. flavones, flavonols, flavanones, flavononols, chalcones and flavan-3-ols), lignins, tocopherols, tannins and phenolic acids present in plants, macroalgae as well as microalgae. Phlorotannins (tannin) is a polyphenolic compounds produced in brown algae and more potent free radical scavenger attibuted to its eight inter-connected rings. (50). They are polymers of phloroglucinols (1,3,5-trihydroxybenzene, and may constitute upto 15% of the dry weight of brown algae (51).

1.2.8. Sterol

Phytosterol is a subgroup of the steroids produced by plants or algae, structurally similar and functionally analogous to cholesterols (pre-dominant sterol in animals). They are one of the key components of the cellular membrane and play important role in regulating membrane fluidity and permeability (52). These are physiological active secondary metabolites which accumulate in cell in response to the stress induced by extreme exposure conditions and have wide biological activities with pharmaceutical and nutritional importance. Microalgae have highly diverse sterol profiles in contrast of higher plant which includes 24-Ethylcholesterol are mostly produced by cyanobacteria (53). 4α-methyl sterols is produced by most of Dinoflagellates (54, 55), epibrassicasterol by Cryptophytes, 24-methylenecholesterol and campesterol by Prasinophyceae. There are of great interest because of its beneficial health effects including immuno-modulatory (56), anti-inflammatory (56), anti-hyper-cholesterolemic (57,58), anti-oxidant (59), anti-cancer (60) and anti-diabetic (61). In 2010, the fortification of foods with phytosterols including dairy products,breakfast cereals, cereal bars and bakery products was authorized by FDA. The European Atherosclerosis Society Consensus Panel approved the utilization of phytosterol-enriched foods among patients with high cholesterol levels (62). Phytosterol contents varied from 8.09 to 19.57 g per kg in different oil seeds with highest content of 32.25g per kg in rice bran, whereas in microalgae phytosterol content ranged from 7 to 34 g per kg. The microalgae, *Pavlovalutheri*, *Tetrasellimis* sp. and *Nannochloropsis* sp. were identified as the top three highest phytosterol producers (0.4%-2.6% dry weight) (63).This indicates microalgae could be good source of phytosterol in various application.

1.2.9. Peptides and Essential Amino acids

Higher organism unable to synthesis essential amino acid and mst be provide throgh diet for growth and several biological activities. Microalgae considered as a viable alternative protein source with all essential amino acid (EAA) and meeting FAO requirements. Many species of microalgae, such as *Chlorella* and *Spirulina* (*Arthrospira* and *Dunaliellasalina* (64), produce higher quantities of proteins and amino acid profile and utilized in food and pharmaceutical industry due to their therapeutic potential in the treatment of various diseases (65). Some proteins, peptides, and amino acids have strong therapeutic effects on health or are necessary for cells and tissues to perform their normal activities. Initial microalgae gained interest as a source of proteins, bio-active peptides and other derivates includes growth factors, enzymes, hormones, and immuno-modulatorsthat can help to replace damaged tissues and contribute health benefits. Microalgal protein can reduce cholesterol levels by increase the production of cholecystokinin involved in reduction of LDL-cholesterol. Some microalgae produces protein that have anti-viral activity for example *Nostoc* produces Cyanovirin protein that have antiviral activities against HIV and influenza virus. *Anabaena* and *Porphyridium* produces the enzyme SOD (superoxide dismutase), which protects against oxidative damages.

1.3. Conclusion

Microalgae are underutilized source of valuable compounds that have health promoting activites. They have great potential for nutraceutical and pharmaceutical application due to its high nutritional value with considerable amount of bio-active compounds like n-3 fatty acid, pigments, polysaccharides, phenolic compounds. Microalgae are required to be explored as future food.

1.4. References

1 Knoll A., Falkowski P.G., Taylor F.J., Schofield O., Raven J.A., Quigg A., Katz M.E., Grzebyk D. Response to Comment on" The Evolution of Modern Eukaryotic Phytoplankton". American Association for the Advancement of Science.

2. Sathasivam R., Radhakrishnan R., Hashem A., Abd_Allah E.F. Microalgae metabolites: A rich source for food and medicine. Saudi Journal of Biological Sciences. 2017 Nov., 3.

3. Chacon-Lee T.L. and Gonzalez-Marino G.E. (2010). Microalgae for "Healthy" Foods-Possibilities and Challenges. Comprehensive Reviews in Food Science and Food Safety. 9: 655–675.

4. Cai T., Park S.Y. and Li Y. (2013). Nutrient recovery from wastewater streams by microalgae: Status and prospects. Renewable and Sustainable Energy Reviews. 19: 360–369.

5. Batista A.P., Ambrosano L., Graca S., Sousa C., Marques P.A., Ribeiro B. (2015). Combining urban wastewater treatment with biohydrogen production—An integrated microalgae-based approach. Bioresource Technology. 184: 230–235.

6. Quintana N., Van der Kooy F., Van de Rhee M.D., Voshol G.P., Verpoorte R. Renewable energy from Cyanobacteria: energy production optimization by metabolic pathway engineering. Applied microbiology and biotechnology. 2011 Aug 1; 91(3): 471-90.

7. Rawat I., Kumar R.R., Mutanda T., Bux F. Dual role of microalgae: phycoremediation of domestic wastewater and biomass production for sustainable biofuels production. Applied energy. 2011 Oct 1; 88(10): 3411-24.

8. Hu C.C., Lin J.T., Lu F.J., Chou F.P., Yang D.J. Determination of carotenoids in Dunaliellasalina cultivated in Taiwan and anti-oxidant capacity of the algal carotenoid extract. Food Chemistry. 2008 Jul 15; 109(2): 439-46.

9. Colla L.M., Reinehr C.O., Reichert C., Costa J.A. Production of biomass and nutraceutical compounds by Spirulinaplatensis under different temperature and nitrogen regimes. Bio-resource technology. 2007 May 1; 98(7): 1489-93.

10. Liu R.H. Health benefits of fruit and vegetables are from additive and synergistic combinations of phytochemicals. The American journal of clinical nutrition. 2003 Sep 1; 78(3): 517S-20S.

11. Croteau R., Kutchan T.M. and Lewis N.G. Natural products (secondary metabolites). Biochemistry and Molecular Biology of Plants. 2000; 24: 1250-1319.

12. Kiuru P., D. Auria M.V., Muller C.D., Tammela P., Vuorela H., Yli-Kauhaluoma J. Exploring marine resources for bio-active compounds. Plantamedica. 2014 Sep; 80(14): 1234-46.

13. Yaakob Z., Ali E., Zainal A., Mohamad M., Takriff M.S. An overview: biomolecules from microalgae for animal feed and aquaculture. Journal of Biological Research-Thessaloniki. 2014 Dec; 21(1): 6.

14. Gouveia L., Marques A.E., Sousa J.M., Moura P., Bandarra N.M. (2010). Microalgae-source of natural bio-active molecules as functional ingredients. Food SciTechnol Bull Funct Foods. 7(2): 21.

15. Simopoulos A.P. (2008). The importance of the omega-6/omega-3 fatty acid ratio in cardiovascular disease and other chronic diseases. Biol. Med. 233: 674-688.

16. Guedes A.C., Amaro H.M., Barbosa C.R., Pereira R.D., Malcata F.X. (2011). Fatty Acid Composition of Several Wild Microalgae and Cyanobacteria, with a focus on Eicosapentaenoic, Docosahexaenoic and Linolenic Acids for Eventual Dietary Uses. Food Research International. 44: 2721-2729.

17. Spolaore P., Joannis-Cassan C., Duran E., Isambert A. Commercial applications of microalgae. Journal of bio-science and bio-engineering. 2006 Feb 1; 101(2): 87-96.

18. Amaro H.M., Guedes A.C., Malcata F.X. Anti-microbial activities of microalgae: an invited review. Science against microbial pathogens: communicating current research and technological advances. 2011; 3: 1272-84.

19. Desbois A.P., Lebl T., Yan L., Smith V.J. Isolation and structural characterization of two anti-bacterial free fatty acids from the marine diatom, Phaeodactylumtricornutum. Applied microbiology and biotechnology. 2008 Dec 1; 81(4): 755-64.

20. Sánchez-Saavedra M, Licea-Navarro A, Bernáldez-Sarabia J. Evaluation of the anti-bacterial activity of different species of phytoplankton. Revista de Biología Marina y Oceanografía. 2010; 45(3).

21. Desbois A.P., Smith V.J. Anti-bacterial free fatty acids: activities, mechanisms of action and biotechnological potential. Applied microbiology and biotechnology. 2010 Feb 1; 85(6): 1629-42.

22. Grima E.M., Belarbi E.H., Fernández F.A., Medina A.R., Chisti Y. Recovery of microalgal biomass and metabolites: process options and economics. Biotechnology advances. 2003 Jan 1; 20(7-8): 491-515.

23. Wen Z.Y., Chen F. Heterotrophic production of eicosapentaenoic acid by microalgae. Biotechnology advances. 2003 Jul 1; 21(4): 273-94.

24. deMorais M.G., Vaz B.D., de Morais E.G., Costa J.A. (2015). Biologically active metabolites synthesized by microalgae. BioMed research international.

25. Amaro H.M., Guedes A.C., Malcata F.X. Anti-microbial activities of microalgae: an invited review. Science against microbial pathogens: communicating current research and technological advances. 2011; 3: 1272-84.

26. Smelcerovic A., Knezevic-Jugovic Z., Petronijevic Z. Microbial polysaccharides and their derivatives as current and prospective pharmaceuticals. Current pharmaceutical design. 2008 Oct 1; 14(29): 3168-95.

27. Thomas N.V., Kim S.K. Potential pharmacological applications of polyphenolic derivatives from marine brown algae. Environmental toxicology and pharmacology. 2011 Nov 1; 32(3): 325-35.

28. Mishra N., Mishra N. Exploring the biologically active metabolites of isochrysisgalbana in pharmaceutical interest: an overview. International journal of pharmaceutical sciences and research. 2018 Jun 1; 9(6): 2162-74.

29. Raposo M.F., de Morais R.M., Bernardo de Morais A.M. Bioactivity and applications of sulphated polysaccharides from marine microalgae. Mar Drugs. 2013; 11(1): 233-52.

30. Rodríguez-Meizoso I., Jaime L., Santoyo S., Cifuentes A., García-Blairsy Reina G., Senorans F.J., Ibáñez E. Pressurized fluid extraction of bio-active compounds from Phormidium species. Journal of agricultural and food chemistry. 2008 May 6; 56(10): 3517-23.

31. Cardozo K.H., Guaratini T., Barros M.P., Falcão V.R., Tonon A.P., Lopes N.P., Campos S., Torres M.A., Souza A.O., Colepicolo P., Pinto E. Metabolites from algae with economical impact. Comparative Biochemistry and Physiology Part C: Toxicology & Pharmacology. 2007 Jul 1; 146(1-2): 60-78.

32. Gong M., Bassi A. Carotenoids from microalgae: a review of recent developments. Biotechnology advances. 2016 Dec 1; 34(8): 1396-412.

33. Ye Z.W., Jiang J.G., Wu G.H. Biosynthesis and regulation of carotenoids in Dunaliella: progresses and prospects. Biotechnology advances. 2008 Jul 1; 26(4): 352-60.

34. Grünewald K., Hagen C. β-carotene is the intermediate exported from the chloroplast during accumulation of secondary carotenoids in Haematococcuspluvialis. Journal of Applied Phycology. 2001 Feb 1; 13(1): 89-93.

35. Grossman P., Niemann L., Schmidt S., Walach H. Mindfulness-based stress reduction and health benefits: A meta-analysis. Journal of psychosomatic research. 2004 Jul 1; 57(1): 35-43.

36. Chuyen H.V., Eun J.B. Marine carotenoids: Bio-activities and potential benefits to human health. Critical reviews in food science and nutrition. 2017 Aug 13; 57(12): 2600-10.

37. Fiedor J., Burda K. Potential role of carotenoids as anti-oxidants in human health and disease. Nutrients. 2014 Jan 27; 6(2): 466-88.

38. Hair Jr J.F., Wolfinbarger M., Money A.H., Samouel P., Page M.J. Essentials of business research methods. Routledge; 2015 Mar 4.

39. Becker W. Microalgae in human and animal nutrition. Handbook of microalgal culture. 2004; 312: 351.

40. Raja R., Hemaiswarya S., Rengasamy R. Exploitation of Dunaliella for β-carotene production. Applied Microbiology and Biotechnology. 2007 Mar 1; 74(3): 517-23.

41. Francavilla M., Trotta P., Luque R. Phytosterols from Dunaliella-tertiolecta and Dunaliellasalina: a potentially novel industrial application. Bio-resource technology. 2010 Jun 1; 101(11): 4144-50.

42. Barsanti L, Gualtieri P. Algae: anatomy, biochemistry and biotechnology. Florida: Taylors R Francis Group, 2006; 301 pp.

43. Phang S.M., Chu W.L., Wong C.Y., Teoh M.L., Peng K., Tan I., Lee H.K. A Checklist of Microalgal Isolates from Ny-Ålesund, Svalbard. Polarnet Technical Report. 2008, 11.

44. Hejazi M.A., Wijffels R.H. Milking of microalgae. TRENDS in Biotechnology. 2004 April 1; 22(4): 189-94.

45. Lorenz R.T., Cysewski G.R. Commercial potential for Haematococcus microalgae as a natural source of astaxanthin. Trends in biotechnology. 2000 April 1; 18(4): 160-7.

46. Riss J., Décordé K., Sutra T., Delage M., Baccou J.C., Jouy N., Brune J.P., Oréal H., Cristol J.P., Rouanet J.M. Phycobiliprotein C-phycocyanin from Spirulinaplatensis is powerfully responsible for reducing oxidative stress and NADPH oxidase expression induced by an atherogenic diet in hamsters. J. of Agricultural and Food Chemistry. 2007 Aug 16; 55(19): 7962-7.

47. Kumar N., Singh S., Patro N., Patro I. (2009). Evaluation of protective efficacy of *Spirulinaplatensis*against collagen-induced arthritis in rats. Inflammopharmacology 2009: 17: 181-190.

48. Romay C.H., Gonzalez R., Ledon N., Remirez D., Rimbau V. C-phycocyanin: a biliprotein with anti-oxidant, anti-inflammatory and neuroprotective effects. Current protein and peptide science. 2003 Jun 1; 4(3): 207-16.

49. Manach C., Williamson G., Morand C., Scalbert A., Rémésy C. Bioavailability and bioefficacy of polyphenols in humans. I. Review of 97 bioavailability studies. The American journal of clinical nutrition. 2005 Jan 1; 81(1): 230S-42S.

50. Hemat, R.A.S. Fat and muscle dysfunction. In R.A.S. Hemat (Eds), Andropathy. Dublin, Ireland: Urotext. 2007; 83-85.

51. Wang T., Jonsdottir R., Liu H., Gu L., Kristinsson H.G., Raghavan S., Olafsdottir G. (2012). Anti-oxidant Capacities of Phlorotannins Extracted from the Brown Algae *Fucusvesiculosus*. J. Agricul. Food. Chem., 60: 5874–5883.

52. Hernandez-Ledesma B., Herrero M. Bioactive Compounds from Marine Foods: Plant and Animal Sources. 1st ed. John Wiley & Sons Ltd; Chichester, UK: 2014, 173–187.

53. Volkman J.K. sterols in micro-organisms. Appl Microbiol Biotechnol. 2003; 60(5): 495–506.

54. Thomson P.G., Wright S.W., Bolch C.J.S., Nichols P.D., Skerratt J.H., McMinn A. Antarctic distribution, pigment and lipid composition, and molecular identification of the brine dinoflagellate *Polarellaglacialis* (Dinophyceae). J Phycol. 2004; 40: 867–73.

55. Giner J.L., Zhao H., Boyer G.L., Satchwell M.F., Andersen R.A. Sterol chemotaxonomy of marine pelagophyte algae. ChemBiodiv. 2009; 6(7): 1111-30.

56. Caroprese M., Albenzio M., Ciliberti M.G., Francavilla M., Sevi A. A mixture of phytosterols from *Dunaliellatertiolecta* affects proliferation of peripheral blood mononuclear cells and cytokine production in sheep. Vet. Immunol. Immunopathol. 2012; 150: 27-35.

57. Rasmussen H.E., Blobaum K.R., Park Y.K., Ehlers S.J., Lu F., Lee J.Y. Lipid extract of *Nostoc commune var. sphaeroides* Kützing, a blue-green alga, inhibits the activation of sterol regulatory element binding proteins in HepG2 Cells. J. Nutr. 2008; 138: 476-481.

58. Chen J., Jiao R., Jiang Y., Bi Y., Chen Z.Y. Algal Sterols are as effective as β-sitosterol in reducing plasma cholesterol concentration. J. Agric. Food Chem. 2014; 62: 675-681.

59. Lee S., Lee Y.S., Jung S.H., Kang S.S., Shin K.H. Anti-oxidant activities of fucosterol from the marine algae *Pelvetiasiliquosa*. Arch. Pharm. Res. 2003; 26: 719-722. doi: 10.1007/BF02976680.

60. Choi J.M., Lee E.O., Lee H.J., Kim K.H., Ahn K.S., Shim B.S., Kim N.I., Song M.C., Baek N.I., Kim S.H. Identification of campesterol from *Chrysanthemum coronarium* L. and its antiangiogenic activities. Phytother. Res. 2007; 21: 954-959.

61. Lee Y.S., Shin K.H., Kim B.K., Lee S. Anti-Diabetic activities of fucosterol from *Pelvetiasiliquosa*. Arch. Pharm. Res. 2004; 27: 1120-1122.

62. Gylling H., Plat J., Turley S., Ginsberg H.N., Ellegård L., Jessup W., Jones P.J., Lütjohann D., Maerz W., Masana L. Plant sterols and plant stanols in the management of dyslipidaemia and prevention of cardiovascular disease. Atherosclerosis. 2014; 232: 346-360.

63. Ahmed F., Zhou W., Schenk P.M. *Pavlovalutheri* is a high-level producer of phytosterols. Algal Res. 2015; 10: 210-217.

64. Amaninezhad P., Emadi H., Emtiazjou M., Hosseinzadeh S.H. Effects of dunaliella microalgae (dunaliellasalina) on different immune indeses (complements c3, c4 and anti-oxidant capacity) in rainbow trout (oncrohynchusmykiss). *Global Journal of Biodiversity Science and Management*, 2013; 3(2): 237-242.

65. Ibañez E., Cifuentes A. Benefits of using algae as natural sources of functional ingredients. Journal of the Science of Food and Agriculture. 2013 Mar 15; 93(4): 703-709.

❑❑❑

Chapter 17

Valorization of Whey

Suchetana Purkayastha

Lecturer, Department of Food Processing Technology, Mirmadan Mohanlal Govt. Polytechnic, Plassey, Nadia, Email:suchetana2008@yahoo.com

ABSTRACT

The dairy industry creates huge quantities of whey as an effluent. The whey protein and lactose must be used in a sustainable way. Milk sugar lactose, major by-product of cheese whey water can be used for the development of various products by using fermentation. Whey proteins are progressively finding utilizations in the food and pharmaceutical industries. Whey is a vital environmental concernment and whey valorization or utilization is necessity. Conversion of lactose to bulk commodities like ethanol and single cell protein are emphasized in this chapter.

1.0 Introduction

Whey is a dairy industry by-product, specifically the watery part that is produced during coagulation of milk protein casein in casein manufacture or in cheese making. It is generated in huge quantities and has the high polluting capacity, therefore, presenting a major environmental hazard. Still, whey contains a lot of milk nutrients, including peptides and functional proteins, lactose, lipids, vitamins, and minerals and for that reason, it has a wide potential source of added value components, demanding the industry to face surplus whey as resource and not as waste [15].Whey contains about 85–95% of milk volume and 55% of nutrients in milk. The most abundant nutrients in whey are lactose, soluble proteins and lipids and mineral salts in minute quantities. They also contain considerable quantities of lactic acids, citric acids, non-protein nitrogen compounds (urea and uric acid) and B vitamins [15,38]. Whey can be classified into two categories depending upon casein precipitation: acid whey (pH<5), generating from production of soft cheeses (such as cottage cheese), and sweet whey (pH 6–7), generating from hard (ripened) cheeses [15,21,38,45,47]. Generally, whey made from rennet-coagulated cheese is low in acidity, but the production of fresh cheeses, such as cottage cheese or paneer or chhana yield medium or acid whey [25].

The worldwide generation of whey is approximated at around 165 million tonnes. Out of which cheese whey contribute at about 95%. The main source of whey in India is from the manufacture of chhena and paneer. In the absence of systematic survey or statistics, the estimated production of whey is at about 5 million tons per annum [17,25]. The valorization was 75% in Europe and less than 50% in the other parts of the world and as a result, a huge quantity of component with potential source as a feed or food is generated as waste [1,25]. Whey after getting waste, becomes a main pollutant, as it causes BOD of 30,000- 50,000 mg/lit and COD of 60,000-80,000 mg/lit. Removing whey also generates a major loss of nutrients and energy sources and has been considered seriously due to its polluting capacity [25].

2.0 Whey Valorization

Whey, a source of biologically active components with special functional and physiological properties provides chances for the food processing industry to make functional food or food that has health advantage.Whey-derived bio-active compounds act as anti-microbial and anti-viral agents, and improve immune defense and bone health, and enhance antioxidative characteristics, and prevent cancer and cardiovascular disease, and increase the efficiency of physically active individuals, among other merits [39]. Whey can be used as raw material for many products [40,43]. Scientists utilize cheese-whey to develop new products formulation [27, 40]. Traditionally, liquid whey can be provided to the farmer for either agricultural bio-fertilizers or for utilizing lactose and proteins for cattle feed. Powdered whey is mainly utilized in animal feeding and small quantities are used as sweeteners. Due to its saline taste, its usage in human foods is not preferred.However, another possible usage of whey is using it as a protein source. It can also be converted into whey protein concentrate (WPC).WPC used as food additive and ingredient. Processing of whey for the recovery of protein has been introduced by the advancements in the membrane.Protein recovery during whey, processing has been enhanced by the developments in membrane separation processes.

However, ultrafiltration and diafiltration techniques are approved separation technique in dairy industries [32,40]that admit their recovery without notable loss of functional attributes. Whey proteins contain about 15 and 22% of milk protein. It contains five major protein fractions with specified molecular weights, viz. bovine serum albumin (BSA), α-lactalbumin (α-La), immunoglobulin-G (IgG), β-lactoglobulin (β-Lg) and proteose peptones (30,40). The product whey protein concentrate (WPC) represents about 50-85% protein on a dry weight basis, while a more refined product whey protein isolate (WPI) represents 90-98% protein and lactose and fat in minute quantities. Whey proteins are high-quality proteins with a protein efficiency ratio (PER) 3.4, more than milk protein casein (2.8) and same as egg albumin [38,40].

Large volumes of a lactose-rich flow, the permeate, are also acquired during processing of whey for the production of whey protein concentrate. Permeate remains one of the main pollutants as it contains lactose, which consists of more than 70% of total whey solid and it is responsible for whey polluting capacity [15,46]. Lactose generated is regained from thepermeate of whey by crystallization [13,15,45].

Another popular use for lactose in whey or whey permeates includes its application as substrate for production of various valuable components by fermentation processes. The traditional examples are single cell protein (SCP) and ethanol production in yeast-based bio practices, though researchers have suggested various alternative bio-products [3,15,34,38,45,46].

2.1 Ethanol Production

Lactose is a carbon source for production of ethanol by using fermentation micro-organisms. [2,9,23]. As the strain of yeast *Saccharomyces cerevisiae* lacks β-D-galactosidase enzyme activity, so lactose cannot be fermented by *Saccharomyces cerevisiae.* However, it can be fermented the hydrolyzed products of lactose i.e. glucose and galactose. On the other hand, *Kluyveromyces fragiis* can ferment lactose directly to ethanol. *Kluyveromyces fragiis* have been studied for utilization of whey, e.g., the effect of multiple substrates in ethanol fermentation from cheese whey. However, fermenter can also be utilized for fermentation with batch, continuous and fed-batch mode. During fed-batch fermentation, feed is given periodically during the process of fermentation and it is one of the methods that is commercially applied. Moreover, fed-batch has many merits compared to batch one, like fed-batch can restrict reducing substrate during fermentation process. As a result of productivity of β-galactosidase enzyme with fed-batch process more than batch system [2, 5, 35,36]. Enhanced ethanol fermentation can be accomplished by controlling process parameters. The parameters like temperature, pH, and substrate concentration affect the fermentation process. Moreover, pH adjustment sometimes becomes tough to implement during fermentation process. Initial lactose concentration of de-proteinated cheese whey shows a great influence on ethanol production by *K. fragilis*. It was found that only the initial lactose concentration, as well as temperature, plays a vital role in the production of ethanol. As substrate concentration is increased product concentration also increases [11]. It can also produce ethanol at high temperature and low pH. Technologists also established that mixed culture immobilized cells of *S. cerevisiae* and *K. marxianus* are favourable in ethanol fermentation from whey. Fermentation with immobilized mixed cultures has been found as one of the excellent processes of fermentation [16]. The immobilized *K. fragiis* cultures show more cell densities and greater ethanol productivities in comparison to free cell suspensions [29,41]. It was established that the immobilization of enzyme does not considerably alter the pH at which it develops maximal activity [18]. The process immobilization has additional merits like reduction in fermentation expenses caused by costly enzymes. However, when co-immobilized with an enzyme, yeast develops special attributes. The literature includes reports on changed metabolisms of micro-organisms in particular micro-environments [20]. The function of the microbial cell can be changed by modifications in water activity, dissolved oxygen concentration or carbon dioxide concentration, the enhanced concentration of substrate and product on phase boundary, pH alteration and the ionic concentration of the culture medium. Immobilized cells are simply handled in continuous operations and separated from product due to reduced necessities for filtration systems [48]. Moreover, there is reduced the risk of microbial hazard due to the greater density of yeast [37].

The continuous process is popularly applied now-a-days. In a continuous process culture medium, feed containing a substrate and necessary sources of other nutrients are continuously pumped into an agitated vessel, where the micro-organisms are active. The product, which is collected from the top of the bio-reactor, carries ethanol, residual sugar, and cells. Fermentation with continuous mode provides numerous merits compared to batch operations, generally resulting in enhanced volumetric productivity and, consequently, smaller bio-reactor volumes and lower investment and operational costs. The major drawback is that yeasts cultivated under anaerobic conditions during a long time diminish their ability to synthesize ethanol. At high dilution rate enabling elevated productivity, the substrate is not totally consumed and yield is decreased [19].

2.1.1 Industrial production of whey ethanol

The Carbery Milk Products (now Carbery Group) from Ireland introduced the process of industrial-scale whey-to-ethanol in the year 1978. The method was considered for production of potable ethanol, however, since 2005, the company has also been providing fuel bio-ethanol to a petrol company in Ireland [10, 15,22]. The Carbery process was then adopted by industries in the United States and New Zealand [15,22]. New Zealand's Anchor Ethanol, generates various ethanol grades (potable ethanol for beverages to anhydrous alcohol for fuels) [15,42].

In the United States, Milbrew method was introduced in the year 1972 to produce single-cell protein (SCP) and bio-ethanol from whey using *K. fragilis* [8,24,34].The yield of ethanol may be manipulated by altering the fermentation conditions, like aeration rate [15,24].

2.2 Single cell Protein Production

Conversion of whey or whey permeate into Single cell protein (SCP) has become one of the excellent remedies and is implemented in most of the countries [28,33,38,44]. Utilization of whey for generation of proteinaceous biomass of yeast has many merits like it is an easy treatment operation and the ultimate removal of the whey is promoted, as pollutant level is considerably declined and lactose is transformed into proteinaceous yeast biomass. SCP is formed using yeasts like *Kluyveromyces*, *Trichosporon* and *Candida* species, as those organisms can metabolize whey lactose [26,44]. *Kluyveromyces marxianus* species have been popularly studied for SCP production from whey and considered as generally regarded as safe (GRAS) microorganisms for food and feed uses [12, 14, 26,44]. There are few shortcomings during whey processing like process economy and contamination problem, lower efficiency of removal of COD and recovery of soluble protein in whey after the process of fermentation (if whey is fermented without removal of protein). SCP as animal feed by using whey is not profit oriented due to its contamination issues and low returns in the economy [4,6,7,40,44]. The SCP production can be facilitated employing different modes of fermentation, like batch, continuous or fed-batch process. However, each method has its own merits and demerits. Moreover, the selection of the mode depends on either treatment or production or both. However, batch process facts are always vital prior running the continuous process or fed-batch process, continuous mode is generally preferred on a commercial scale [31,44].

3.0 Conclusion

Many methods have been introduced on whey utilization, these techniques revealed that in bio-ethanol, Single cell protein and also in whey protein concentrate or isolate production we don't require any additional raw material cost and gradually, we will be able to reduce the pollution problem developed by the disposal of whey.

4.0. Acknowledgement

I would like to express my gratitude to Prof. Chiranjib Bhattacharjee of Jadavpur University, Kolkata and Dr. Manuj Kumar Hazarika, Associate Professor of Tezpur University for the support and help.

5.0 Reference

1. Aneja, R.P., Mathur, B.N., Chandan, R.C. & Banerjee, A.K. (2002). *Technology of indian milk products: handbook on process technology modernization for professionals, entrepreneurs and scientists*. Dairy India Yearbook.
2. Ariyanti, D., Aini, A.P. & Pinundi, D.S. (2014). Optimization of ethanol production from whey through fed-batch fermentation using Kluyveromycesmarxianus. *Energy Procedia*, 47: 108-112.
3. Audic J.L., Chaufer B., Daufin G. (2003). Non-food applications of milk components and dairy coproducts: a review. *Lait*. 83: 417–38.
4. Chatzipaschali, A.A. & Stamatis, A.G. (2012). Biotechnological utilization with a focus on anaerobic treatment of cheese whey: current status and prospects. *Energies*. 5(9): 3492-3525.

5. Cheng, N.G., Hasan, M., Kumoro, A.C., Ling, C.F., danTham, M. (2009). Production of Ethanol by Fed-batch Fermentation. *J Science and Technology Pertanika*. 17(2): 399-408.

6. Çinar, Ö., Hasar, H. & Kinaci, C. (2006). Modelling of submerged membrane bio-reactor treating cheese whey wastewater by artificial neural network. *Journal of biotechnology*. 123(2): 204-209.

7. Cristiani-Urbina, E., Netzahuatl-Muñoz, A.R., Manriquez-Rojas, F.J., Juárez-Ramýìrez, C., Ruiz-Ordaz, N. & Galýìndez-Mayer, J. (2000). Batch and fed-batch cultures for the treatment of whey with mixed yeast cultures. *Process Biochemistry*. 35(7): 649-657.

8. Detroy R.W., Julian G. Biomass conversion: fermentation chemicals and fuels. Crit Rev.

9. Domingues, L., Guimares, P.M.R., dan Oliveira, C. (2010). Metabolic engineering of Saccharomyces cerevisiae for lactose/Whey fermentation. J Bioengineered Bugs. 1(3): 164-171.

10. Doyle A. Another step in biofuel supply. Irish Farmers Journal Interactive, (2005). (availableonline bhttp://www.farmersjournal.ie /2005/1008/farmmanagement/crops/index

11. Dragone, G., Mussatto, S.I., e Silva, J.B.A. & Teixeira, J.A. (2011). Optimal fermentation conditions for maximizing the ethanol production by Kluyveromycesfragilis from cheese whey powder. *Biomass and bioenergy*. 35(5): 1977-1982.

12. Fonseca, G.G., Heinzle, E., Wittmann, C. & Gombert, A.K. (2008). The yeast Kluyveromycesmarxianus and its biotechnological potential. *Applied microbiology and biotechnology*. 79(3): 339-354.

13. GänzleM.G., Haase G., Jelen P. (2008). Lactose: crystallization, hydrolysis and value-added derivatives.*Int Dairy J*. 18: 685–694.

14. Grba, S., Stehlik-Tomas, V., Stanzer, D., Vahcic, N. & Skrlin, A. (2002). Selection of yeast strain *Kluyveromycesmarxianus* for alcohol and biomass production on whey. *Chemical and biochemical engineering quarterly*. 16(1): 13-16.

15. Guimarães, P.M., Teixeira, J.A. & Domingues, L. (2010). Fermentation of lactose to bio-ethanol by yeasts as part of integrated solutions for the valorisation of cheese whey. *Biotechnology advances*. 28(3): 375-384.

16. Guo, X., Zhou, J. & Xiao, D. (2010). Improved ethanol production by mixed immobilized cells of *Kluyveromycesmarxianus* and *Saccharomyces cerevisiae* from cheese whey powder solution fermentation. *Applied biochemistry and biotechnology*. 160(2): 532-538.

17. Gupta, V.K. (2008). Overview of production, processing and utilization of dairy by-products. *Technological Advances in the Utilization of Dairy By-products. Course Compendium, NDRI, Karnal.* 22: 1-7.

18. Haider, T. & Husain, Q. (2009). Hydrolysis of milk/whey lactose by â galactosidase: A comparative study of stirred batch process and packed bed reactor prepared with calcium alginate entrapped enzyme. *Chemical Engineering and Processing: Process Intensification.* 48(1): 576-580.

19. Jain, A. & Chaurasia, S.P. (2014). Bio-ethanol production in membrane bio-reactor (MBR) system: a review. *Int. J. Environ. Res. Dev*, 4: 387-394.

20. Jirkù, V. (1999). Whole cell immobilization as a means of enhancing ethanol tolerance. *Journal of industrial microbiology & biotechnology,* *22*(3): 147-151.

21. Kosikowski, F.V. (1979). Whey Utilization and Whey Products. *Journal of Dairy Science.* 62(7): 1149-1160.

22. Ling K.C. Whey to ethanol: a biofuel role for dairy co-operatives? Washington DC: USDA Rural Development; 2008 (available online bhttp://www.rurdev.usda.gov/RBS/

23. Lukondeh, T., Ashbolt, N.J., Rogers, P.L. (2005). Fed-batch fermentation for production of Kluyveromycesmarxianus FII 510700 cultivated on a lactose-based medium. *J Industrial Microbiology and Biotechnology.* 32(7): 284-288.

24. Lyons T.P., Cunningham J.D. (1980). Fuel alcohol from whey. *Am Dairy Rev.* 42: 42A-E.

25. Macwan, S.R., Dabhi, B.K., Parmar, S.C. & Aparnathi, K.D. (2016). Whey and its utilization. *Int. J. Curr. Microbiol. App. Sci*, 5(8): 134-155.

26. Mansour, M.H., Ghaly, A.E., Ben-Hassan, R.M., &Nassar, M.A. (1993). Modelling batch production of single cell protein from cheese whey. *Applied Biochemistry and biotechnology*, 43(1).

27. Marwaha, S.S. & Kennedy, J.F. (1988). Whey—pollution problem and potential utilization. *International Journal of Food Science & Technology.* 23(4): 323-336.

28. Mawson, A.J. (1994). Bioconversions for whey utilization and waste abatement. *Bioresource Technology.* 47(3): 195-203.

29. Melzoch, K., Rychtera, M. & Hábová, V. (1994). Effect of immobilization upon the properties and behaviour of Saccharomyces cerevisiae cells. *Journal of Biotechnology*. 32(1): 59-65.

30. Merina, U., Bernsteina, S., Bloch-Damtib, A., Yagilb, R., Van Creveldb, C., Lindnera, P., Gollop, N. (2001). A Comparative Study of Milk Serum Proteins in Camel (Camelusdromedarius) and Bovine Colostrum. Livestock Production Science. 67: 297–301.

31. Michael L., Shuler F.K. (2002). Bioprocess engineering: basic concepts. Prentice Hall: Upper Saddle River, New York Microbiol 1982(10): 203–28.

32. Ounis, W.B., Gauthier, S.F., Turgeon, S.L., Roufik, S. & Pouliot, Y. (2008). Separation of minor protein components from whey protein isolates by heparin affinity chromatography. *International dairy journal*. 18(10-11): 1043-1050.

33. Panesar, P.S., & Kennedy, J.F. (2012). Biotechnological approaches for the value addition of whey. *Critical reviews in biotechnology*. 32(4): 327-348.

34. Pesta G., Meyer-Pittroff R., Russ W. Utilization of whey. In: Oreopoulou V., Russ W., editors. Utilization of by-products and treatment of waste in the food industry. *Springer*.

35. Rech R. danAyub, M.A.Z. (2006). Fed-batch Bioreactor Process with Recombinant Saccharomyces cerevisiae Growing on Cheese Whey. Brazilian J. Chemical Engineering. 23(04): 435-442.

36. Rech, R. danAyub, M.A.Z. (2007). Simplified feeding strategies for fed-batch cultivation of *Kluyveromycesmarxianus* in cheese Whey. *Journal of Elsevier*. 42(5): 873–877.

37. Shindo, S., Takata, S., Taguchi, H. & Yoshimura, N. (2001). Development of novel carrier using natural zeolite and continuous ethanol fermentation with immobilized *Saccharomyces cerevisiae* in a bioreactor. *Biotechnology Letters*, 23(24).

38. Siso, M.G. (1996). The biotechnological utilization of cheese whey: a review. *Bioresource Technology*. 57(1): 1-11.

39. Solak, B.B. & Akin, N. (2012). Health benefits of whey protein: a review. *Journal of Food Science and Engineering*. 2(3): 129.

40. Spalatelu C. (2012). Biotechnological valorisation of whey. *Innov Romanian Food Biotechnol* 10: 1–8.

41. Staniszewski, M., Kujawski, W. & Lewandowska, M. (2009). Semi-continuous ethanol production in bio-reactor from whey with co-

immobilized enzyme and yeast cells followed by pervaporative recovery of product–Kinetic model predictions considering glucose repression. *Journal of food engineering*. 91(2): 240-249.

42. Thiele J.H. Estimate of the energy potential for fuel ethanol from putrescible waste in New Zealand. Dunedin: Waste Solutions Ltd; 2005. (available online bhttp://www.bioenergy.org.nz/documents/liquid biofuels/energy-potential-for-fuel-ethanolfrom-putrescible-waste-in-nz-report-05.pdfN last visited: 2nd June 2009).

43. Türkmenoðlu, S. (2006). *Organic acids production from cheese-whey* (Doctoral dissertation, Middle East Technical University).

44. Yadav, J.S.S., Bezawada, J., Elharche, S., Yan, S., Tyagi, R.D., & Surampalli, R.Y. (2014). Simultaneous single-cell protein production and COD removal with characterization of residual protein and intermediate metabolites during whey fermentation by K. marxianus. *Bioprocess and biosystems engineering*. 37(6): 1017-1029.

45. Yang, S.T. & Silva, E.M. (1995). Novel products and new technologies for use of a familiar carbohydrate, milk lactose. *Journal of Dairy Science*. 78(11): 2541-2562.

46. Zadow J.G. (1984). Lactose-properties and uses. *J Dairy Sci*. 67: 2654–79.

47. Zall, R.R. (1984). Trends in whey fractionation and utilization, a global perspective. *Journal of Dairy Science*, 67(11): 2621-2629.

48. Zhang, S., IMADA, K. & OHASHI, T. (1998). Preparation of Whey Syrup by Immobilized β-Galactosidase. *Food Science and Technology International, Tokyo*. 4(1): 36-39.

□□□

Chapter 18

Edible Films and Coatings - Classification, Properties and their Role in Preservation

[1]K. Shunmugapriya and [2]P.Vennila

[1] Department of Food science and Nutrition, Home Science College and Research Institute, Madurai, [2] Professor, Post Harvest Technology Centre, Tamil Nadu Agricultural University, Coimbatore, Corresponding author: priyakannan44 @gmail.com

ABSTRACT

Edible coatings and films are emerging packaging technology and it provides excellent protection to raw materials and final food products. It became an alternative for petrochemical based packaging material which are biodegradable and requires high cost, man power for the production of these materials. Since, edible coatings and film are prepared from easily available natural resources, it is low cost and does not cause any environmental pollution. Starch, protein, waxes from both animal and plant source can be utilized for the production of edible coating and films. At some instances, to enhance the functionality of edible films and coatings, anti-oxidants, anti-microbials, vitamins, minerals, essential oil, bacteriocins, plasticizer can be incorporated in the polymer solution itself. These coating solution are easily imposed on food products by dipping, panning, spraying and films are prepared by solvent casting and extrusion methods. These edible films and coatings can be applied to wide range of food products, such as fruits, vegetable, meat and meat products, confectionary items, eggs, etc., where they are offering mechanical, water and moisture barrier property.

1. Introduction

Food packaging plays a significant role in food technology, aims to protect and preserve all kind of food products and their raw materials. Apart from this, it also protects the products from oxidative detoriation, photo-sensitive oxidation and microbial spoilage. Earlier days, dried leaves were used as packaging material for various items but due to some constraints, these traditional packaging systems were replaced by petrochemical based plastics, such as polyolefin, polyesters, polyamides, etc. due to their availability in large quantities at low cost and favourable functionality characteristics. But these petrochemical plastics are totally non-biodegradable, and therefore, lead to environmental pollution, which pose serious ecological problems (1) Of late, a drastic changes was observed in packaging technology and this shift was imposed by the growing environmental awareness by all to look for packaging films and processes, which are biodegradable. Thus, the concept of biodegradability enjoys both user-friendly and eco-friendly attributes. Though a bit expensive, biopackaging is tomorrow's need for packaging especially for a few value added food products.

1.1. Edible films and coatings

An edible film or coating could be defined simply as a thin and continuous layer of edible material that is usually placed on food surfaces for preservation or quality purposes. The terms 'edible film' and 'edible coating' are often used interchangeably; however, edible films are regarded as stand-alone films, as they are formed before use. Such films can be used as covers, wraps or separation layers (2). On the other hand, edible coatings are formed directly on the food surface. In this case, coatings are regarded typically as a part of the final product.

Materials used to form edible films and coatings include bio-polymers, such as proteins or carbohydrates as well as lipids. Edible films provide replacement and/or fortification of natural layers to prevent moisture losses, while selectively allowing for controlled exchange of important gases, such as oxygen, carbon dioxide, and ethylene, which are involved in respiration processes. A film or coating can also provide surface sterility and prevent loss of other important components. Generally, its thickness is less than 0.3 mm (3).

1.2. Properties of Edible coating

The properties of edible coating depend primarily on molecular structure rather than molecular size and chemical constitution. Specific requirements for edible films and coatings are (4):

i) The coating should be water-resistant so that it remains intact and covers a product adequately, when applied.

ii) It should not deplete oxygen or build up excessive carbon dioxide. A minimum of 1–3% oxygen is required around a commodity to avoid a shift from aerobic to anaerobic respiration.

iii) It should be easily emulsifiable, non-sticky or should not be tacky, and have efficient drying performance.

iv) It should never interfere with the quality of fresh fruit or vegetable and not impart undesirable order.

v) It should be translucent to opaque but not like glass and capable to tolerate slight pressure.

1.3. Functionality of Edible Films

1. Mechanical properties

Tensile strength (TS), elongation at break (EB), elastic modulus EM), compression strength, puncture strength, stiffness, tearing strength, burst strength, abrasion resistance, adhesion force, folding endurance, etc. are parameters that are related to the film's mechanical properties and chemical structure.

Generally, protein films have lower TS than most polysaccharide films and petroleum-based films. Protein and carbohydrate films are often brittle as a result of the strong cohesive bond. So, plasticizers are commonly required to increase film flexibility by reducing the intermolecular H-bonding along polymer chains and increased intermolecular spacing. Both type and amount of plasticizers affect the interactions between polymer molecules. Increasing plasticizer concentration in the formula results in decreased TS and EM, and increased film elongation.

When natural extracts are added, Changes in the mechanical properties of bio-polymer-based films will occur. Green tea water extract (GTE) slightly increases the mechanical resistance and extensibility of chitosan-based films but decreases the TS and EB of both agar and agar–gelatin films, negatively affecting the film's mechanical properties.

2. Moisture and gas barrier

For food applications, the oxygen (O_2), carbon dioxide (CO_2), and water vapor (WVP) permeability of biomaterials should be considered. Due to the hydrophilic nature of films, its barrier properties are dependent on the surrounding relative humidity conditions (5). For this reason, tests should be carried out at constant storage conditions, % RH, and temperature. Increasing the polarity, hydrogen bond forces, and crystallinity decreases the free volume of the film, enhances the cohesion, and reduces the flexibility, thereby increasing the barrier properties of the polymer (6).

Overall, hydrocolloid-based films have quite high WVP compared to films of edible waxes and plastics. At high relative humidity (RH) and high plasticizer concentration, WVP of hydrophilic films increase because of their substantial polarity. On the other hand, lipids or other hydrophobic compounds are often used to make moisture barrier coatings or enhance the moisture-barrier properties of hydrocolloid based films, because of their low water affinity, low polarity, and dense-structured molecular matrixes. To prevent or diminish the dehydration of foods, films used as packaging or coatings must control the transfer of humidity from the product to the environment; hence, the WVP of these films must be as low as possible (7).Films made from proteins and carbohydrates are excellent barriers to oxygen, because of their tightly packed, ordered hydrogen bonded network structure (8).

3. Light Barrier Properties

Transparent packaging induces the oxidation and degradation of nutritional compounds, because light acts as a catalyst for these processes. Therefore, opaque packaging and packaging containing specific compounds that absorbs light in the UV–V is spectrum have been developed to prevent these reactions. Plant extracts are commonly used to provide colour and opacity to polymers (9), and as a result, films containing plant extracts are less transparent than films without plant extracts. Thus, the incorporation of plant extracts in films provides an adequate barrier to light, which is also important for preventing the degradation of ascorbic acid and, consequently, browning (10).

4. Carrier properties

During the raw material blending process, active compounds can be added to edible films and coating solutions. These include anti-oxidants, anti-microbial agents, flavoring, pigments and nutrients. Anti-microbial edible packaging is a promising application that can be used alone or synergistically in combination with other preservation, such as refrigeration, modified atmosphere packaging or irradiation to improve microbial stability (11).

1.4. Film Forming Material

1. Polysaccharides based materials

Edible coatings can be made from a variety of polysaccharides. The polymers most commonly available are extracted from marine and agricultural plants and animals. Such coatings have been used to retard moisture loss of some foods during short-term storage. However, polysaccharides, being hydrophilic in nature, do not function well as physical moisture barriers. The method by which they retard moisture loss is by acting as a sacrificial moisture barrier to the atmosphere, so that the moisture content of the coated food can be maintained (12).

1.1 *Starch and derivatives*

Starch, the storage polysaccharide of cereals, legumes, and tubers, is a renewable and widely available raw material. Starches contain two types of glucose polymers: a linear chain glucose polymer termed as amylose and a branched polymer of glucose termed as amylopectin (13). Films produced from high amylose cornstarch (71% amylose) had no detectable oxygen permeability at RH levels less than 100% (14). This was true for both unplasticized and plasticized (16% glycerol) films. Branched structure of amylopectin generally leads to films with poor mechanical properties (decreased tensile strength and elongation).

Amylose and hydroxyl propyl amylose films have been formed as protective edible coatings on foods and encapsulating agents, to provide an O_2 or lipid barrier and to improve appearance, texture, and handling. Edible starch films and coatings are commonly used in bakery, confectionary, batters, and meat products.

1.2 *Cellulose and derivatives*

Cellulose ethers are polymer substances obtained by partial substitution of hydroxyl groups in cellulose by ether functions. In general, cellulose derivatives possess excellent film-forming property. The most common commercially produced cellulose derivatives are carboxymethyl cellulose (CMC), methyl cellulose (MC), hydroxypropyl cellulose (HPC), and hydroxypropylmethyl cellulose (HPMC). CMC coatings have shown the capabilities for helping retain the original firmness and crispness of apples, berries, peaches, celery, lettuce, and carrots, when used in a dry coating process, preserving important flavour components of some fresh fruits and vegetables and reducing oxygen uptake without increasing carbon dioxide level in the internal environment of coated apples and pears by simulating a controlled atmosphere environment.

1.3 *Chitin*

Chitin, is a commercial by-product or a waste from crab, shrimp, and crawfish processing industries. It has anti-microbial activity by penetrating into cells, binding to DNA and subsequent inhibition of RNA and protein synthesis (15).

1.4 *Pectin*

Pectin with a degree of esterification (DE) above 50% are labelled high-methoxyl pectin (HMP) and below 50% are termed low-methoxyl pectin (LMP). The differences in methyl ester content and DE affect solubility and gelation properties of pectin (16). HMP forms gels with sugar and acid, especially in jams and jellies. LMP, produced from chemical de-esterification, forms gels in the presence of divalent cations. Calcium cations bridge adjacent LMP chains via ionic interactions and with inter chain H-bonding yield a 3D-gel network.

Edible pectin film can be formed by evaporating water from pectin gel. Although pectinate coatings are not adequate moisture barriers, they can retard water loss from enrobed food by acting as a sacrificing agent, when moisture evaporates from their gel matrix rather than dehydrating the food significantly. Pectin coatings have been investigated for their ability to retard moisture loss and lipid migration and improve handling and appearance of foods.

1.5 *Seaweed extracts*

Alginates are the major structural polysaccharides of brown seaweed known as Phaeophyceae. Alginate gel coating can act as a sacrificing agent, where moisture is lost from the coating before the food significantly dehydrates. Alginates are the salts of alginic acid, which is a linear copolymer of D-mannuronic and L-guluronic acid monomers. Alginate coating formation is based on the ability of alginates to react with divalent and trivalent cations, such as calcium, ferrum, or magnesium, which are added as gelling agents (17). Calcium-alginate coatings were found to improve the quality of fruits and vegetables, such as reducing shrinkage, oxidative rancidity, moisture migration, oil absorption, and sealing-in volatile flavours, improving appearance and colour, and reducing weight loss of fresh mushrooms in comparison with uncoated ones (18).

Carrageenan is a complex mixture of at least five different water-soluble galactose polymers designated as λ , κ , ι, μ, and ν-carrageenan. Gelation of ι and κ–carrageenan occurs in the presence of monovalents or divalent cations. Carrageenan film formation includes this gelation mechanism during moderate drying, leading to a three-dimensional network formed by polysaccharide double helices and to a solid film after solvent evaporation (19).

Other gums, including exudate gums (gum arabic or acacia gum and gum karaya) and microbial fermentation gums (xanthan gum), have also been studied as coating materials for fruits and vegetables. Xanthan gum provides uniform coatings with good cling and improved adhesion in wet batters. Gum arabic has been used for coating pecan nut halves to eliminate moist and oily appearance. It acts as an inhibitor of after-cooking darkening of potatoes (20).

2. Protein based material

Protein film-forming materials are derived from many different animal and plant sources. Proteins are macromolecules, hetero-polymers with specific amino acids as their monomer units. The structure of a protein matrix can be modified by heat, pressure or other actions in order to further control the physical and mechanical properties of the resulting films and coatings. Several globular proteins, including wheat gluten, corn zein, soy protein, and whey protein, have been investigated for their film properties. Protein films are expected to be good oxygen barriers at low relative humidity.

2.1 *Zein*

Zein is the key storage protein of corn and comprises approximately 45% to 50% of the proteins in corn. The ability of zein and its resins to form tough, glossy, and hydrophobic grease-proof coatings and their resistance to microbial attack have been of commercial interest. Zein-based coatings have water vapour permeability is higher than that of LDPE. Zein-based coatings have been applied to nuts and fresh and dried fruits, often as a substitute for shellac coatings. Zein coatings were able to retard ripening of tomatoes to maintain the original firmness and colour of broccoli florets (21) and to reduce the growth of *Listeria monocytogenes* on cooked sweet corn. Compared to commercial shellac coatings, zein coatings are favourable for gloss and other quality characteristics on apples (22)

2.2 *Soy protein*

Soy protein concentrate (SPC) or soy protein isolate (SPI) is extracted from defatted protein meal. SP coatings generally exhibit poor moisture resistance and water vapour barrier properties due to the inherent hydrophilicity of the protein and the addition of hydrophilic plasticizers (23). The high oxygen-barrier capability of SPI coatings has led to their applications as microencapsulating agents of flavours and pharmaceuticals, or in coatings of fruits, vegetables, and cheese, where the SP coatings are able to preserve freshness of apple slices and to retard the senescence process of kiwi fruit (24).

2.3 *Wheat gluten*

Wheat gluten is a general term for water-insoluble proteins of wheat flour. Edible films can be formed by drying aqueous ethanol solution of wheat gluten. Cleavage of native disulfide bonds during heating of film-forming solutions and then formation of new disulfide bonds during film drying are believed to be important to the formation of wheat gluten films structure along with hydrogen and hydrophobic bonds.

2.4 *Casein*

Caseins represent about 80% of the total milk proteins and can form films from aqueous solutions without further treatment based on their random-coil nature and the ability to form extensive intermolecular hydrogen, electro-static, and hydrophobic bonds, resulting in an increase of inter-chain cohesion. In general, casein films have an opaque appearance.

2.5 *Whey protein*

Whey protein is the protein that remains soluble after casein has been precipitated at pH 4.6 and 20^{0}C. It is rich in β -lactoglobulin. Whey proteins are globular and heat labile in nature. Whey protein isolate (WPI)-based edible films are water insoluble, have good mechanical properties and are excellent oxygen, lipid and aroma barriers. Formulation of WPI-based films must include low-molecular weight plasticizers to enhance their flexibility. Due to their inherent emulsifier capacity, these films have been used in a variety of lipid-based composite films (25).

2.6 *Egg white protein*

Formation of egg white protein films involves de-naturation of egg white protein in aqueous solution, followed by the casting process in alkaline medium. The egg white solutions were denatured at 45°C for 20 minutes. pH range of 10.5 to 12 was necessary to obtain smooth and homogeneous egg white films, and pH above 12 should be avoided as the film-forming solution becomes too viscous. Egg white protein films are clear and transparent and have properties similar to other proteins. To improve mechanical and barrier properties, different chemical cross- cross-linking agents, such as transglutaminase and dialdehyde starch, have been added to egg white protein solutions upon heat denaturation. Egg albumen has been studied for its potential to retain moisture inside raisins in cereal/raisin mixtures.

3. Lipids based coating

Lipid based coatings have been used for over 800 years. Historically, uses of lipid-based coatings included waxing of fruits as well as coating confectionery products. Lipid coatings are mainly used for their hydrophobic properties, which make them good barriers to moisture loss. In addition to preventing water loss, lipid-based coatings have been used to reduce respiration, thereby extending shelf-life and to improve appearance by generating a shine on fruits and vegetables. Lipid-based coatings can be made from a wide array of lipid substances including acetylated mono-glycerides, natural waxes and surfactants. The most effective lipid substances are paraffin wax and bees-wax.

3.1 Waxes

Waxes are esters of long chain aliphatic acids with long-chain aliphatic alcohols. Since waxes are highly hydrophobic, they are insoluble in bulk water, but soluble in typical organic solvents. Wax coatings are substantially more resistant to moisture transport than other lipid or non-lipid edible coatings and therefore, they are used worldwide as the most efficient barriers to water vapour transfer (26).

3.1.1 *Carnauba wax*

Carnauba wax is an exudate of palm tree leaves from the Tree of Life (*Copernica cerifera*). Carnauba wax is considered a GRAS (generally recognized as safe) substance and is permitted for use in coatings for fresh fruits and vegetables, in chewing gum, in confections, and in sauces with no limitations other than good manufacturing practice.

3.1.2 *Paraffin wax*

Paraffin wax is derived from the wax distillate fraction of crude petroleum. Paraffin waxes are permitted for use as protective coatings for raw fruit and vegetables and for cheese. They can also be used in chewing gum base, as a defoamer, and as a component in microencapsulation of spice-flavoring substances. Paraffin wax is frequently used to coat root crops, such as rutabaga, turnip, and yucca. In some cases, the stems of sugarcane are also coated. To adhere well to the surface, vegetables to be coated with paraffin waxes must be clean. Because molten waxes are applied to vegetable surfaces well above the boiling point of water, any free water on the commodity will form steam pockets and will flake during subsequent handling. For this reason, the product must be completely dry before application of paraffin.

3.1.3 *Other waxes*

Bee-wax (white wax) is produced from honeybees. Bees-wax is considered a GRAS substance and is allowed for direct use in food with some limitations. Maximum allowed levels are 0.065% in chewing gum, 0.005% in confections and frostings, 0.04% in hard candy, 0.1% in soft candy, and 0.002% in all other categories.

Candelilla is obtained from candelilla plant. Candelilla wax is considered a GRAS substance and is allowed for certain food uses including chewing gum and hard candy with no limitations other than good manufacturing practices.

Waxes are divided into three types, which are as follows:

Animal waxes: Bees-wax, spermaceti wax, shellac wax, lanolin wax, Chinese insect wax.

Vegetable waxes: Carnauba wax, candelilla wax, bayberry wax, rosin wood wax, sugarcane wax, palm wax, esparto wax, cotton seed wax, oricury wax, rice bran wax, Japan wax, waxol.

Mineral and synthetic waxes: Montan wax, paraffin wax, synthetic wax, microcrystalline wax.

3.2 Resins

Resins included in edible coatings are wood rosin and coumarone indene, both of which are used for coating citrus fruits. A shellac resin is a secretion by the insect *Laccifer lacca* and is composed of a complex mixture of aliphatic alicyclic hydroxyl acid polymers. This resin is soluble in alcohols and in alkaline solutions. Shellac is not generally regarded as safe (GRAS) substance, as it is only permitted as an indirect food additive in food coatings and adhesives. It is mostly used in coatings for the pharmaceutical industry.

4. Composite material

Composite film/coating can be categorized as a bi-layer or a stable emulsion. For bi-layer composite films/coatings, lipid generally forms an additional layer over the polysaccharide or protein layer, while the lipid in the emulsion composite films/coatings is dispersed and entrapped in the matrix of protein or polysaccharide. The amphiphilic character of proteins enables proteins to stabilize the protein - lipid emulsions through the balance between forces, primarily electrostatic and hydrophobic. Polysaccharides stabilize emulsions by strongly attaching to the surface of the lipid and significantly protruding into the continuous phase to form a polymeric layer or a network of appreciable thickness. In many cases, addition of emulsifier is required to improve emulsion stability.

In general, bi-layer films/coatings are more effective water vapour barrier than emulsion films/coatings due to the existence of a continuous hydrophobic phase in the matrix, and their moisture-barrier property can be improved by increasing the degree of lipid saturation and chain length of fatty acids.

For emulsion composite films/coatings, the type of lipid, location, volume fraction, polymorphic phase, and drying conditions significantly impact moisture-barriers property. The improved moisture-barrier properties of composite coatings have made them promising candidates for coating fresh and minimally processed fruits and vegetables.

Incorporation of functional ingredients into coating matrix for enhancing functionality

One of the unique functions of edible coatings is the capability to incorporate functional ingredients into the matrix to enhance its functionality. This may include:

1. Improving basic coating functionality, such as plasticizers for improving mechanical properties and emulsifiers for stabilizing composite coatings and improving coating adhesion; and
2. Improving quality, stability, and safety of coated foods by incorporating anti-oxidants, anti-microbial agents, nutraceuticals, flavours, and/or colour agents.

1. Plasticizers

A plasticizer, in most cases, is required for making edible coatings, especially for polysaccharide-and protein-based coatings since the structure of such coatings is often brittle and stiff due to extensive interactions between polymer molecules.

Glycerol, acetylated monoglyceride, polyethylene glycol, and sucrose are common plasticizers incorporated into the polymeric coating matrix for decreasing glass transition temperature of the polymers and increasing coating flexibility. Plasticizers are usually hygroscopic that attracts water molecules. Water can also function as plasticizer, but easily lost due to dehydration at a low relative humidity environment. In addition to improving mechanical properties, the plasticizer also affects the resistance of coatings to the permeation of vapours and gases, where the hydrophilic plasticizers usually increase the water vapour permeability of the coatings.

2. Emulsifiers

Emulsifiers are surface-active agents of an amphiphilic nature and are able to reduce the surface tension of water - lipid or water -

air interface. Emulsifiers are essential for the formation of protein or polysaccharide coatings containing lipid emulsion particles. Emulsifiers also modify surface energy to control adhesion and wettability of the coating surfaces (25).

3. Antimicrobials

Incorporation of both natural and synthetic anti-microbial agents into various edible packaging has been developed as an effective alternative for controlling the growth of micro-organisms. Benzoic acid as preservative is more active against yeasts and moulds than bacteria. Sorbic acid and its salts are effective in pH range 3.0-6.5 against a broad spectrum of yeast and moulds and lactic acid bacteria. Acetic, lactic, propionic, and fumaric acids also can be used in coatings and contribute to anti-microbial activity.

Chitosan's antimicrobial activity is most effective against yeasts and moulds, followed by gram-positive bacteria and gram-negative bacteria. Its mechanism of anti-microbial action was proposed to be a deleterious leakage of microbial proteinaceous and intercellular components as a result of the interaction between positively charged chitosan and the negatively charged microbial cell membrane. Chitosan also chelates trace metals, thus preventing microbial growth and toxin production.

4. Plant extracts

Essential oil extracts from plants, e.g., grapefruit seed, cinnamon, all spice, clove, thyme, rosemary, onion, garlic, radish, mustard, horseradish, and oregano, are rich in phenolic compounds, such as flavonoids and phenolic acids, which exhibit a wide range of biological effects, including anti-oxidant and anti-microbial activity (27). These naturally occurring anti-microbial agents can be added into foods without labelling as anti-microbial agents or preservatives (28). Carvacrol is GRAS and used as a flavouring agent in baked goods, sweets, ice cream, beverages, and chewing gum. Among the complex constituents of citrus essential oils, the terpene, citral is known to have strong anti-fungal properties. In addition, cinnamon oil and its active compound (cinnamaldehyde) also have been tested for their inhibitory activity against *E. coli*. An apple skin extract added to an apple-based film inhibited the growth of *Listeria monocytogenes* (29). Grape seed and grape pomace extracts inactivated the Shiga toxin produced by *E. coli* O157:H7 (30).

5. Bacteriocins

Bacteriocins are protein-containing macromolecules produced by various bacteria and possessing different anti-bacterial spectra, modes of action, and chemical properties. They are generally heat stable, hypoallergenic, and readily degraded by proteolytic enzymes in the human intestinal tract. Numerous bacteriocins have been characterized, such as colicins (*Escherichia coli*), lacticin (*Lactococcus lactis*), pediocins (*Pediococcus acidilactici*), and nisin (*Lactococcus lactis*). Nisin remains the most commercially important bacteriocin because of its history of safe use and documented effectiveness against gram-positive pathogenic and spoilage bacteria. Nisin interacts with the sulphur-containing compounds in bacterial membranes, disrupting their semi-permeable function and causing cell lysis (31).

6. Antioxidants and anti-browning agents

Anti-oxidants can be added into the coating matrix to protect against oxidative rancidity, degradation, and discolouration of certain foods. Nuts were coated with pectinate, pectate, and zein coatings containing BHA, BHT, and citric acid to prevent rancidity and maintain their texture. Ascorbic acid was incorporated into edible coatings to reduce enzymatic browning in sliced apples and potatoes. Xanthan gum coatings mixed with α -tocopherol enhanced nutritional quality and improved the surface colour of peeled baby carrots.

Anti-oxidant constituents of fruit, such as phenolic compounds and ascorbic acid are related to enzymatic browning. Phenolic compounds are oxidized to highly unstable quinones that are later polymerized to brown, red, and black pigments. Ascorbic acid is extensively used to avoid enzymatic browning of fruit due to the reduction of the o-quinones, generated by the action of the polyphenol oxidase (PPO) enzymes, back to the phenolic substrates.

Application of edible films and coatings on fruits and vegetables

The purpose of using edible films and coatings for fruits and vegetable is to retard transfer of gas, vapour and volatiles, thus providing food with a modified atmosphere that decreases respiration and senescence, reduces aroma loss, retains moisture and delays colour changes throughout storage.

Edible films/coatings decrease water vapour transmission rate by forming a barrier on the fruit or vegetable surface. This barrier prevents texture decay, since water is essential for preservation of cell turgor. Metabolic alterations that can cause accelerated rate of senescence due to water loss can also be avoided with their use. Ability of films and coatings to function as barriers to water vapor relies on external conditions, which include (i) temperature and relative humidity, (ii) characteristics of commodity, such as type of product, variety, maturity and water activity, and (iii) characteristics of film, such as composition, concentration of solids, viscosity, chemical structure, polymer morphology, degree of cross-linking, solvents used in casting the film, and type of plasticizer used. In case of minimally-processed fruits, there is usually very high water activity present at the surface, making it difficult to develop a coating that delays water loss, since capacity of films to work as barriers to water vapour decreases as water activity increases in the commodity (or relative humidity of the environment).

When coating or film is applied to a commodity, a modified atmosphere is developed. Reduction of internal oxygen and increase of carbon dioxide in the commodity will in some cases delay softening.

Application on meat and poultry

The mixture of hydrocolloids (alginates, gums, cellulose derivatives, etc.) and acids (lactic or acetic acids) or anti-microbial agents laid out on meat pieces reduces the growth of micro-organisms such as *Listeria monocytogenes*.

Study showed that the plantago major seed mucilage (PMSM) containing dill essential oil enhanced the shelf-life of beef by preventing lipid oxidation and microbial spoilage. The results of the study indicated that the effect of PMSM+ 1.5%Dill wrapping on beef is sufficient to retain their good quality characteristics and extend the shelf life during storage at 4æ%C (32).

Study reported that combination of chitosan coating and orange peel essential oil was effective in extending the shelf-life of fresh shrimps upto 15 days. The composite chitosan coating is a transparent layer from a sensorial point of view and prevented black spot development. Moreover, the use of different coating material combined orange peel essential oils will also contribute on preventing melanosis and prolonging the shelf-life of shrimp (33).

Advantages of Edible Coatings

1. Improves external appearance by giving additional shine to the fruit surface.
2. Reduces weight loss and keep the fruit firm, so that its fresh look can be maintained.
3. Reduces rate of respiration and ethylene production, thus delaying senescence.
4. Prevents fruits and vegetables against chilling injuries and storage disorders.
5. Act as barrier to free gas exchange.
6. Provides a carrier for post harvest chemical treatments.
7. Encapsulates aroma compounds, antioxidants, pigments, ions that stop browning reactions and nutritional substances, such as vitamins.

Limitations

1. Thick coating on fruits and vegetables surface becomes an undesirable barrier between the external and internal atmosphere and restricts exchange of respiratory gases (CO_2 and O_2).
2. Use of coatings results disorders like core flush, flesh breakdown, accumulation of ethanol, and alcoholic off-flavours, which is due to modification of internal atmosphere.
3. The addition of nutraceutical compounds to edible coatings may impart bitter taste, astringent or off-flavour.

Conclusion

Synthetic polymers are gradually being replaced by biodegradable materials especially those derived from replenish able, natural resources. Use of such edible packagings will open up potential economic benefits to farmers and agricultural processors. Bilayer and multicomponent films resembling synthetic packaging materials with excellent barrier and mechanical properties need to be developed. The tax levied on the shipment of packaging material can be saved by use of edible coatings and films. Hence, the usage of edible coating and films as packaging material should be improved in order to maintain the environment.

References

1. Tharanathan R.N. (2003). Biodegradable films and composite coatings: past, present and future. Trends in Food Science & Technology. 14: 71-78.

2. Krochta J.M. (2002). Protein as raw materials for films and coatings: definitions, current status, and opportunities. In Protein-Based Films and Coatings, ed. Gennadios, 1:1–40. New York: CRC Press.

3. Pavlath A.E., Wong D.S.W., Kumosinski T.F. (1993). New Coatings for cut fruits and vegetables. Chem tech 23: 36-40.

4. Arvanitoyannis I., Gorris L.G.M. (1999). Edible and Biodegradable Polymeric Materials for Food Packaging or Coating. In: Processing Foods: Quality Optimization and Process Assessment. pp. 357–371. CRC Press, Boca Raton, Florida.

5. Lopez-de-Dicastillo C., Gomez-Estaca J., Catala R., Gavara R., Hernandez-Munoz P. (2012). Active antioxidant packaging films: development and effect on lipid stability of brined sardines. Food Chem 131: 1376-1384.

6. Mali S., Grossmann M.V., Garcýa M., Martino M.N., Zaritzky N.E. (2006). Effects of controlled storage on thermal, mechanical and barrier properties of plasticized films from different starch sources. J Food Eng 75: 453-460.

7. Ma X., Chang P., Yu J. (2008). Properties of biodegradable thermoplastic pea starch / carboxymethyl cellulose and pea starch/ microcrystalline cellulose composites. Carbohydr Polym 72: 369-375.

8. Yang L., Paulson A.T. (2000). Effects of lipids on mechanical and moisture barrier properties of edible gellan film. Food Research International. 33: 571-578.

9. Hong S., Park J., Kim D. (2000). Anti-microbial and physical properties of food packaging films incorporated with some natural compounds. Food Sci Biotechnol 9: 38-42.

10. Wang S., Marcone M.F., Barbut S., Lim L.T. (2012). Review: Fortification of dietary bio-polymers-based packaging material with bioactive plant extracts. Food Res Int 49: 80-91.

11. Kang H.J., Jo C., Kwon J.H., Kim J.H., Chung H.J., Byun M.W. (2007). Effect of pectin-based edible coating containing green tea powder on the quality of irradiated pork patty. Food Control 18: 430-433.

12. Kester J.J., Fennema O.R. (1986). Edible films and coatings: A review. Food Technol 48: 47–59.

13. Rodriguez M., Oses J., Ziani K., Mate J.I. (2006). Combined effect of plasticizers and surfactants on the physical properties of starch based edible films. Food Res. Int. 39: 840-846.

14. Mark, A.M., Roth, W.B., Mehltretter, C.L. & Rist, C.E. (1966). Oxygen permeability of amylo maize starch films. *Food Technol.* 20: 75-80.

15. Kumar, MNVR, (2000). A review of chitin and chitosan applications. React. Funct. Polym. 46: 1-27.

16. Baldwin E.A. (2007). Surface treatments and edible coatings in food preservation. In Handbook of Food Preservation ed. MS Rahman, 21: 478-508. Boca Raton, FL: CRC Press.

17. Cha D.S., Chinnan M.S. (2004). Bi-polymer-based anti-microbial packaging: A review. *Crit. Rev. Food Sci. Nutr.* 44: 223-237.

18. Hershko V., Nussinovitch A. (1998). Relationships between hydrocolloid coating and mushroom structure. J Agric Food Chem 46: 88-97.

19. Karbowiak T., Debeaufort F., Voilley A. (2007). Influence of thermal process on structure and functional properties of emulsion-based edible films. Food Hydrocolloid. 21: 879-888.

20. Mazza G., Qi H. (1991). Control of after cooking darkening in potatoes with edible film-forming products and calcium chloride. *J Agric Food Chem.* 39: 2163-2166.

21. Park H.J. & Chinnan M.S. (1990). Properties of edible coatings for fruits and vegetables. Am Soc Agric Eng 90: 10-11

22. Bai J., Baldwin E.A., Hagenmaier, R.H. (2002). Alternatives to shellac coatings provide comparablegloss, internal gas modification, and quality for 'Delicious' apple fruit. Hort Science 37: 559-63.

23. Rhim J.W., Gennadios A., Handa A., Weller C.L., Hanna M.A. (2000). Solubility, tensile, and colour properties of modified soy protein isolate films. *J Agric Food Chem* 48: 4937-41.

24. Xu S., Chen X., Sun D.W. (2001). Preservation of kiwifruit coated with an edible film at ambient temperature. *J Food Eng* 50: 211-216.

25. Lin S.Y.D. Krochta J.M. (2005). Whey protein coating efficiency on surfactant-modified hydrophobic surfaces. J Agric Food Chem 53: 5018-5023.

26. Lin, D. and Zhao, Y. (2007). Innovations in the development and application of Edible coatings for fresh and minimally processed Fruits and Vegetables. Comprehensive Reviews Food Sci. Food Safety, 6(3): 60-75.

27. Oussalah M., Caillet S., Salmieri S., Saucier, L., Lacroix M. (2004). Anti-microbial and anti-oxidant effects of milk protein-based film containing essential oils for the preservation of whole beef muscle. J. Agric. Food Chem. 52: 5598–605.

28. Suppakul P., Miltz J., Sonneveld K., Bigger S.W. (2003). Active packaging technologies with an emphasis on anti-microbial packaging and its applications. *J. Food Sci.* 68: 408-20.

29. Du W., Olsen C.W., Avena-Bustillos R.J., McHugh T.H., Levin C.E., Friedman M. (2009). Effects of all spice, cinnamon, and clove bud essential oils in edible apple films on physical properties and anti-microbial activities. Journal of Food Science 74: 372-378.

30. Quinones B., Massey S., Friedman M., Swimley M.S., Teter K. (2009). Novel cell-based method to detect Shiga Toxin from *E. coli* O157:H7 and inhibitors of toxin activity. Applied Environmental Microbiology 75: 1410-1416.

31. Thomas L.V., Clarkson M.R., Delves-Broughton J. (2000). Nisin. In Natural Food Anti-microbial Systems, ed. AS Naidu, pp. 463-524. New York: CRC Press.

32. Behbahani, Shahidi, Yazdi, Mortazavi, Mohebbi. (2017). Use of Plantago major seed mucilage as a novel edible coating incorporated with *Anethum graveolens* essential oil on shelf-life extension of beef in refrigerated storage. International Journal of Biological Macromolecules 94: 515–526.

33. Yunus Alparslan, Taçnur Baygar (2017). Effect of chitosan film coating combined with orange peel essential oil on the shelf-life of deep water pink shrimp. Food Bioprocess Technol. DOI 10.1007/s11947-017-1862.

□□□

Chapter 19

Biodegradable Packaging

Niyati Jain[1], Arghya Mani[2] and Vijay Bahadur[1]

[1] Department of Horticulture, SHUATS, Allahabad, India

[2] Department of Post Harvest Technology, BCKV, Mohanpur, Nadia, Email: sweetunicky19@gmail.com

---

ABSTRACT

This chapter provides an overview of the development, use and future trends of biodegradable packaging, types of bio-polymers, the properties of biodegradable packaging, the concept of biodegradability, biodegradable technology and packaging advantages and disadvantages, environmental benefits of biodegradable packaging products and factor affecting biodegradable rate. In the opening statement, described the concept of packaging in general and how biodegradable packaging developed through history. The chapter on bio-polymers shows that all types of bio-polymers are used for the production of bio-packaging and how we share it. The next two chapters relate to the properties of biodegradable materials and distribution of bio-packaging shapes. A special place in the work occupies a section of biodegradation and composting, which clearly illustrates the mechanism of biodegradation but also clarifies the concept compostable and biodegradable plastics. The next two chapters deal with the technology of production of biodegradable packaging and a brief analysis of the advantages and disadvantages biodegradable packaging. The last two chapters explain the environmental benefits of biodegradable packaging products and factors affecting biodegradable rate.

Keywords: Biodegradable packaging, polymers, degradation, environmental benefits.

1. Introduction

Packaging is any product that is used to hold, protection, handling, delivery and presentation of goods, from raw materials to finished products, from producer to consumer. Packaging is usually divided according to the basic raw material of which is produced according to the type of packaging material can be divided into metal, glass, polymer, paper cardboard, wood, textile, multi-layered, ceramic and other types. Food packaging must meet a number of conditions, such as legislation, safety and many other conditions as well as functionality since, it is required to be innovative, easy to use and attractive design. One of the main tasks of packaging in the food industry is to protect the product of chemical, mechanical and microbiological impact, and also allows the freshness of the product and keeps all its nutritional value. The key point in food packaging is that the packaging is an integral part of the production, preservation, storage, distribution, and at the present time an integral part of the preparation of foods. The properties of the food product are only possible to maintain proper selection of appropriate packaging and packing process[1,2].

1.1. Development of Bio-plastic

Today, at the beginning of the 21st century, great importance is given to products from renewable sources, for their positive impact on nature. Generally, increasing awareness among consumers worldwide to conventional plastic products, although very useful, creating huge damage to the environment, water resources, and the entire ecosystem. Accumulation of plastic in the environment, reduction of arable land, wear oil wells, releasing gases during cineration have prompted efforts to develop biodegradable packaging/ plastics[3].

The largest sector in the demand for bio-packaging is the food industry. The rapid development of the industry has led to problems with non-degradable packaging, but it takes time, work and patience while reorienting them to bio-packaging (bio-plastic)[4]. In addition to efforts to find a replacement for plastic supports the development and cardboard packaging produced only from renewable sources[5,6]. Bio-polymers Biomaterials (bio-polymers) are polymers produced from renewable sources. Bio-polymers are manufactured from plant raw materials, in the first place, but in recent times and of animal. Their main feature is their biodegradability. Classified in many ways, such as, chemical structure, origin, methods of synthesis, cost-effectiveness, application, etc.[7]

Polymers from renewable resources are different from natural polymers because their synthesis is induced intentionally. Conventional polymers are not biodegradable because of long chains of molecules that are too big and too well-connected to each other to make them able to separate the micro-organisms to break down.

Unlike conventional, polymers made from natural plant materials from wheat, potato or corn starches have molecules that are easily microbiologically degradable. For 1 kg of bio-plastics should be 1 to 2 kg of maize and 5 to 10 kg of potatoes, which means that 500 000 tonnes of bio-plastics per year requires 50,000 to 100,000 hectares of soil. At the same time, it means the destruction of large areas of forest/ rainforest to cultivated plants for the production of biodegradable materials.[8]

1.2. Classification of biopolymers

The traditional way of dealing with biodegradable packaging materials is divided into three generations based on historical development.

A. First generation

The first generation of material was used for shopping bags, consisting of synthetic polymers, such as low density polyethylene (LDPE-low density polyethylene) with a pro portion of 5–15% starch fillers and pro-oxidizing and auto-oxidative additives. Later these materials decompose or bio-fragment into smaller molecules that are not biodegradable. Such materials have created a very bad image of bio materials especially for consumers who were convinced that they played in terms of biodegradability[9].

Low density polyethylene-LDPE produced in 1933 by Imperial Chemical Industries (ICI) using high pressure process via free radical polymerization. Its production uses the same methods today. It is estimated that about 5.7 wt% of LDPE can be re-cycled. LDPE is in the range defined by the density of 0.910 to 0.940 g/cm3. Non-reactive at room temperature, except in the action of strong oxidizing agents and some solvents cause swelling. Excellent resistance to acids, alcohols, esters, and a base, followed by resistance to various aldehydes, ketones and vegetable oils, and low is resistant to halogen hydrocarbons. They are stable upto a temperature of 80 °C. It is produced in transparent or opaque variations, and is quite flexible and tough, but also fragile. It is used for general purposes (packaging for juices) or for industrial purposes (corrosion resistant materials, welding machine, etc.[10]

B. Second generation

The second generation of biomaterials comprises a mixture of pre-gelatinized starch (40–70 %) and low density polyethylene (LDPE) with the addition of the hydrophilic copolymer, such as ethylene acrylic acid, polyvinyl alcohol and vinyl acetate, which are used to compact. Complete degradation of starch takes 40 days and the degradation of the whole of the above-mentioned film lasts 2–3 years.

C. Third generation

The third generation of the material fully consists of biomaterials and can be divided into three main categories according to the origin and production methods:

1. Polymers extracted/isolated directly from biomass
2. Polymers produced by classical chemical synthesis and bio-monomers
3. Polymers obtained directly from natural or genetically modified organisms [11].

Third generation bio-polymers those are interesting for this study can be divided into three main categories according to their origin and method of production.

1.3. Polymers extracted/isolated directly from biomass

This category of biopolymers is most present on the market. Polymers of this category are obtained from plants, marine and domestic animals. Examples are polysaccharides, such as cellulose, chitin and starch, whey protein, casein, collagen, soy protein, myofibrillar proteins of animal muscle, etc., can be used alone or as a mixture with synthetic polyesters, such as polylactic acid (PLA). The most prevalent category that is used in food packaging is cellulose-based paper. There appears also re-generated cellulose film (cellophane paper) and cellulose acetate. Hemi-cellulose, the second most abundant plant polymer in the world, is in its infancy as a biomaterial research for food packaging[12].

A. Starch

From raw materials annually renewable, based on starch are most used. Green plants, such as potatoes, corn, wheat and rice are the raw materials for the production of bio-polymers. Starch as the main component in them, it is potentially the most acceptable biodegradable polymer material due to its low cost, availability, and because it is produced from renewable sources. How is not a thermoplastic material, is used in mixtures with non-biodegradable polymers and biodegradable materials. The nature of the crystal grains in the form of 15–100 microns in diameter after extraction, the crystal structure is distorted to him pressing, heat, mechanical work and plasticizers, such as water, glycerol and other polyols to make the thermoplastic starch[13]. Plastic-coated starch (known as thermoplastic starch or TPS) is usually obtained by destroying or plastic coating of native starch with water using the thermo-mechanical energy in a continuous extrusion process. TPS can be manufactured in the same way as traditional plastics, but the sensitivity to water vapour and low mechanical properties render it useless for many applications. TPS achieves equilibrium properties after a few weeks[14].

Mixing starch with aliphatic polyesters improves its workability and biodegradation. The combination of starch with a water soluble polymer, such as polyvinyl chloride (PVC) is used for the production of starch film [15].

This generation of plastics made from starch is completely biodegradable. From biodegradable plastics based on starch are made of different bags and sacks, rigid packaging, such as hot-formed trays and containers, as well as products for filling gaps in packages. This material successfully replaces polystyrene and polyethylene in many applications [16].

B. Chitin and Chitosan

Chitin is the second most abundant polysaccharide after cellulose and differs from it only by the OH group. The most abundant in the cell walls of the skin of insects, the shells of shellfish and insects, and can be found in the cell walls of some fungi. Chitosan is actually deacetylated derivative of chitin [17]. Both are applied to produce various biodegradable films for packaging, and the largest they use as an edible coating to prolong the shelf-life of fresh fruits and vegetables [18]. Chitosan has a very poor mechanical properties and resistance to water.

More recently used composite films formed by mixing chitosan and starch showing good properties when it comes to water vapour and mechanical properties. Chitin and chitosan have good anti-microbial properties to a variety of fungi, yeasts and bacteria found in food and thus enable good use as materials that produce biodegradable packaging for food packaging, in particular, are important for prolonging the shelf-life of foods [19].

C. Other plant proteins

The two most common vegetable proteins used in the production of biodegradable packaging are chickpeas and isolated soy protein. Other proteins that are used include those extracted from wheat, pistachios, peas and sunflower [20]. Other polymers based on proteins such as albumin, casein, fibrinogen, silk and elastin are taken into account due to its biodegradability as a raw material for the production of biodegradable packaging. Unfortunately, they have not found wide use because they are difficult to process, do not melt without decompression, and difficult to mix with other polymers due to incompatibility and processing them is expensive unlike other polysaccharides [21].

D. Soybeans

Films made from isolated soy protein are very sensitive to moisture and is not strong. Adding stearic acid of about 25 % to improve properties of tensile and thermal properties and reduced moisture sensitivity [22]. More recently it was discovered that the soy protein together with glycerol, gellan gum, or K-carrageenan suitable for the production of biodegradable/edible soybean based packaging containers (trays) [23]. Always in the production of such packaging must pay attention to the barrier properties to moisture because most hydrophilic in nature. Work on the properties of these materials is demanded because they have not yet come across a great application in production but is thought to want in the future. Collagen covers (intestine) for now remain the only widely applied product based on proteins. It is believed that the film based on a protein found extensive application as edible films in future also [24].

E. Polymers produced by conventional chemical synthesis of bio-monomers

It is possible to get a large range bio-polyesters by chemical synthesis. In theory, all previous packaging materials can be replaced by new types derived from renewable monomers, but the question of economic viability [25, 26] . The most famous of these groups bio-polymer is Poly Lactic Acid (PLA-polylactide). PLA is biodegradable thermoplastic linear polyester, for its properties similar to polystyrene. The raw material for obtaining the lactic acid is obtained by fermentation of glucose or starch from another source. As a source of carbohydrate may be used corn, wheat or alternatively whey and molasses [27].

There are several routes to useful industrial (or high molecular weight) PLA [28].

There are the two principal monomers lactic acid and lactide. The most common way to the PLA is a ring opening polymerization of lactide with various metal catalysts (typically tin (II)-ethyl hexanoate) in solution or as a suspension. The reaction metal catalyst often causes racemization of PLA, reducing stereo specificity with respect to the starting material.

The second time the PLA is a direct condensation of lactic acid monomer. This procedure should be carried out at a temperature lower than 200°C because above this temperature in the formation of low molecular weight materials. Direct condensation is carried out in steps, where the lactic acid in the first oligomerizes PLA oligomers. Thereafter, the polycondensation is carried out in solution or the melt where the short oligomeric units combine to give a high molecular weight polymer chain. Removing the water using vacuum or azeotropic desti lation favours to polycondensation during transesterifi cation. Even higher molecular weight can be obtained by careful crystallization of the crude polymer from the melt. The carboxylic acids and alcohols, the last group are concentrated in the amorphous area of the solid polymer can be reacted.

Polymerization of the mixture of L- and D-lactide typically leads to the synthesis of poly-DL-lactide (PDLLA), which is amorphous. The use of a catalyst can result in PLA exhibiting crystallinity. In addition to lactic acid and lactide is used and lactic acid O-karoboksihidrid [29]. In aerobically degrades completely over the lactic acid into water and carbon dioxide, and the biodegradation favourable conditions for 3–4 weeks [30]. PLA is mainly processed into thermoformed pads and containers for packing and serving food, films, transparencies and bottles and other packaging blown, but also mixed with other materials to improve their. PLA excellent water vapor-permeable which is important in the packaging of fresh food, which is necessary for the water vapour which evaporates quickly while reducing disturbance of the packaging. Is largely processed into thermoformed pads and containers for packing and serving food, films, transparencies and bottles and other packaging blown, but also mixed with other materials to improve their.

An example of the good properties of the variety of raspberry "Polana" which during storage maintains a stable concentration of anthocyanins in all combinations of packaging with PLA thickness of 25 or 40 microns (polypropylene boxes in PLA bags and cardboard boxes in the PLA bags) Unlike polypropylene box with holes and cardboard boxes. PLA bags thickness 25 micrometers maintain optimal gas composition (10 to 20% CO_2 and 5 to 10% O_2) for storing raspberries [31].

F. Polymers obtained directly from natural or genetically modified organisms

Many bacteria accumulate these polymers as a source of energy and as a carbon reserve. This group includes Poly Hydroxyl Alkanoates (PHAs) and bacterial cellulose. PHAs are polyesters that are part of the living organisms structure, hydrophobic and insoluble in water. Their characteristics are most associated with the properties of the monomer building blocks of which a wide variety of different bio-polymer can be synthesized by microbial fermentation. Enzyme PHA-polymerase catalyzes the reaction of polymerization HA in the PHA within the cell. The overall biochemical pathway of synthesis carried out in the cell in a series of enzyme reactions.

PHA, which synthesizes type Alcaligenes, Azotobacter, Bacillus, Halobacterium, Rhizobium and many other micro-organisms can be produced in large quantities biotechnologicaly, renewable substrate, using fermentation and known physical and chemical processes extracted from biomass after production. Depending on the bacteria and the carbon source, the poly hydroxyl alkanoate may be manufactured from rigid brittle to plastic to rubber-like polymer. Have similar properties, such as propylene and polyethylene, elastic and the thermoplastic (retained upon cooling forms) [32].

The most common is the use of derivative poly hydroxyl butyrate labeled PHB. PHB is biodegradable polyester linear prepared by bacterial fermentation of sugar or lipid. It can be used for food packaging, cosmetics and pharmaceutical products, as well as in agriculture. The aerobic conditions are completely degraded into water and carbon dioxide. Biodegradation in favorable conditions takes 5–6 weeks [33].

Apart from renewable, biodegradable plastics can be produced from synthetic polymers by using bacteria. The bacterium Pseudomonas putida converts styrene monomer in the poly hydroxyl alkanoate (PHA), biodegradable plastic which has a wide range of applications. PHA is water in soluble, biodegradable material and compostable whose improvement works intensively before its commercialization [34].

1.4. Properties of biodegradable materials

Materials were based on the need to be useful in the food packaging industry so as to their physical and mechanical properties enable their eligibility and the application of a certain degree, but it also applies largely to prices.

A. Barrier properties

Poor barrier properties (especially humidity resistance) of the traditional and most widely used bio-materials (paper, cellulose films, and cellophane) are all known and is, therefore, necessary to mix these materials with synthetic polymers to achieve the desired barrier properties for packaging of many foodstuffs. Bio-materials made of polysaccharides having poor barrier properties, when it comes to water vapour and other polar substances in a large proportion of the humidity, but at low or middle portion of humidity create good properties to oxygen and other non-polar substances, such as various flavours and oils. Moisture vapour transmission rate was prepared from the starch material is 4–6 times higher than conventional materials made from synthetic polymers. Materials made of arabinoxylan as barley a low permeability as regards oxygen and CO2 and a high permeability in the case of water vapour (the fluorinated materials are less surface hydrophilic such or yet to be finalized).

Some of barrier properties (Tab. 1) where the bio-material and oil derived materials eg., PLA (polylactic acid) has a moisture vapour transmission rate 3–5 times greater than that of PET (polietilentetraftalat), LDPE (low density polyethylene), HDPE (high density polyethylene) and OPS (oriented polystyrene). PLA has improved barrier properties to oxygen from PS (polystyrene), but not as well as PET.

PHA (polihidroksialkonati) has similar moisture vapor transmission rate as well as materials made from petroleum. PHB (polyhydroxyalkanoates) has better barrier properties to oxygen from the PET and PP (polypropylene), and adequate barrier properties, when it comes to fat and fragrances for products with a short shelf life. Barrier properties of gases in most bio-materials depend on the ambient humidity, or PLA and PHA are exceptions.

Table 1: The barrier properties of polymers to bio-based and those derived from oil [34].

Polymers	Transfer rate of oxygen ml. m-2 day-1 at 0 % (relative humidity)	Moisture vapor transmission rate (g m-2 day-1 at 100 % relative humidity)	Temperature (°C)	Material thickness (mm)
OPLA(Oriented polylactic acid)	56.33	15.30	22	4.6
PLA(Polylactic acid)	200	66	23	0.1
PLA-M(Polylactic acid average molecular weight)	-	210 (at 90 % relative humidity)	25	0.25
PLLA + SiOx(Polylactic acid medium molecular weight silicic compounds + hydrocarbons)	84-99	34-40 (at 37.8 °C)	23	0.1
PHB(Polyhydroxybutyrate)	183	1.16	30	1
PHBV(Poly-hydroxybutyrate and hydroxy valerate)	-	1.39	30	1
PET(Polyethyleneterephthalate)	9.44	3.48	22	4.6
OPS(Oriented polystyrene)	532	5.18	22	4.6
LDPE(Low density polyethylene)	-	7.9	38 (at 90 % relative humidity)	0.75
LDPE+ 5% starch	-	36.85	38 (at 90 % relative humidity)	0.75

B. Mechanical properties

The mechanical properties of most organic material similar to the materials derived from petroleum. For example, properties of the PLA are defined by molecular weight of the polymer chain structure (linear with respect to the branched), the degree of crystallization, etc. Orientation PLA improves the mechanical strength and heat stability, a different molecular weight and crystallization result of soft and elastic to hard and high strength materials. The amorphous and poorly crystallized PLA has a transparent, shiny surface; a highly crystalline PLA has an opaque surface. In Table-1 is available to the melting temperature is 130–180°C, and in the glassy form exceeds already at a temperature of 40–70°C. The physical properties of PHA copolymer depend of the composition and molecular structure of the copolymer. PHB is generally hard, highly crystalline thermoplastic polymer which most resembles the isolactic PP because of its mechanical properties. As such polymer PHB generally rigid and brittle, the introduction of the HV (hidoksivaleratne subunit) copolymers improves his mechanical pro perties so that it reduces the level of crystallization and melting temperature resulting in a decrease or increase in hardness toughness and resistance to impact. It is evident that a variety of PHA used in various applications due to its properties and a melting temperature which is 50–180°C.

Experimental methods try to determine which materials have superior properties and are carried out and also, the mixing of different types to obtain a better packing material.

1.5. Current restrictions

The main problem was the most-used for food packaging are their properties, processing and price. In particular, brittleness, low temperatures at which creates distortion, low resistance during processing (excluding PHA polymers) and their barrier properties, particularly to water vapour, limiting their use. In recent years, the price drops and they certainly will decline in the future and with time should increase process optimization and efficiency of plants for the production of such material. The limited availability of raw materials is still one of the most pressing problems that hinder the development of such materials. However, it is unlikely that will be enough to PLA and to meet the needs of the food industry for some time.

1.6. Methods for improving functionality

It is necessary to develop new techniques and processes to improve the barrier properties of bio-packaging. For example, the addition of bio-nanocomposite material shows that the improved mechanical properties of bio-materials and coating SiOx compounds with PLA material reduces moisture vapour transmission rate of 60 %. There are many applications of different techniques that improve different properties, when it comes to these materials or to achieve this and applied in the future requires more effort and research that will help to these materials even more products and use [34].

A. Bio-packaging forms

Biodegradable packaging is produced in several different forms to adapt to the requirements for packaging and storage of various products, currently the most biodegradable gels, films, bags, boxes with lids and trays.

B. Biodegradable gel

The gels are commonly used to prevent microbial contamination, such as a hydrogel, the hydrogel chemical and polymeric network (IPN) [35]. In lettuce, for example, impregnating gel no visible positive effects on maintaining the quality and content of pectic substances while the fruits of Solanum muricatum, protective gel positive effect on maintaining the beta-carotene [36]. Radish gel coat starch-based proved effective to maintain the content of pectin while the same is not performed well for content Glucosinolate. It has been found that a combination of hydrogels for various polymeric materials reduce the service life of certain fruits, probably due to migration of water from the surrounding area [37]. White extruded ginseng extract has good potential to maintain the concentration of anti-oxidants, if used together with biodegradable stretch film [38].

C. Biodegradable films

Biodegradable films are designed with the intention of replacing the polyethylene film used for different purposes, from various industrial films, packaging products to the bag for the collection of organic waste. Such materials have better properties than traditional non-degradable plastics. They are resistant to moisture, warm organic materials for a period of several weeks or even months without changes in physical properties. This allows greater flexibility composting program. Good as a replacement for current films used in storage, transport and packaging of the product and are completely biodegradable. In addition, do not contain polyethylene, do not leave residues after composting and are made from renewable bio-materials (polyester derived from corn dextrose). A comparative study of the permeability of the biodegradable film for oxygen and carbon dioxide as a form of packaging for the fruit of tomatoes showed that films with low permeability negatively affected the quality of the fruit. However, when the permeability of the biodegradable films is into line with the respiration of the fruit, the prevention of contamination by micro-organisms and insects achieved a positive effect on the durability and quality [39].

Compared with polyphenol foil, biodegradable film permeability is significantly decreased. Two kinds of experimental films have been applied to freshly chopped pineapple and melon and observed for their influence on the microbiological quality control of the fruit during storage at 10 °C. The types of films that were used in this study are commercial plastic stretch film and experimental methyl-cellulose film that includes vanilla as a natural anti-microbial agent. Fresh sliced fruit, without any foil wrapping was used as a control. Methyl-cellulose film had inhibitory effect against Escherichia coli, and the yeast was reduced was recorded. Methyl cellulose films with vanillin increased the intensity of the yellow colour with pineapple. Pineapple, which was guarded in an ordinary commercial plastic film had a larger amount of ethanol. However, with pieces of pineapple coated biodegradable film with vanillin recorded a decrease of ascorbic acid by 90% [40].

D. Biodegradable bags

The production technology is based on the use of bio-materials, namely polyester obtained from dextrose corn as the main raw material. Thus, was obtained 100% biodegradable packaging, which is the influence of microorganisms in conditions, such as normal composting degraded organic material to carbon dioxide and water. By using natural raw materials, are not diminished physical properties of this product, but on the contrary, they are improved. Biodegradable bags are strong, flexible, resistant to breakage and damage, and resistant to moisture and temperature changes because of its raw material composition, the largest application with the food industry. Tests show that these bags are safe packaging, and can be used for the storage and packaging of food products. The addition of certain additives, the use of bags extends to other industrial branches. Prednost application of biodegradable bags is not only the functionality but also the fulfillment of all the rules of environmental protection. Upon completion of use bags as packaging, the bags, when disposing off the land or compost decomposable to carbon dioxide and water, over a period of several weeks, and at the same time does not diminish the value of the resulting compost, which makes this container substantially different from previously known polyethylene [41].

E. Biodegradable boxes

Box with a cover made of bi-oriented polystyrene, produced from corn. Such packaging is biodegradable after 47 days, depending on the conditions, and the process does not release harmful substances into the environment. It is crystal clear, allowing better visibility of contents, and is resistant to grease and withstands temperatures from -60°C to +80°C. PLA containers significantly better maintain fruit quality blueberries from standard ventilated switching containers at temperatures between 10 and 23°C [42].

F. Trays for fruits and vegetables

It was found that salad, and sliced broccoli, tomatoes, sweet corn and blueberries can be successfully kept in biodegradable trays of pulp wrapped in foil packaging of poured caprolactone. Such patches are resistant to moisture but brittle. Correspond to the products during freezing them does not change their structural properties well [43].

G. Biodegradable packaging with silver

Silver is used in the fight against infection and deterioration even in ancient Greece and Rome. In the 19th century, botanist Von Nagel found that small concentrations or silver particles have an anti-bacterial effect. Silver is used today in the food packaging system. Food longer maintains the quality texture, improve its storage capacity and maintain food security addition of silver in the packaging. Silver as an element in the production of biodegradable packaging has anti-bacterial role. It damages the cell walls, cell membranes and cytoplasm of bacteria. More recent studies have found that silver also affects the replication of DNA. Silver may be incorporated in the bio-polymer in the form of pure silver, silver coated, or in the form of micro particles. Biodegradable film with silver is preferred over any other packaging materials because of its suppleness and exceptional adhesion ability to fruit. It is used for packaging fresh fruit and vegetables, as well as for storage and transportation of food. The product meets the requirements on sanitary objects that come into direct contact with food [44, 45].

H. Biodegradable Technology

Extrusion, thermo-forming, casting, baking, blow moulding, injection molding, laminating, coating and many other techniques are the main methods in the processing of plastics currently used for industrial production of plastic packaging fed, but mostly it LDPE (low density polyethylene), HDPE (high density polyethylene), PP (polypropylene), PS (polystyrene), PET (polietilentetraftalat), etc. the good fact is that, renewable at bio-osnovi plastics generally show good adaptation to many of the aforementioned methods to be used for plastics processing, and require little or no adjustment. However, some species proved to have limitations properties, when it comes to processing, so for example, have poor mechanical and thermal properties making the material rigid, dry and stiff, but on the other hand have a poor performance, when it comes to gases and moisture. In order to remove shortcomings in their properties need to overcome their own shortcomings before commercial use of packaging on bio-osnovi, for example, to overcome the brittleness used biodegradable plasticizers. The plasticizers include glycol and other low molecular weight polyhydroxy component, polyethers and urea.

All in all, as far as the differences in technology and production bioplatike ordinary plastic, the only difference is that during the production of plastic on bio-osnovi using various accessories (additive) to enhance its properties during processing. PLA is a hybrid material that has similar properties as synthetic plastics (PET, PP) and very good processing standard equipment used for processing synthetic plastic. In particular PLA films that develop techniques, such as injection moulding and thermo-forming very good barrier characteristics (section 3.1). At present, the most represented film bags and containers for organic foods. Services that distribute food are increasingly using cups, pots, containers, laminated or extrusion coated PLA materials on the market more accessible.

The aliphatic polyester co-polymer PHBV and PHB (poly-hydroxybutyrate and hydroxyvalerate) are commercially important as bio-ambalazna/biodegradable plastic and very well quoted, when it comes to the food packaging industry. PHB is a very good thermoplastic material with a high degree of crystallization, while PHA with medium chain length longer behaves as an elastic material for which has a low melting point and a low degree of crystallization. But an important feature of PHA material is that it has low water vapour permeability, making it similar to LDPE. PHA is formed technique of blowing, extrusion and injection moulding into forms, such as films, bottles and containers for food. It is also very used in the biomedical industry (is non-toxic and biocompatible in humans). PCBs are also used for implants, bone plates, sutures and surgical. PHB transparencies above 130°C allow a wide range of use and industrial design. It is important to add that the PCB has a low permeability and breaks down without any residual residue. When conditions rashlad-uvanja and freezing PHB shows slightly worse performance than the PPA, or at high temperatures its properties are much stronger and more stable than the PPA [46].

1.7. Environmental benefits of biodegradable packaging products

Biodegradable packaging products refers to products that break down naturally into organic agents from exposure to heat, moisture and micro-organisms existing by nature. For example, most paper and textile products are considered biodegradable but plastics made from petroleum sources are not. In fact, oil based packaging can take upto 1000 years to biodegrade naturally, especially plastic. What is more, deadly toxins and particles released over time frequently harm wildlife, pollute rivers and produce landfill sites that mark our beautiful landscape.

The environmental benefits of using biodegradable packaging products are vast. In fact, developing an ethical supply chain that embraces biodegradable packaging can generate extensive business advantages. Eco-friendly responsibility is not in conflict with profitability. It is just the opposite. A sustainable business can help you build your brand, attain new contracts and protect your reputation.

1.8. Functionality

Eco-friendly packaging's have just as much convenience and functionality as their counterparts. In truth, biodegradable packaging can be constructed from food grade materials, and come in various sizes, shapes and thicknesses. As well, the packaging easily stacks up for stress-free storage. And to fit the needs of your business, biodegradable packaging comes in cartons, die-cuts and shelf ready displays.

A. No allergens or toxins

Since eco-friendly packaging is made from biodegradable material, many are non-toxic and allergy free. By and large, they are safe for consumers that a sensitivity to allergens. As well, biodegradable packaging contains no chemicals that release toxic and poisonous wastes during decomposition.

B. Energy conservation

Because biodegradable products break down easily, it takes far less energy to recycle. In production, Eco-friendly packaging requires half as much energy to produce compared to non-degradable products.

C. Compostable

Compostable products break down into smaller compounds with the aid of micro-organisms and specific conditions. As a general rule, it decomposes in a particular timeframe within an environment that contains moisture, warmth and air. The result is non-toxic compost that can enhance soil and support plant life. Compostable products are easily confused with biodegradable products, but there is a big difference between the two. Biodegradable products return to nature, disintegrate or disappear entirely. Compostable products break down into usable compost in a safe and timely manner.

D. Marketing Benefits

The benefits of utilizing biodegradable packaging encompass more than protection and well-being of the planet. It can also add great value to retail and manufacturing businesses.

Increasingly, consumers are purchasing only groceries, household products and even restaurant foods that use only biodegradable packaging products. It is very advantageous for a business owner to advertise their use of sustainable packaging. Not only does it enhance sales but also it increases potential clientele. Plus, using biodegradable packaging expands brand recognition and builds loyalty with environmentally aware consumers.

1.9. Advantages and disadvantages of bio-packaging

By studying the available literature can be found arguments for both positive and negative characteristics of bio-packaging. The good is the fact that all of the research leading to the fact that the disadvantages of bio-packaging and to eliminate its production continues to grow. There are many different opinions, different experts who express their opinions on bio-packaging as something good or as something that will become a reality. Below are separated some of the advantages and disadvantages of bio-packaging (Tables 2 and 3).

Table 2: Seven advantages and disadvantages of biodegradable packaging

S. no.	Advantages	Disadvantages
1.	Renewal	The lack of arable land (future)
2.	Good for the environment	Comostability
3.	Less energy to produce	A single bad properties
4.	Easier recycling	He awareness of people
5.	It is not toxic	Installations for the production
6.	Reduced dependence on oil	Processing plants
7.	Reduced emissions Co_2	Short-lived

Table 3: The advantages of biodegradable bags of paper and PE bag

S. No.	Properties	Biodegradable bag	Paper bag	PE bag
1.	Biodegradable	+	+	-
2.	Compostable	+	+	-
3.	Resistance to tearing	+	-	+
4.	Waterproof	+	-	+
5.	Fat resistant	+	-	+
6.	Can be welded	+	-	+
7.	Can be printed	+	+	+
8.	High resistant to melting	+		+

1.10. Factors affecting biodegradation rate

In practice, almost all chemical compounds and materials are subject to biodegradation processes. The significance, however, is in the relative rates of such processes, such as days, weeks, years or centuries. A number of factors determine the rate at which this degradation of organic compounds occurs [47]. Salient factors include light, water and oxygen. Temperature is also important because chemical reactions proceed more quickly at higher temperatures. The degradation rate of many organic compounds is limited by their bioavailability. Compounds must be released into solution before organisms can degrade them [48].

Biodegradability can be measured in a number of ways. Respirometry tests can be used for aerobic microbes. First one places a solid waste sample in a container with micro-organisms and soil, and then aerates the mixture. Over the course of several days, micro-organisms digest the sample bit by bit and produce carbon dioxide –

the resulting amount of CO_2 serves as an indicator of degradation. Biodegradability can also be measured by anaerobic microbes and the amount of methane or alloy that they are able to produce. In formal scientific literature, the process is termed bio-remediation [49].

Table 4: Approximated time for compounds to biodegrade in a marine environment

Product	Time to biodegrade
Paper towel	2–4 weeks
Newspaper	6 weeks
Apple core	2 months
Cardboard box	2 months
Wax coated milk carton	3 months
Cotton gloves	1–5 months
Wool gloves	1 year
Plywood	1–3 years
Painted wooden sticks	13 years
Plastic bags	10–20 years
Tin cans	50 years
Disposable diapers	50–100 years
Plastic bottle	100 years
Aluminium cans	200 years
Glass bottles	Undetermined

2. References

1. Stricevic, N. (1982). Suvremena ambalaza I & II, Skoolska knjiga, Zagreb, Croatia.
2. Anita L., karlo Z., Stanislava T., Anita M.B. and Marija L. (2017). Biodegradable packaging in the food industry. *Journal of Food Safety and Food Quality* 68, heft 2; Seiten 23-52.
3. Mohatny A.K., Misra M., Drzal L.T., Selke S.E., Harte B.R. and Hinrichsen G. (2005). Natural fibers, bio-polymers & bio-composites: an introduction, in Mohatny AK. Natural Fibers, Bio-polymers and Bio-composites, *CRC Press*, Chapter 1.
4. Platt D.K. (2006). Biodegradable Polymers Market Report, Shawnbury, UK. *Rapa Technology Ltd.*
5. Kolybaba M., Tabil L.G., Panigrahil S., Crerar W.J., Powell T. and Wang B. (2003). Biodegradable Polymers: Past, Present, and Future,

American Society of Agricultural Engineers annual Meeting, Fargo, North Dakata, USA.

6. Narayan R. (2006). Bio-based and biodegradable polymer materials: rationale, drivers and technology exemplars, in Khemani K and Scholz C. (Eds). Degradable Polymers and Materials: Principles and Practice, *American Chemical Society;* 939: 282–306.
7. Davidovic A. and Savic A. (2010). Microbial production of biodegradable polymer. *Technological Acta;* 3: 3–13.
8. Goodship V. and Ogar E.O. (2004). Polymer Processing with Supercritical Fluids. *RAPRA Review;* Rep., 15(8).
9. Chiellini E. (2008). Environmentally compatible food packaging. *Wood head publishing limited,* Cambridge England; 8–10.
10. Malpass D. (2010). Introduction to Industrial Polyethylene: Properties, Catalysts, and Processes, *John Wiley and Sons.*
11. Chiellini E. (2008). Environmentally compatible food packaging. *Wood head publishing limited,* Cambridge England (a); 8–10.
12. Grondahl M., Gustafsson A. & Gatenholm P. (2006). Gas-phase surface flourination of arabinoxlyan films, *Macro-molecules;* 39: 2718–2712.
13. Bastioli C. (2005). Starch-based technology, Handbook of Biodegradable Polymers, Shawbury, UK, *Rapa Technology Ltd.,* 257–286.
14. Averous L. and Baqquillon N. (2004). Bio-composite based on plasticizes starch: thermal and mechanical behaviours, *Carbo-hydroPolism.* 56: 111–122.
15. Adeodato Vieira M.G., Altenhofen da Silva M., Oliveira dos Santos L. and Beppu M.M. (2011). Natural-based plasticizers and biopolymer films: A review, *European Polymer Journal.* 47(3): 254–263.
16. Rustogy R. and Chandra R. (1998). Biodegradable polymers. Prog. Polym. Sci., 1273–1335.
17. Clarnival A.M. and Halleux J. (2005). Classification of biodegradable polymers. Biodegradable polymers for Industrial Application, Cambridge, UK, *Wood head Publishing Ltd.,* 3–31.
18. Zhao Y. and Mc Daniel M. (2005). Sensory quality of foods associated with edible film and coating systems and shelf-life extension, Innovations in Food Packaging, San Diego, *California, Elsevier Ltd.,* 434–453.

19. Chiellini E. (2008). Environmentally compatible food packaging. *Wood head publishing limited*, Cambridge England, 8–10.

20. Dean K. and Yu L. (2005). Biodegradable protein-nanoparticles composites, Biodegradable Polymers for Industrial Application, Cambridge, UK, *Wood head Publishing Ltd.*, 289–309.

21. Battacharya M., Reis L., Correlo V. and Boesel L. (2005). Material properties of biodegradable polymers, Biodegradable Polymers for Industrial Application, Cambridge, UK, *Wood head Publishing. Ltd.*, 336–356.

22. Lodha P. and Nteravali A.N. (2005). Thermal and mechanical properties of environmental-friendly grees plastics from stearic acid modified-soy protein isolate. *Industrial Crops Products*, 21: 49–64.

23. Mohareb E. and Mittal G.S. (2007). Formulation and process conditions for biodegradable/edible soy-based packaging trays, *Packaging and Technological Science.*, 20: 1–15.

24. Guilbert S. and Cuq B. (2007). Material formed from proteins. Rapa Technology. 339–374.

25. Kobayashi S. (2010). Lipase-catalyzed polyester synthesis - A green polymer chemistry, *Proceedings of Japan Academy, Ser. B. Physical and Biological Science,* 86(4).

26. Berezina N., Maria Martelli S.M. (2014). Bio-based Polymers and Materials, in Renewable Resources for Bio-refineries, Chapter 1, pp. 1–28 DOI: 10.1039/9781782620181-00001.

27. Wackett L.P. (2008). Polylactic acid (PLA) An annotated selection of World Wide Web sites relevant to the topics in Environmental Microbiology, *Microbial Biotechnology*, 1(5): 432–433.

28. Woo S.I., Kim B.O., Jun H.S. and Ho Nam Chang H.N. (1995). Polymerization of aqueous lactic acid to prepare high molecular weight poly (lactic acid) by chain-extending with hexa-methylene diisocyanate, *Polymer Bulletin*, 35(4): 415–421.

29. Kricheldorf H.R. & Jonte, J.M. (1983). New polymer syntheses, Polymer Bulletin, 9: 6–7.

30. Mahalik N.P. and Nambiar A.N. (2010). Trends in food packaging and manufacturing systems and technology, *Trends of Food Science and Technology*, 21: 117–128.

31. Seglia D., Krasanova I., Heidemane G., Kampuse S., Dukalska L. and Kampus K. (2009). Packaging technology influence on the shelf-life extension of fresh raspberries, *Acta Horticulture*, 877: 433–440.

32. Zivkovic N. (2009). Polyhydroxy alkanoates, Green Chemistry (seminar). *Food Technology Zagreb*, (in Croatian).

33. Botana A., Mollo M., Eisenberg P. and Torres Sanches R.M. (2010). Effect of modified montmorillonite on biodegradable PHB nano-composites, *Bio-resource Technology*, 4: 263–270.

34. Chiellini E. (2008b). Environmentally compatible food packaging. *Wood Head Publishing Limited,* Cambridge England, 8–10.

35. Farrisa S., Schaich K.M., Liu L.S., Piergiovanni L. and Yamb K.L. (2009). Development of polyion-complex hydrogels as an alternative approach for the production of bio-based polymers for food packaging applications, *Trends of Food Science and Technology*, 20: 316–332.

36. Schreiner M., Huyskens-Keil S., Krumbein A., Prono-Widayat H. and Ludders P. (2003). Effect of film packaging and surface coating on primary and secondary plant compounds in fruit and vegetable products, *Journal of Food Engineering*, 56: 237–240.

37. Garcia E. and Barrett D.M. (2002). Preservative treatments for fresh cut fruits and vegetables, *Science, Technology and Market,* University of California.

38. Rico D., Martin-Diana A.B., Barat J.M. and Barry-Ryan C. (2007). Extending and measuring the quality of fresh-cut fruit and vegetables, *Science and Technology*, 18: 373–386.

39. Muratore G., Del Nobile M.A., Buonocore G.G., Lanza C.M. and Asmundo C.N. (2005). The influence of using biodegradable packaging films on the quality decay kinetic of plum tomato (Pomodorino Datterino), *International Journal of Food Engineering*. 567: 393-399.

40. Sungsuwan J., Rattanapanone N., Rachanapun P. (2008). Effect of chitosan/methyl cellulose films on microbial and quality discharacteristics of fresh-cut cantaloupe and pineapple. *Post harvest Biology and Technology*, 49: 403–410.

41. Nampoothiri K.M., Nair N.R. and John R.P. (2010). An overview of the recent developments in polylactide (PLA) research, *Bio - resource Technology*, 101(22): 8493–8501.

42. Alemar E., Samsudin H., Auras R., Harte B. and Rubino M. (2008). Post harvest shelf-life extension of blueberries using a biodegradable package. *Food Chemistry*, 110: 120–127.

43. Makino Y. and Hirata T. (1996). Modified atmosphere packaging of fresh produce with a biodegradable laminate of chitosan cellulose and polycaprolactone, *Post harvest Biology and Technology*, 10: 247-254.

44. Tokic ´I., Fruk G. and Jemric´ T. (2011). Biorazgradiva ambalaža za cÇuvanje voc ´a i drugih hortikulturnih proizvoda: materijali, svojstva i uc Çinak na kakvoc´u. *Journal of Central European Agriculture*. 12: 231.

45. Otoni C.G., Espitia P.J.P., (2016). Avena-Bustillose RJ and McHughe TH. Trends in anti-microbial food packaging systems.

46. Chiellini E. (2008c). Environmentally compatible food packaging. Wood head publishing limited, Cambridge England, 8–10.

47. Sims, G.K. (1991). The effects of sorption on the bio-availability of pesticide, Landon: springer verleg. Pp. 119-137.

48. Sims G.K. and Cupple A.M. (1999). Factors controlling degradation of pesticides in soil. *Pesticide science*. 55: 598-601.

49. Measuring biodegradability "the University of Waikato", June 19, 2008.

□□□

Chapter 20

3D Printing : Advancement in Food Formulation

***Hradesh Rajput[1], Deepika Goswami[2], Sandeep GM Nigam[1], Rekha Rani[1] and Pratistha Srivastav[1]**

[1]Warner College of Dairy Technology, Sam Higginbottom University of Agriculture Technology and Science, Allahabad, [2]Central Institute of Post Harvest Engineering and Technology, Ludhiana, Punjab, *Email: hrdesh802@gmail.com

ABSTRACT

The use of 3-Dimensional (3D) printing, also known as additive manufacturing (AM), technology in food sector has a great potential to fabricate 3D constructs with complex geometries, elaborated textures and tailored nutritional contents. For this reason, 3D technology is driving major innovations in food industry. Here, we review the use of 3D printing techniques to design food materials. Our discussions bring a new insight into how essential food material properties behave during application of 3D printing techniques. The fabrication of food products is one of the emergent uses of 3D printing technologies. The news media are important sources for constructing and reproducing the meanings and narratives of emerging technologies, such as 3D food printing and introducing them to publics. Moreover, this article makes an attempt at an introductory survey of 3D printers commercially available on the market and different technologies are used for the development of 3D food. Additionally, advantages of 3D food printing are indicated. The article is an overview of the major food components that may be printed now and the components which could be used in near future.

Keywords: 3D food, printer, additive, technology and material.

1. Introduction

3D printing, also known as additive manufacturing, was developed in the 1980's as a process used to make three dimensional objects. Additive manufacturing creates parts from the ground up by fusing together layers of material. Its counterpart, subtractive manufacturing, begins with material and removes excess until only the desired shape remains. The worlds of food science and micro-technology have collided, and today, we are reeling from their cosmic impact. Modern 3D printing technology can exceed our wildest expectations by transforming our ability to prepare and consume foods. In fact, current food exploration has the ability to realize seemingly ludicrous ideas, such as printing pizza for space missions to Mars [1].

Quite simply, 3D printing is "additive manufacturing," or slowly building layer upon layer. Imagine printing an essay from your word processor program. After connecting the printer to the computer, the inkjet contains black ink, and row by row, prints each line of your report until it is complete. The process of printing food is similar, but instead of ink, the printer must be outfitted to work with foods, such as sugar, starches, and/or proteins. In essence, printing food involves carefully layering tiny semi-liquefied food particles on top of each other to create novel processed foods [2]. However, printing something like a pizza is much easier said than done. Maintaining the pressure at the right level involves calibrating the printer to apply the appropriate amount of force to the food [3]. This concept is extremely important for yeast dough, which expands as it prints. Particles must also be extremely small; even crushed herb flakes will clog the printer. In addition, controlling the texture of the food takes some trial and error for the perfect balance; for example, thickening agents, such as xanthan gums are added to tomato sauce.

Lastly, the printer will not cook the food. Either the printed pizza is baked elsewhere, or the printer must be outfitted with a localized laser so that the pizza will bake as it prints.From restaurants to spaceships, many places can benefit from the versatility of printing food. Restaurants and food manufacturers can create countless diverse food products from a small number of ingredients. In addition, food in conjunction with software has created many new concepts, such as the idea of digital pantries. Digital pantries store online recipes made from a limited number of semi-liquefied ingredients; these recipes can then be recreated by others with a push of a button. 3D printers can also receive information about your health and use these data to print foods accordingly.

At the Cornell Creative Machines Lab, one printer created 2 identically shaped cookies of different caloric intake for people with very different health histories. As a result, 3D printing can customize nutrition by making the same food for different people based on their personal biometrics [4]. With this capability, even places with limited access to food can benefit from 3D printing. Food can be sent to Third World countries and prepared from pre-packaged cartridges, which have longer shelf-lives than fresh fruits and vegetables and can be customized with essential nutrients and specific dietary requirements [4].Eventually, with advanced 3D food printing technology, it may be possible to create more attractive meals for those who travel to remote places for extended periods of time. For example, scientists making geological and environmental contributions may be in need of 3D printed food, when studying in isolated areas. In addition, NASA hopes that 3D printing can be utilized to help feed astronauts on longer space missions, such as flights to Mars. 3D printing would allow for relatively simple cooking in space as well as longer shelf-life [1].

From the lab to your kitchen, printers, such as the ChefJetTM are rolling out to consumers this year. However, 3D printing of food is mainly a novelty; current printing technology is limited by cost, time, and amount of production on a large scale. Baking a single cookie can take an hour, and today's commercial printers usually cost upwards of $4000 [5]. Thus, it is important to note that 3D printing is not a replacement for conventional food manufacturing; rather, it is a way to customize and optimize foods for mass production by efficiently making foods with improved textures, flavours, and appearances. For example, scientists are attempting to revolutionize the meat industry by using 3D printing technology to create tissues and organ structures [6]. With all the different types of foods, textures, and shapes to experiment with, there are truly no creative limits to what 3D printing can offer the food industry.

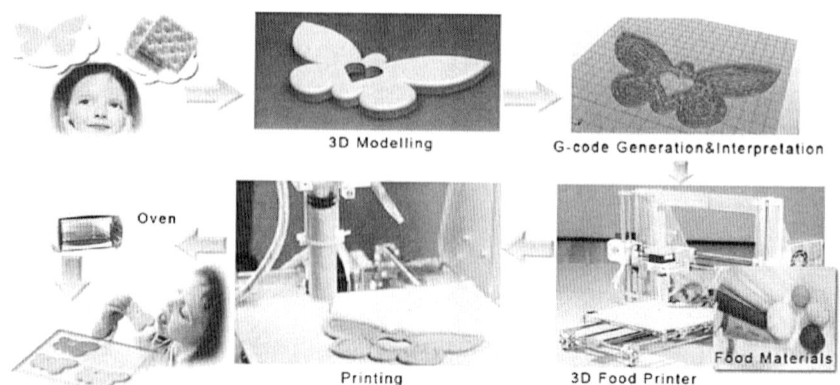

From Idea to 3D Printed Cookie

The message is clear: there is a place for the union of both food and technology. Much progress has already been achieved, but the future of 3D food printing now depends on the desire to make these imaginations into realities. The possibilities and applications are truly endless for creating kitchen magic with 3D printing. The world is your oyster ... just print it!

1.1. History of 3D Printing

The earliest 3D printing technologies first became visible in the late 1980's, at which time they were called Rapid Prototyping (RP) technologies. This is because the processes were originally conceived as a fast and more cost-effective method for creating prototypes for product development within industry. As an interesting aside, the very first patent application for RP technology was filed by a Dr. Kodama, in Japan, in May 1980. Unfortunately for Dr. Kodama, the full patent specification was subsequently not filed before the one year deadline after the application, which is particularly disastrous considering that he was a patent lawyer! In real terms, however, the origins of 3D printing can be traced back to 1986, when the first patent was issued for stereo lithography apparatus (SLA). This patent belonged to one Charles (Chuck) Hull, who first invented his SLA machine in 1983. Hull went on to co-found 3D Systems Corporation one of the largest and most prolific organizations operating in the 3D printing sector today.

a) 1984–86, Charles Hull invents 3D printing and coins the term "Stereo Lithography".

b) 1992, First 3D printer built by 3D Systems.
c) 1999, First application of 3D printing in the medical field - creating the human bladder.
d) 2000, Miniature human kidney created through 3D printing.
e) 2006, The Selective Laser Sintering machine - printing multiple materials & fields.
f) 2009, First usable prosthetic leg - this opens the door for customized products using 3D printing.
g) 2011, 3D printers start offering 14k gold as printable material.
h) 2012, Alternative 3D printing process were introduced at the entry level of the market
i) 2013, StratasysacquiresMakerbot

Throughout the 1990's and early 2000's a host of new technologies continued to be introduced, still focused wholly on industrial applications and while they were still largely processes for prototyping applications, R&D was also being conducted by the more advanced technology providers for specific tooling, casting and direct manufacturing applications. This saw the emergence of new terminology, namely Rapid Tooling (RT), Rapid Casting and Rapid Manufacturing (RM), respectively.

In terms of commercial operations

a) Sanders Prototype (later Solidscape) and ZCorporation were set up in 1996
b) Arcam was established in 1997
c) Objet Geometries launched in 1998
d) MCP Technologies introduced the SLM technology in 2000
e) EnvisionTec was founded in 2002,
f) ExOne was established in 2005 as a spinoff from the Extrude Hone.

These companies all served to swell the ranks of Western companies operating across a global market. The terminology had also evolved with a proliferation of manufacturing applications and the accepted umbrella term for all of the processes was Additive Manufacturing (AM).

In 2007, the market saw the first system under $10,000 from 3D Systems, but this never quite hit the mark that it was supposed to. This was partly due to the system itself, but also other market influences. The holy grail at that time was to get a 3D printer under $5000 this was seen by many industry insiders, users and commentators as the key to opening up 3D printing technology to a much wider audience. Dr. Bowyer conceived the RepRap concept of an open source, self-replicating 3D printer as early as 2004, and the seed was germinated in the following years with some heavy slog from his team at Bath, most notably Vik Oliver and Rhys Jones, who developed the concept through to working prototypes of a 3D printer using the deposition process. 2007 was the year the shoots started to show through and this embryonic, open source 3D printing movement started to gain visibility.

But it wasn't until January 2009 that the first commercially available 3D printer in kit form and based on the RepRap concept was offered for sale. This was the BfBRapMan 3D printer. Closely followed by Makerbot Industries in April, the same year, the founders of which were heavily involved in the development of RepRap until they departed from the Open Source philosophy following extensive investment. Since 2009, a host of similar deposition printers have emerged with marginal unique selling points (USPs) and they continue to do so. The interesting dichotomy here is hat, while the RepRap phenomenon has given rise to a whole new sector of commercial, entry-level 3D printers, the ethos of the RepRap community is all about Open Source developments for 3D printing and keeping commercialization at bay.

2012 was the year that alternative 3D printing processes were introduced at the entry level of the market. The B9 Creator (utilizing DLP technology) came first in June, followed by the Form 1 (utilizing stereolithography) in December. Both were launched via the funding site Kickstarter and both enjoyed huge success. As a result of the market divergence, significant advances at the industrial level with capabilities and applications, dramatic increase in awareness and uptake across a growing maker movement, 2012 was also the year that many different mainstream media channels picked up on the technology. 2013 was a year of significant growth and consolidation. One of the most notable moves was the acquisition of Makerbot by Stratasys.

1.2. Different Technology used for 3D Food Printing

Food is one emerging application (and/or 3D printing material) that is getting people very excited and has the potential to truly take the technology into the mainstream. After all, we will all, always, need to eat! 3D printing is emerging as a new way of preparing and presenting food. Initial forays into 3D printing food were with chocolate and sugar, and these developments have continued a pace with specific 3D printers hitting themarket. Some other early experiments with food including the 3D printing of "meat" at the cellular protein level. More recently pasta is another food group that is being researched for 3D printing food.

1.2.1. Selective Laser Sintering (SLS)

Selective Laser Sintering (SLS) applies a laser, as a power source, to sinter powder particles. A solid structure is built by directing the laser at points pre-determined by a 3D model. The laser fuses together specific areas of the particulate powder bed by scanning cross-sections. Afterwards, the powder bed is lowered by one layer thickness, a new layer of particles is deposited on top, and the process is repeated until the object is finalised. This method can be applied to generate multiple layers of food matrix each layer containing different food material components [7]. The interaction between the laser beam and the particles used in SLS is important to define the feasibility and quality of any SLS process [8].

The selection of the laser has a relevant influence on the fusion of powder for two main reasons:

1. The laser absorptivity of materials relies on the laser wavelength.
2. The mechanism for powder densification is affected by the input laser energy density (Gu, 2012).

Laser Sintering / Laser Melting

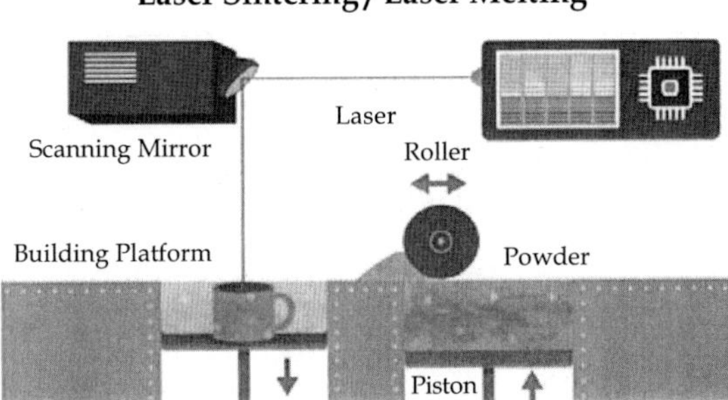

Although SLS has been extensively used to sinter metals, this technology was also adapted for some food designs. Based on SLS technology, using an infrared laser that heats and sinter the material, TNO has incorporated nutritional value and flavours to powder-based 3D objects. The principle relies on the powder binding as a result of the melting fat and/or sugar of the composition [7, 9]. The Candy-Fab (*CandyFab Project*, 2007)applied a selective low-velocity stream of hot air to sinter and melt a bed of sugar. The fabrication powder bed is heated to just below the material melting point to minimize thermal distortion and facilitate fusion among layers. Selective sintering offers more freedom to build complex food items in a short time without post-processing. It is suitable for sugar and fat-based materials with relatively low melting points. However, the fabrication operation is complicated as many variables are involved.

1.2.2. Powder Bed Printing (PBP)

In standard binder jetting technology, each powder layer is distributed evenly across the fabrication platform, and liquid binder sprays to bind two consecutive layers of powder [10]. The powder material is usually stabilized by spraying water mist to minimize the disturbance caused by binder dispensing. In edible 3D printing project, [11]utilized sugars and starch mixtures as a powder material and a Z Corporation powder/binder 3D printer as a fabrication air was applied to melt chocolate and force it out of the chambers. A "3D Food-Inks Printer" which dispenses 3D colour images on extruded base material may also fall into this category [12], while a post-processing step was applied to fuse the layers together. Platform Sugar Lab (*3D Systems, 2013*)used sugar and different flavour binders to fabricate complex sculptural cakes for weddings and other special events.

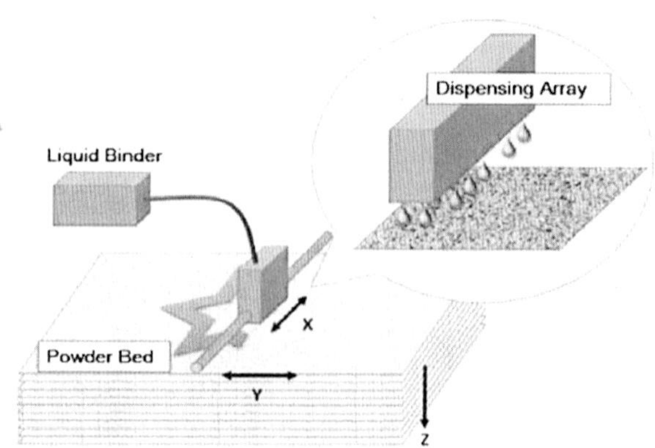

This fabrication adopted 3D Systems' Colour Jet Printing technology, and the material and fabrication process met all the requirements of food safety. Binder Jetting offers advantages, such as faster fabrication, ability to build complex structures, and low ingredients cost. But the fabricated products suffer from rough surface finish and high sugar content, and the machine cost is high. Post-processing may be required, such as curing at higher temperature to strengthen the bonding among layers.

1.2.3. Fused Deposition Modelling (FDM)

Also known as hot melt extrusion, this method uses a single nozzle head to extrude melted material, typically plastic, layer by layer onto a build platform according to the 3D data that has been supplied to the printer. One of the most familiar printers in this space, the Maker Bot, uses the FDM method. This method developed by Crump [13, 14] and trademarked by Stratasys Inc [15]. In this method, a moving nozzle is used to extrude a hot-melt filament polymer as a continuous melted threat, fusing it to the preceding layer on cooling. While FDM is primarily used for prototyping plastics, the technology has been adapted to 3D food printing for few years.Schematic diagram of 3D food extrusion processes which can be considered an adaptation of FDM into the sector of food design. In extrusion processes, formulated ink composed of food ingredients is loaded in a cylinder (extruder).

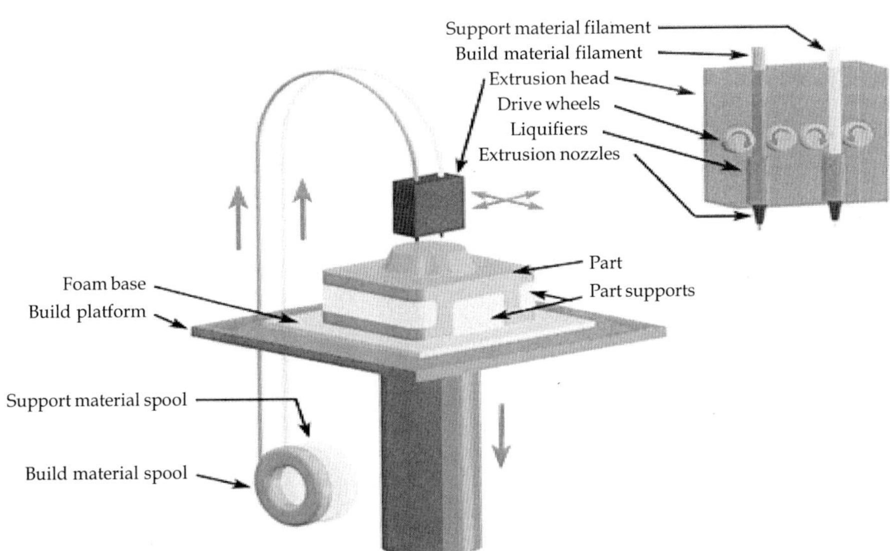

The edible ink is extruded out of the nozzle by the force produced by an acting hydraulic piston. The consecutive deposition of layers is undertaken by directing the cylinder at points pre-determined by a 3D model. Depending on the materials used in extrusion processes, the binding mechanisms may happen by the accommodation of layers controlled by the rheological properties of the materials, solidification upon cooling or hydrogel-forming extrusion.

1.2.4. Inkjet 3D printing (IJP)

Inkjet Printing (IJP) technology relies on the fundamental of accumulation of droplets of material deposited on-demand by inkjet printing nozzles, as depicted by [8]. Inkjet printers generally operate using thermal or piezoelectric heads. In a thermal inkjet printer, the print head is electrically heated to generate pulses of pressure that push droplets from the nozzle. Piezoelectric inkjet printers contain a piezoelectric crystal inside the print head which creates an acoustic wave to separate the liquid into droplets at even intervals. Employing a voltage to a piezoelectric material arouses a prompt change in shape, which in succession produces the pressure necessary to eject droplets from the nozzle [16].

The technology developed by [17]**Grood and Grood, 2011** for dispensing a liquid onto layers can be classified as drop-on-demand deposition [17, 18]. This technology was commercialized by the name of FoodJet printing and uses an array of pneumatic membrane nozzle-jets, which layers tiny drops onto a moving object. The drops together shapes a digital image in the format of a graphical decoration, surface fill or cavity deposition [19]. Inkjet printers generally handle low viscosity materials; therefore, it does not find application on the construction of complex food structure. Typical deposited materials are: chocolate, liquid dough, sugar icing, meat paste, cheese, jams, gels, etc.

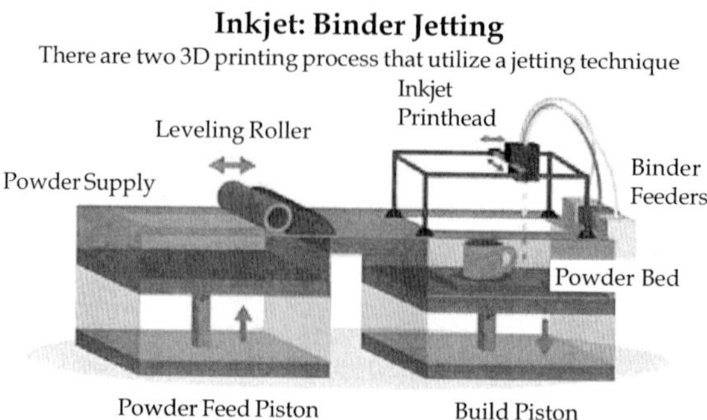

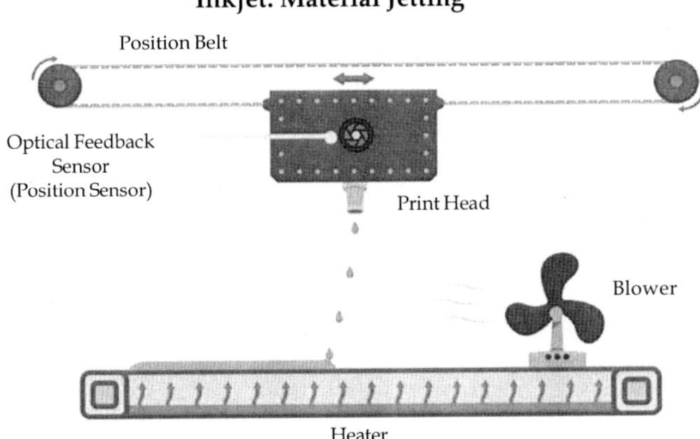

Inkjet food printing dispenses multi-material streams/droplets from a syringe-type multi-channel printhead in a drop-on-demand way and creates 3D edible food products, such as cookies, cakes, or pastries. It involves pre-patterning food items at multiple layers of processing. The De Grood Innovations' FoodJet Printer used pneumatic membrane nozzle-jets to drop on-demand materials onto pizza bases, biscuits, and cupcakes. The ejected stream/droplets fall under gravity, impact on the substrate, and dry through solvent evaporation. The drops can form a two and half dimensional digital image as a decoration or surface fill. Compared with other methods, 3DP is an economical and innovative way for mass customization in food fabrication. The quality of fabricated food items depends on the process and planning rather than people's skill. This technology can easily and accurately perform fabrication based on customer demands.

1.3. Major Food Components

Food components for 3D food printing are solid or semi-solid for extrusion or in the form of powders. Chocolate is an interesting, commonly used material, creating great opportunities in the market of personalized products. The 3D chocolate printing technology was developed at the University of Exeter and is a quite well known material [20, 21]. Frosting, cheese [22] and ice cream [23] are beside chocolate other food components used for 3D printing, which do not require any finishing processing. Components for food printing are divided into three categories:

- **Natively printable materials:** there are mainly materials directly extruded from a syringe, such as chocolate, icing, hydrogel, cheese, hummus, pasta dough, butter, jelly, etc. However, these include also the materials in the form of powder, such as sugar, starch, etc.
- **Non-printable traditional food materials:**these food like- meat, rice, vegetables and fruits. In order to make them suitable for extrusion used in 3D printing, the hydrocolloids are added to these. Xanthan gum and gelatine are examples of commonly used hydrocolloids.
- **Alternative ingredients:** Insect powder may be an alternative additive, which was studied as a protein source to replace traditional meat [24].

People [22] believe that printable food material must be compatible with traditional techniques of cooking. Traditional foods like turkey, celery and scallops after grinding and modifying by additives can be successfully 3D printed and subjected to various processing steps. The viscosity, consistency and solidifying are the properties that determined the printability of food components.

1.4. 3D Food Printing in India

Chocobot is the first made in India commercial chocolate 3D printer

As the hotbed of 3D is still solidly located in the west and China, we are always very happy to hear about 3D printing businesses popping up elsewhere. And in the respect, India is definitely up-and-coming, as a remarkable number of start-ups have been appearing in that subcontinent. However, few have yet been as successful as the Bangalore-based Global 3D labs. While that company has had quite a bit of success over the past year, they are already back for more as they are about to release India's first commercially available chocolate. 3D printer: the chocobot, a multi-functional tool capable of directly 3D printing dark, white, milk and other chocolates.

To give a little background, Global 3D Labs was founded by entrepreneurs, who saw the distinct lack of 3D printing innovation in india and an overly strong reliance on foreign export. Gopal Krishna, AviralKedia (both from MIT, Manipal), Manish Amin, ShreyasKudva (both from MITE, Manglore), therefore, set out to develop their own India-made 3D printer, which would grow to become the pramaan 3D printer, a study and decent plug-and play machine with SD card support. The company has since grown out to an eight-member team has been very successful in its home country. Specializing in 3D printers for professionals, startups, researchers and students, they are frequently seen throughout the 3D printing community offering seminars, custom solutions and more. 'we have supplied 3D printer across India and have reputed clients like Indian railway, Bremer Engineering India, Innovation & Knowledge Park, Central Manufacturing Technology India, National Chemical Laboratories and more,' they tell us.

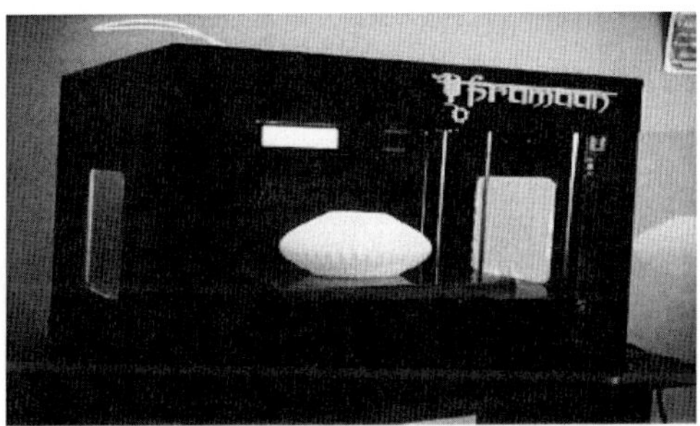

The Pramaan V2 3D Printer

The Pramaan Mini 3D Printer

The upcoming Pramaan V3

Aside from their upcoming chocobot, they currently have a repertoire of four products: the pramaan V2 the V3 the pramaan mini DIY kits. The pramaan V3 is our latest FDM 3D printer which comes equipped with a 25x25x25 cm3 build volume. It comes with the octoprint ability which can be used to 3D print over WiFi and the internet,' they reveal. It also sport as on board camera which can be used to get a live feed of the print. Pramaan V3 comes with a filament sensor as well, which notifies the user when the filament is entangled or about to get over. Connectivity options include Cloud connectivity, SD card, WiFi and LCD screen.' The machine is also ready for release.

Following that success and establishing an India only manufacturing process-the team looked for some new options to expand their market. 'After coming up with pramaan V2 we thought of experimenting with food printing. And what batter food article than chocolateÃ So after months of experimentation and R&D, we come up with chocobot,' the team tell us. 'we have designed it in such a way that we, ourselves would like to purchase and use it.'And indeed, it looks like a remarkable 3D printer. While it won't be fully unveiled until next week (at the 3D Print UAE Exhibition and Conferences in Dubai), the Indian team already revealed that is capable of 3D printing just any form of chocolate through a series of different syringes. 'Not only can it print directly with chocolate, but also print using polycarbonates which can then used as a mould with liquid chocolate. The liquid chocolate can then be frozen in the mould and the mould can when be removed. It boasts of a build volume of 18x18x18 cm3 and has a sturdy and robust build,' they reveal. With an operating temperature of 34 degrees centigrade, modular head options and a 60 ml syringe, you can do some serious chocolate 3D printing. Aside from that, it largely operates through the same principles as their Pramaan 3D printer and with the same software, though obviously extruding something vastly different. They estimate that it takes about 40 minutes to complete one plate of chocolates. We love 3D printing and everybody love food,' GopalKrishna explains. Expected to be released next week, only one thing remains to be revealed: the price tag. With taxes of 14.5%, the total price comes upto Rs. 91600 definitely interesting enough to draw some attention in India and beyond.

1.5. 3D Printing Food in Space

Space food product development, Food tasting session in the Habitability and Environmental Factors Office at NASA's Johnson Space Center. NASA and a Texas company are exploring the possibility of using a "3D printer" on deep space missions in a way where the "D" would stand for dining. NASA has awarded a Small Business Innovation Research (SBIR) Phase I contract to Systems and Materials Research Consultancy of Austin, Texas to study the feasibility of using additive manufacturing, better known as 3D printing, for making food in space. Systems and Materials Research Consultancy will conduct a study for the development of a 3D printed food system for long duration space missions. Phase I SBIR proposals are very early stage concepts that may or may not mature into actual systems. This food printing technology may result in a phase II study, which still will be several years from being tested on an actual space flight. program is interested in developing methods that will provide food to meet safety, acceptability, variety, and nutritional stability requirements for long exploration missions, while using the least amount of spacecraft resources and crew time. The current food system wouldn't meet the nutritional needs and five year shelf-life required for a mission to Mars or other long duration missions. Because refrigeration and freezing require significant spacecraft resources, current NASA provisions consist solely of individually pre-packaged shelf stable foods, processed with technologies that degrade the micronutrients in the foods.

NASA is funding this phase I six month $125,000 study on 3D printing of foods to determine the capability of this technology to enable nutrient stability and provide a variety of foods from shelf stable ingredients, while minimizing crew time and waste. NASA selected this proposal because the research team, subcontractors and consultants included premier food rheology and flavour expertise that would be required for a novel product development system. The work plan for this feasibility study also was well laid out and the technology offers the potential to meet some of the food requirements using basic food components for long duration missions.NASA recognizes in space and additive manufacturing offers the potential for new mission opportunities, whether "printing" food, tools or entire spacecraft. Additive manufacturing offers opportunities to get the best fit, Printing: Food in Space Arrival of fresh food and supplies delivered on the International Space Station,form and delivery systems of materials for deep space travel. This is why NASA is a leading partner in the president's National Network for Manufacturing Innovation and the Advanced Manufacturing Initiative. 3D printing is just one of the many transformation technologies that NASA is investing in to create the new knowledge and capabilities needed to enable future space missions while benefitting life here on the Earth.

1.6. Today uses of 3D Printing

3D printing has long played a role in model and prototype creation. But more and more it is being used to produce end-products like jewellery, sculptures, and shoes.Thanks to advances in 3D printed metal, automotive companies are now looking at the potential of 3D printing to fulfill after sale parts on demand, rather than holding large inventories. As an early adopter of 3D printing for product development and prototyping, the aerospace industry is following suit, now experimenting with non-critical part production.In the marketing industry, 3D printing is being used in both playful and functional ways. Barilla recently challenged customers to design a unique 3D pasta design, while McDonald's is rumoured to be creating a 3D printing toys kiosk to print on-demand Happy Meal parts.

1.7. Some of the Latest Development with Respect to 3D Printed Food Include

a) **Chocolate** - International chocolate producers Hershey has partnered with a company called 3D Systems to develop new, innovative and customized chocolate designs using 3D printing technology.

b) **Pizza** - NASA is exploring the use of 3D printers to produce pizza in outer-space. The benefit of this technology here is that the machine will be able to use powdered ingredients with a longer shelf life, combining them with water just prior to printing. The printer would be capable of printing first dough, then a tomato paste layer, followed by cheese and a protein layer.

c) **Pasta -** This has been explored by several individuals and includes products from corn pasta to stuffed ravioli. Of course, cooking of the pasta is required after printing, but just think that with the press of a few buttons, a gourmet meal customized to your preferences (such as salt, flavour, etc.) can be prepared whilst you finish other work.

d) **Meat products** - Though this hasn't been officially produced yet, the technology exists and the possibility is very real through 3D bio-printing using a biopsy of the animal tissue. One of the leads in this area is a U.S. based company called Modern Meadows. 3D printed meat cannot only reduce and in some cases eliminate the vast resources required to cultivate natural meat, but also introduce a new food source to vegetarians or vegans who have a stand against the treatment of animals in the industry as well as improve or modify the flavour, texture and taste of the meat.

There exists local as well as foreign regulations, standards, and guidelines that detail the design of food equipment as well as the constituents of food products, e.g. Public Health, USFDA, National Sanitation Foundation (NSF), Codex Alimentarius, WHO, etc. Provided that the printer and its raw materials follow these principles and that effective cleaning and sanitation methods are developed and practiced, then the safety of 3D printed food would be like any other food product on the market. Quality, nutritional content and personal preferences, however, are of course, an entirely different story.As the technology advances, it may become necessary for governing agencies to create or modify food safety laws and regulations to accommodate these new products that we are almost certain to see on grocery shelves and food service establishments in the very near future.

1.8. Consumer Benefits

Companies are not the only ones experimenting with 3D printing. As the technology moves from maker to mainstream, more and more people are playing with it at work, at school and at the library. According to OITP Perspectives, a publication by the American Library Association, more than 250 libraries in the U.S. offer the use of 3D printers to patrons. Home use is growing as well. A 2014 report by research agency OnePoll found that one in three Americans would consider buying a 3D printer for their home.While consumer applications of 3D printing spanned across dozens of applications one-of-a-kind gifts, prototypes, decorations, toys, jewellery, functional fixes the purposes behind 3D printing all ladder up to 5 key benefits:

1. Thoughtful Moments: I can capture the memories and connect to others.
2. Playful Creation: I can bring my imagination to life.
3. Design Control: I am in the driver's seat to design my products.
4. Confident Decisions: I can rest assured that I've gotten it right
5. Immediate Access: I have what I want, when I want it. These consumer benefits are core to what differentiates 3D printing from other processes and why it appeals to consumers and companies alike.

1.9. Company benefits

3D printing is transforming the way companies do everyday business. When the benefits of 3D printing are leveraged, the technology enables companies to reach new audiences, connect with customers, and capitalize on the efficiencies the technology provides. While 3D printing has the potential to add value to businesses in numerous ways, there are unique technological benefits that make it stand apart from other processes:

1.10. Accessibility

Software: A key challenge in 3D printing is creating the file in the right format for print without 3D modeling skills. Today, the ability to build files is a specialized skill of engineers and designers. Software companies like AutoCAD are building intuitive software to make 3D file design and creation accessible to everyone. Programs like Tinkercad by AutoCAD are leveraging intuitive features like drag and drop.

Hardware: As more companies create and evolve the 3D printer, the technology continues to become cheaper and better in quality. The average cost of a home 3D printer ranges from \$500-\$2500. At an enterprise level, multiple models are available ranging in level of fidelity and price.

1.11. The future of 3D printing

3D printing is a nascent technology with big promise. While there are still many hurdles to ensuring the technology is affordable and practical, the rate at which the technology is evolving is rapid. Hurdles, such as cost, speed, quality and accessibility are already being addressed today. The message is clear, there is a place for the union of both food and technology. Much progress has already been achieved, but the future of 3D food printing now depends on the desire to make these imaginations into realities. The possibilities and applications are truly endless for creating kitchen magic with 3D printing. The world is your oyster ... just print it!

2. Summary

3D food printing has demonstrated its capability of making personalized chocolates or producing simple homogeneous snacks. Currently, these applications are still primitive with limited internal structures and monotonous textures. It is necessary to develop a systematic way to investigate recipes,platform design, printing technologies, and their influences on food fabrication. Sustenance printing innovations apply computerized advances to control nourishment structures and materials. This flexibility, connected to household cooking or providing food benefit, will permit proficient conveyance of high caliber, naturally arranged sustenance things to shoppers. With the advancement of an open electronic media interface, nourishment printers may frame nature of arranged machines that can arrange new fixings, plan most loved sustenance on request, and even team up with specialists to create more advantageous eating methodologies. 3D printing is one mechanized sustenance readiness technique, which as of now conveys upgraded masterfulness and could be utilized to convey these information driven joys. With time, more developments in this innovation will assist humanity with large scale manufacturing of sustenance and medical advantages.

The future of 3D printing will be developed in the coming years. Undoubtedly, food printing can have many advantages, but whether the market is ready for such a big change and the technology will grow fast enough are the questions. Can the 3D printed food replace traditional meals? Certainly not today, but it may already be an interesting alternative to traditional meals, as well as an edible decoration limited only by our imagination, difficult or impossible to prepare in other ways. In the future, ready, healthy meal, tailored totheir individual needs, will be waiting, when coming home.It will be prepared by a previously programmed devicethanks to 3D printing technology, or even the printers willbe programmed remotely via mobile phone, from office or on the way home. Furthermore, no food will be wasted; a portion will be one, fresh and prepared especially for us, at a time, when the need arises.

3. References

1. Terfansky M., Thangavelu B., Fritz B., Khoshnevis B. (2013). 3D Printing of Food for Space Missions. AIAA SPACE 2013 Conference and Exposition; San Diego, CA; September 10–12, 2013. San Diego, CA: University of Southern California.
2. Lam CXF, Mo X.M., Teoh S.H., Hutmacher D.W. (2012). Scaffold development using 3D printing with a starch-based polymer. Mater SciEng 20(1–2): 49–56.
3. Lipson H., Kurman M. (2012). "Fabricated: The New World of 3D Printing." Indianapolis, IN: John Wiley and Sons. p 75–77.
4. Lipson H. (2014). Personal communication. (information from an interview with Dr.Hod Lipson at Cornell University on March 15, 2014).
5. Bloomberg Businessweek Technology, (2014). Food Science - A Guide to All the Food That's Fit to 3D Print (So Far). Available from: http://www.businessweek.com/articles-28/all-the-food-thatsfit
6. Jakab K., Norotte C., Marga F., Murphy K., Vunjak-Novakovic G., Forgacs G. (2010). Tissue engineering by self-assembly and bioprinting of living cells. Biofabrication 2: 1–14.
7. Diaz, J.V., Van Bommel, K.J.C., Noort, M.W., Henket, J., Brier, P., (2014b). Preparing Edible Product, Preferably Food Product Including Bakery Product, and Confectionary Product, Involves Providing Edible Powder Composition, and Subjecting Composition to Selective Laser Sintering. Nederlandse Org ToegepastNatuurwetensch (Nede-C).
8. Kruth, J.P. (2007). Consolidation phenomena in laser and powder-bed based layered manufacturing. CIRP Ann. 56(2): 730e759.
9. Diaz, J.V., Van, B.K.J.C., Noort, M.W.J., Henket, J., Bri€er, P. (2014a). Method for the Production of Edible Objects Using Sls and Food Products. Google Patents.

10. Sachs E., Cima M. and Cornie J. (1990). Three dimensional printing: rapid tooling and prototypes di-rectly from a CAD model. *CIRP Annals-Manufacturing Technology*, 39(1): 201–204.

11. Southerland D., Walters P. and Huson D. (2011). Edible 3D Printing, in NIP & Digital Fabrication Conference, Minnesota, USA.

12. Golding M., Archer R., Gupta G., *et al.* (2011). Design and development of a 3-D food printer, in In Proceedings of NZIFST 2011 Conference, Rotorua, New Zealand.

13. Crump, S.S. (1991). Fast, Precise, Safe Prototypes with FDM, American Society of Mechanical Engineers. Production Engineering Division (Publication) PED, pp. 53e60.

14. Crump, S.S. (1992). Apparatus and Method for Creating Three-dimensional Objects. Google Patents.

15. Batchelder, J.S. (2012). Additive Manufacturing System for Printing a Chocolate Confection Comprises at Least One Controller to Receive Instructions and to Relay Commands; Platen; Print Head; and Cartridge. STRATASYS INC (STRANon-standard) STRATSYS INC (STRA-Non-standard).

16. Murphy, S.V., Atala, A. (2014). 3D bio-printing of tissues and organs. Nat. Biotech. 32(8): 773e785.

17. Grood, J.P.W., Grood, P.J. (2011). Method and Device for Dispensing a Liquid. Google Patents.

18. Grood, J.P.W., Grood, P.J., Tillie, L.W.M. (2013). Method and Device for Dispensing a Liquid. Google Patents.

19. FoodJet, (2015). FoodJet Printing Systems - Web Page.

20. Jia F., *et al.* (2016). Technological Forecasting & Social Change, 102: 202–213.

21. Hao L. *et al.* (2010). Virtual and Physical Prototyping, 5(2): 57-64.

22. Lipton J., *et al.* (2010). http://sffsymposium.engr.utexas.edu/Manuscripts/2010/2010-68-Lipton.pdf (last checked on Jan. 14th, 2016).

23. Yang F., Zhang M., Bhandari B. (2015). Critical Reviews in Food Science and Nutrition, DOI: 10.1080/10408398.2015.1094732

24. Sun J., *et al.* (2015). International Journal of Bioprinting, 1(1): 27–38.

□□□

Chapter 21

Food Secuity, Safety and Quality Laws in India

K. Rama Krishna*,GajananGundewadi and Pushpendar Kumar

Division of Food Science and Post Harvest Technology, IARI, New Delhi-110012, *Corresponding author E-mail: ramakrishna.karri001@gmail.com

1. Introduction

Food security, as defined by the United Nations' Committee on World Food Security, is the condition in which all people, at all times, have physical, social and economic access to sufficient safe and nutritious food that meets their dietary needs and food preferences for an active and healthy life. Food security has been a major concern in India. According to UN-India, there are nearly 195 million undernourished people in India, which is a quarter of the world's hunger burden. Also, roughly 43% children in India are chronically undernourished[1]. India ranks 74 out of 113 major countries in terms of food security index[2].

Over the coming decades, a changing climate, growing global population, rising food prices, and environmental stressors will have significant yet highly uncertain impacts on food security.Adaptation strategies and policy responses to global change, including options for handling water allocation, land-use patterns, food trade, post-harvest food processing, and food prices and safety are urgently needed. These policy responses will be vital to improve the living conditions of farmers and rural populations across the country. Economic growth is only sustainable if India have food security with country-owned and country-driven food security strategies. Otherwise, there will be obstacles and additional costs to global, regional, and country-level economic growth. Food security needs to encompass women and other vulnerable and disadvantages groups.

1.1. Pillars of Food Security

The WHO states that there are negative one pillars that determine food security: food availability, food access, and food use[15]. The FAO adds a fourth pillar: the stability of the first three dimensions of food security over time[13]. In 2009, the World Summit on Food Security stated that the "four pillars of food security are availability, access, utilization, and stability[14]".

1. Availability

Growth in food production has been greater than population growth. Food per person increased since 1961. Data source: Food and Agriculture Organization. Food availability relates to the supply of food through production, distribution, and exchange. Food production is determined by a variety of factors including land ownership and use; soil management; crop selection, breeding, and management; livestock breeding and management; and harvesting. Crop production can be affected by changes in rainfall and temperatures. The use of land, water, and energy to grow food often competes with other uses, which can affect food production. Land used for agriculture can be used for urbanization or lost to desertification, salinization, and soil erosion due to unsustainable agricultural practices. Crop production is not required for a country to achieve food security. Nations don't have to have the natural resources required to produce crops in order to achieve food security, as seen in the examples of Japan and Singapore.

Because food consumers outnumber producers in every country, food must be distributed to different regions or nations. Food distribution involves the storage, processing, transport, packaging, and marketing of food. Food-chain infrastructure and storage technologies on farms can also affect the amount of food wasted in the distribution process. Poor transport infrastructure can increase the price of supplying water and fertilizer as well as the price of moving food to national and global markets. Around the world, few individuals or households are continuously self-reliant for food. This creates the need for a bartering, exchange, or cash economy to acquire food. The exchange of food requires efficient trading systems and market institutions, which can affect food security. Per capita world food supplies are more than adequate to provide food security to all, and thus, food accessibility is a greater barrier to achieving food security.

2. Access

Goats are an important part of the solution to global food security because they are fairly low-maintenance and easy to raise and farm. Food access refers to the affordability and allocation of food, as well as the preferences of individuals and households. The UN Committee on Economic, Social, and Cultural Rights noted that the causes of hunger and malnutrition are often not a scarcity of food but an inability to access available food, usually due to poverty. Poverty can limit access to food, and can also increase how vulnerable an individual or household is to food price spikes. Access depends on whether the household has enough income to purchase food at prevailing prices or has sufficient land and other resources to grow its own food. Households with enough resources can overcome unstable harvests and local food shortages and maintain their access to food.

There are two distinct types of access to food: direct access, in which a household produces food using human and material resources, and economic access, in which a household purchases food produced elsewhere. Location can affect access to food and which type of access a family will rely on. The assets of a household, including income, land, products of labour, inheritances, and gifts can determine a household's access to food. However, the ability to access sufficient food may not lead to the purchase of food over other materials and services. Demographics and education levels of members of the household as well as the gender of the household head determine the preferences of the household, which influences the type of food that are purchased. A household's access to enough and nutritious food may not assure adequate food intake of all household members, as intra-household food allocation may not sufficiently meet the requirements of each member of the household. The USDA adds that access to food must be available in socially acceptable ways, without, for example, resorting to emergency food supplies, scavenging, stealing, or other coping strategies.

3. Utilization

The next pillar of food security is food utilization, which refers to the metabolism of food by individuals. Once food is obtained by a household, a variety of factors affect the quantity and quality of food that reaches members of the household. In order to achieve food security, the food ingested must be safe and must be enough to meet the physiological requirements of each individual. Food safety affects food utilization, and can be affected by the preparation, processing, and cooking of food in the community and household. Nutritional values of the household determine food choice, and whether food meets cultural preferences is important to utilization in terms of psychological and social well-being. Access to healthcare is another determinant of food utilization, since the health of individuals controls how the food is metabolized. For example, intestinal parasites can take nutrients from the body and decrease food utilization. Sanitation can also decrease the occurrence and spread of diseases that can affect food utilization. Education about nutrition and food preparation can affect food utilization and improve this pillar of food security.

4. Stability

Food stability refers to the ability to obtain food over time. Food insecurity can be transitory, seasonal, or chronic. In transitory food insecurity, food may be unavailable during certain periods of time. At the food production level, natural disasters and drought result in crop failure and decreased food availability. Civil conflicts can also decrease access to food. Instability in markets resulting in food-price spikes can cause transitory food insecurity. Other factors that can temporarily cause food insecurity are loss of employment or productivity, which can be caused by illness. Seasonal food insecurity can result from the regular pattern of growing seasons in food production.

Chronic (or permanent) food insecurity is defined as the long-term, persistent lack of adequate food. In this case, households are constantly at risk of being unable to acquire food to meet the needs of all members. Chronic and transitory food insecurity are linked, since the re-occurrence of transitory food security can make households more vulnerable to chronic food insecurity.

1.2. Laws

In order to provide the Right to food to every citizen of the country, the Parliament of India, enacted a legislation in 2013 known as the National Food Security Act, 2013. Also, called as the Right to Food Act, this Act seeks to provide subsidized food grains to approximately two thirds of India's 1.2 billion population.[3] It was signed into law on 12 September, 2013, retroactive to 5 July, 2013[4] [5].

1.3. Schemes

A. Central Initiatives

The National Food Security Act, 2013 (NFSA 2013) converts into legal entitlements for existing food security programmes of the Government of India. It includes the Midday Meal Scheme, Integrated Child Development Services scheme and the Public Distribution System. In 2017-18, over ₹ 1500 billion (7.6% of the government's total expenditure) have been allocated to provide food subsidies under the Targeted Public Distribution System (TPDS)[6]. The NFSA, 2013 also recognizes maternity entitlements. Pregnant women, lactating mothers, and certain categories of children are eligible for daily free cereals.

State Initiatives

- Government of Tamil Nadu has launched 'AmmaUnavagam' (Mother's canteen), or more commonly known as Amma canteen[7].
- The State of Uttar Pradesh in 2013 passed a food bill. Food that is going to be wasted from parties will be preserved and distributed to poor and needy people.[citation needed].
- The Chhattisgarh Food Security Act, 2012 law was enacted by the Chhattisgarh Government. It was passed on 21 December 2012, by the State Assembly unopposed to ensure "access to adequate quantity of food and other requirements of good nutrition to the people of the State, at affordable prices, at all times to live a life of dignity[8]."

In the country like India, food availability is not that reliable. Their population right now is 1.3 billion and by 2030, it is estimated that their population will increase to 1.5 billion. The challenge to produce more and more for the growing population is becoming increasingly hard for a country of its size and economic growth. Since the land in India is a shrinking resource for agriculture, the production rate for agriculture needs to be higher per unit of land and irrigation water. Over 60% of the Indian population depends on the agriculture for their daily meals. India produces around 100 million tonnes of rice every year. To access food in India can be considered as not as easy as Australia. It is really hard to get hands on fresh fruit and vegetables. While there might be enough food for the whole population of India, many families and especially children in India don't have access to food because of financial problems. Thus, this is the cause of millions (43%) of malnourished children around India.

B. Technological solutions

India needs to concentrate on methods to improve the availability and affordability of protein rich food products using the latest environmental friendly technology without the need of additional land and water[9]. Bio gas or natural gas or methane produced from farm/ agro/crop/domestic waste can also be used in addition to mined natural gas for producing protein rich cattle/fish/poultry/pet animal feed economically by cultivating *Methylococcuscapsulatus* bacteria culture in a decentralized manner near to the rural / consumption areas with tiny land and water foot print[10 11 12].

C. Food safety and Quality laws

Economic growth, rising incomes, and urbanization have influenced Indians' eating habits. There is increased demand for greater variety in food choices, and Indians are becoming more concerned about food quality and safety. The 2015 withdrawal of Maggi noodles-a top ramen brand in India-has drawn further attention to the food safety situation. Meanwhile, monitoring by government agencies and NGOs has raised awareness and demand for safe food[16].

According to 2015 global estimates of foodborne diseases, "the risk of foodborne diseases is most severe in low-and middle-income countries, linked to preparing food with unsafe water; poor hygiene and inadequate conditions in food production and storage; lower levels of literacy and education; and insufficient food safety legislation or implementation of such legislation"[17].In India, food-borne diseases can be erratic and often go unreported, but in a nationwide 2006 study 13.2% of households reported food-borne illness[18].

As agricultural development is considered a means of fostering rural growth and reducing poverty, India needs to hurdle a number of policy, regulatory, infrastructural, and institutional obstacles and to produce food that meets basic quality and safety standards/ requirements.

Food Safety and Standards Authority of India (FSSAI) is an autonomous body established under the Ministry of Health &

FamilyWelfare, Government of India[22]. The FSSAI has been established under the Food Safety and Standards Act, 2006 which is a consolidating statute related to food safety and regulation in India. FSSAI is responsible for protecting and promoting public health through the regulation and supervision of food safety.

The FSSAI has its headquarters at New Delhi. The authority also has 6 regional offices located in Delhi, Guwahati, Mumbai, Kolkata, Cochin, and Chennai[24]. 14 referral laboratories notified by FSSAI, 72 State/UT laboratories located throughout India and 112 laboratories are NABL accredited private laboratories notified by FSSAI[25 26].

FSSAI was established by Government of India on 5 September, 2008 under Food Safety and Standards Act, 2006. The FSSAI consists of a chairperson & 22 members. The FSSAI is responsible for setting standards for food so that there is one body to deal with and no confusion in the minds of consumers, traders, manufacturers, and investors[2728]. Ministry of Health & Family Welfare, Government of India is the Administrative Ministry of Food Safety and Standards Authority of India[29 30].

The following are the statutory powers that the FSS Act, 2006 gives to the Food Safety and Standards Authority of India (FSSAI)[31]:

1. Framing of regulations to lay down food safety standards
2. Laying down guidelines for accreditation of laboratories for food testing
3. Providing scientific advice and technical support to the Central Government
4. Contributing to the development of international technical standards in food
5. Collecting and collating data regarding food consumption, contamination, emerging risks etc.
6. Disseminating information and promoting awareness about food safety and nutrition in India

D. Location

FSSAI is located in 5 regions[32]:

1. Northern Region – With head office at New Delhi

2. Eastern Region
3. North-Eastern Region
4. Western region
5. Southern Region

E. Regulatory framework

The Food Safety and Standards Authority of India is a statutory body under Food Safety and Standards Act, 2006. The Food Safety and Standards Act (FSS), 2006 is the primary law for regulation of food products. This act also sets up the formulation and enforcement of food safety standards in India. The FSSAI appoints food safety authorities on the state level[33].

The FSSAI functions under the administrative control of the Ministry of Health and Family Welfare[34]. The main aim of FSSAI is to:

1. Lay down science-based standards for articles of food
2. To regulate manufacture, storage, distribution, sale and import of food
3. To facilitate food safety

The FSS Act is a bucket for all the older laws, rules and regulations for food safety. The FSS Act took 8 older Acts into one umbrella:

1. Prevention of Food Adulteration Act, 1954[35]
2. Fruit Products Order, 1955[36]
3. Meat Food Products Order, 1973[37]
4. Vegetable Oil Products (Control) Order, 1947
5. Edible Oils Packaging (Regulation) Order 1988[38]
6. Solvent Extracted Oil, De- Oiled Meal and Edible Flour (Control) Order, 1967[39].
7. Milk and Milk Products Order, 1992[40].

F. Departments

1. Import Division
2. International Co-operation
3. Regulatory Compliance Division (RCD)

4. Food Safety Management System (FSMS) Division
5. Risk Assessment and R&D Division (RARD)
6. Information Education Communication (IEC) Division
7. Regulation and Codex Division
8. Quality Assurance/ Lab Division
9. HR Division
10. Standards Division

1.4. Research and Quality Assurance

A. Research

FSSAI has set certain guidelines for food safety research. The Research and Development division is responsible for research with the following objectives:

1. Generate new knowledge that would help in continuously updating and upgrading food safety standards which are compatible with international organizations
2. Carry out evidence based studies for improving or building policies.

B. Quality Assurance

FSSAI has been mandated to perform various functions related to quality and standards of food. These functions in addition to others include "Laying down procedure and guidelines for notification of the accredited laboratories as per ISO17025[41]." The FSSAI notified laboratories that are classified as:

1. FSSAI notified NABL accredited labs- 112[42]
2. State Labs-72
3. Referral Labs-14[43]

C. Standards

Standards framed by FSSAI are prescribed under Food Safety and Standards (Food Product Standards and Food Additives) Regulation, 2011, Food Safety and Standards (Packaging and Labelling) Regulation, 2011 and Food Safety and Standards (Contaminants, Toxins, and Residues) Regulations, 2011.

D. The FSSAI has prescribed standards for following food products:

1. Dairy products and analogues[44]
2. Fats, oils and fat emulsions
3. Fruits and vegetable products
4. Cereal and cereal products
5. Meat and meat products
6. Fish and fish products
7. Sweets & confectionery
8. Sweetening agents including honey
9. Salt, spices, condiments and related products
10. Beverages, (other than dairy and fruits & vegetables based)
11. Other food product and ingredients
12. Proprietary food
13. Irradiation of food

The development of standards is a dynamic process based on the latest developments in food science, food consumption pattern, new food products and additives, changes in the processing technology leading to changed specifications, advancements in food analytical methods, and identification of new risks or other regulatory options.

Formulation of standards of any article of food under the Food Safety and Standards Act, 2006, involves several stages. After consideration by the Food Authority, the draft standard is published (Draft notified), for inviting stakeholder comments. Since India is a signatory to the WTO-SPS Committee, Draft Standard is also notified in WTO. Thereafter, taking into account the comments received from the stakeholders, the Standard is finalized and notified in Gazette of India, and implemented[45].

E. Consumer outreach

Consumers can connect to FSSAI through various channels or call Toll free Number 1800112100. Recently, an online platform called 'Food Safety Voice' has been launched which helps consumers to register their complaints and feedbacks about food safety issues related to adulterated food, unsafe food, substandard food, labelling defects in food and misleading claims & advertisements related to various food products. A GAMA portal for concerns regarding misleading claims and advertisements too is operated[46].

E. Applicable FSSAI License

FSSAI issues three types of license based on nature of food business and turnover:

1. Registration: For Turnover less than ₹ 12 Lakh
2. State License: For Turnover between ₹ 12 Lakh to ₹ 20 Crore
3. Central License: For Turnover above ₹ 20 Crore

Other criteria like the location of the business, number of retail stores, etc. is needed while evaluating the nature of license applicable[47].

In spite of the decade-old transition from the previous food safety laws into the FSSA's integrated standards and regulations, there remain overlapping and residual/pre-existing standards maintained by other regulatory bodies. Clarity is needed, if all stakeholders are to conform to FSSA regulations.

2. References

1. Nutrition and Food Security - UN India. UN India. Retrieved 5 January, (2018).
2. India: Global Food Security Index. Retrieved 17 Jan, (2018).
3. National Food Security Act (PDF). PRS Legislative Research.
4. Press Information Bureau. Retrieved 5 January, (2018).
5. Food Security Act To Be Implemented From July 5. BTVI.in. Retrieved 5 January, (2018).
6. Expenditure Budget, Union Budget 2017-18 (PDF).
7. Tamil Nadu's Amma canteen concept catches on in other states, Live Mint, 27 March, (2017).
8. Chhattisgarh passes own food security bill - Hindustan Times. 11 January 2013. Retrieved 5 January, (2018).
9. Food made from natural gas will soon feed farm animals – and us. Retrieved 31 January, (2018).
10. BioProtein Production (PDF). Retrieved 31 January, (2018).
11. New venture selects Cargill's Tennessee site to produce Calysta FeedKind® Protein. Retrieved 31 January, (2018).

12. Assessment of environmental impact of FeedKind protein (PDF). Retrieved 20 June, (2017).
13. FAO Agricultural and Development Economics Division (June 2006). "Food Security (PDF) (2). Retrieved June 8, (2012).
14. FAO (2009). Declaration of the World Food Summit on Food Security (PDF). Rome: Food and Agriculture Organization of the United Nations.
15. WHO. "Food Security".Retrieved 15 October, (2013).
16. Sulaiman V., R., N.J. Kalaivani, J. Handoo, T.S. Vamsidhar Reddy, K. Dorai & A. Hall, (2011). "Organised Retailing of Fresh Fruit and Vegetables: Opportunities for Putting Research Into Use." Discussion Paper 12, April, Research into Use Programme (RIU), Department of International Development (DFID), London. https://assets.publishing.service.gov.uk/media/57a08ac2ed915d622 c0008bf/riu-retail-rasheed.pdf
17. WHO (World Health Organization), (2016). "WHO's First Ever Global Estimates of Foodborne Diseases Find Children under 5 Account for Almost One Third of Deaths." Media Centre, WHO. http://www.who.int/mediacentre/news/releases/2015/foodborne-disease-estimates/en/
18. Kohli, C. & S. Garg (2015). "Food Safety in India: An Unfinished Agenda." MAMC Journal of Medical Sciences 1(3): 131–135.
19. Enforcement of FSS Act (PDF). FSSAI. Retrieved 2 April, (2012).
20. Authority of FSSAI. FSSAI.Retrieved 2 April, (2012).
21. Former Union Health Minister.
22. Food Safety and Standards Authority of India (FSSAI). Food Safety and Standards Authority of India, Government of India. Retrieved 2 April, (2012).
23. Section 5(3) of the Food Safety and Standards Act, (2006).
24. Regional Laboratories. FSSAI.Retrieved 2 April, (2012).
25. Referral Laboratories (PDF). FSSAI.Retrieved 2 April, (2012).
26. List of Laboratories. FSSAI.Retrieved 2 April, (2012).
27. FAQ's on FSSA, (2006) (PDF). Food Safety and Standards Authority of India, Government of India. Retrieved 10 October, 2016.
28. FSSAI calls for EoI for nomination of a scientific committee and panels. FnBnews.com. 7 May, (2016).

29. FSSAI to formulate new policies for food safety. India Today. 18 February, (2016).
30. Drug & Food Division (FSSAI/CDSCO). Ministry of Health & Family Welfare.
31. Food safety in India - Regulatory framework and challenges (PDF). swaniti.in. Retrieved 10 October, (2016).
32. FSSAI Head Office. FSSAI.
33. Ministry of Law and Justice (PDF). FSSAI.
34. FSSAI to formulate new policies for food safety. India Today. 8 February, (2016).
35. The Prevention of Food Adulteration Act, 1954. India Kanoon.
36. Fruit Products Order, 1955 (PDF). Consumer Advice. 8 February, (2016).
37. Meat Food Products Order, 1973. agmarknet.nic.in.
38. The Vegetable Oil Products (Regulation) Order – 1998. Crirec. 17 January, (2011).
39. The Edible Oils Packaging (Regulation) Order, 1998. Crirec. 17 January, (2011).
40. Milk and Milk Product Order, 1992. Department of Animal Husbandry, Dairying & Fisheries.
41. Addresses and contact details of NABL accredited food testing laboratories for analysis of food sample taken under FSS Regulation-for 1 Year" (PDF). fssai.gov.in.
42. FSSAI Notified NABL Accredited Food Testing Laboratories. Food Safety Helpline.Retrieved 28 June, (2014).
43. Food Safety and Standard Authority of India - Order (PDF). Food Safety Helpline.Retrieved 28 June, (2014).
44. Pawan Kumar Agarwal appointed CEO, FSSAI. Business Standard. 22 December, (2015).
45. FSSAI appoints new members for scientific panels. Nuffoods Spectrum. Retrieved 13 February, (2015).
46. Grievances Against Misleading Advertisements. gama.gov.in.
47. FSSAI Registration online in india - Enterslice. enterslice.com. Retrieved 2017-07-18.

❑❑❑